3

Springer Series on Chemical Sensors and Biosensors

Methods and Applications

Series Editor: O. S. Wolfbeis

Springer Series on Chemical Sensors and Biosensors

Series Editor: O. S. Wolfbeis

Recently Published and Forthcoming Volumes

Surface Plasmon Resonance
Volume Editor: Homola J.
Vol. 4, 2006

Frontiers in Chemical Sensors
Novel Principles and Techniques
Volume Editors: Orellana G., Moreno-Bondi M. C.
Vol. 3, 2005

Ultrathin Electrochemical Chemo- and Biosensors
Technology and Performance
Volume Editor: Mirsky V. M.
Vol. 2, 2004

Optical Sensors
Industrial, Environmental and Diagnostic Applications
Volume Editors: Narayanaswamy R., Wolfbeis O. S.
Vol. 1, 2003

Frontiers in Chemical Sensors

Novel Principles and Techniques

Volume Editors:
Guillermo Orellana · Maria C. Moreno-Bondi

With contributions by

J. Alonso · Y. Amao · L. Basabe-Desmonts · S. Busche
V. Cadarso · C. M. Charlton · M. Crego-Calama · C. Domínguez
I. Garcés · D. Garcia-Fresnadillo · G. Gauglitz · R. Horváth
M. Kasper · U. J. Krull · N. B. Larsen · A. Llobera
M. D. Marazuela · M. Massey · B. Mizaikoff · M. C. Moreno-Bondi
M. Nakagawa · G. Orellana · H. C. Pedersen · P. A. E. Piunno
M. Puyol · D. N. Reinhoudt · Z. Rosenzweig · I. Salinas
M. Schäferling · N. Skivesen · K. Suzuki · Y. Suzuki · Y. Takeuchi
N. T. K. Thanh · B. T. Thompson · A. Vernhet · F. Villuendas
N. Yamashita · R. S. Zimmerman

Springer

Chemical sensors and chemical biosensors are becoming more and more indispensable tools in life science, medicine, chemistry and biotechnology. The series covers exciting sensor-related aspects of chemistry, biochemistry, thin film and interface techniques, physics, including opto-electronics, measure- ment sciences and signal processing. The single volumes of the series focus on selected topics and will be edited by selected volume editors. The Springer Series on Chemical Sensors and Biosensors aims publish state-of-the-art arti- cles that can serve as invaluable tools for both practitioners and researchers active in this highly interdisciplinary field. The carefully edited collection of papers in each volume will give continuous inspiration for new research and will point to existing new trends and brand new applications.

Library of Congress Control Number: 2005929413

ISSN 1612-7617
ISBN-10 3-540-27756-0 Springer Berlin Heidelberg New York
ISBN-13 978-3-540-27756-9 Springer Berlin Heidelberg New York
DOI 10.1007/b96933

Springer is a part of Springer Science+Business Media

springeronline.com

Cover design: *Design & Production* GmbH, Heidelberg
Typesetting and Production: LE-TEX Jelonek, Schmidt & Vöckler GbR, Leipzig

Printed on acid-free paper 52/3100 YL – 5 4 3 2 1 0

Series Editor

Prof. Otto S. Wolfbeis

Institute of Analytical Chemistry
University of Regensburg
Chemo- and Biosensors
93040 Regensburg, Germany
otto.wolfbeis@chemie.uni-regensburg.de

Volume Editors

Prof. Dr. Guillermo Orellana

Lab. of Applied Photochemistry
Universidad Complutense
28040 Madrid, Spain
orellana@quim.ucm.es

Prof. Dr. Maria C. Moreno-Bondi

Dept. of Analytical Chemistry
Universidad Complutense
28040 Madrid, Spain
mcmbondi@quim.ucm.es

Springer Series on Chemical Sensors and Biosensors Also Available Electronically

For all customers who have a standing order to Springer Series on Chemical Sensors and Biosensors, we offer the electronic version via SpringerLink free of charge. Please contact your librarian who can receive a password or free access to the full articles by registering at:

springerlink.com

If you do not have a subscription, you can still view the tables of contents of the volumes and the abstract of each article by going to the SpringerLink Homepage, clicking on "Browse by Online Libraries", then "Chemical Sciences", and finally choose Springer Series on Chemical Sensors and Biosensors.

You will find information about the

- Editorial Board
- Aims and Scope
- Instructions for Authors
- Sample Contribution

at springeronline.com using the search function.

Preface

With their similarity to the organs of the most advanced creatures that inhabit the Earth, sensors are regarded as being the "*senses of electronics*": artificial eyes and ears that are capable of seeing and hearing beyond the range of human perception; electronic noses and tongues that can recognise odours and flavours without a lifetime training; touch that is able not only to feel the texture and temperature of the materials but even to discern their *chemical* composition. Among the world of chemical sensors, *optical* devices (sometimes termed "optodes", from the Greek "the optical way") have reached a prominent place in those areas where the features of light and of the light-matter interaction show their advantage: contactless or long-distance interrogation, detection sensitivity, analyte selectivity, absence of electrical interference or risks, and lack of analyte consumption, to name just a few. The introduction of optical fibres and integrated optics has added more value to such sensing since now light can be confined and readily carried to difficult-to-reach locations, higher information density can be transported, indicator dyes can be immobilised at the distal end or the evanescent field for unique chemical and biochemical sensing (including multiplexed and distributed measurements), optical sensors can now be subject to mass production and novel sensing schemes have been established (interferometric, surface plasmon resonance, fluorescence energy transfer, supramolecular recognition ...).

This third volume of Springer Series on Chemical Sensors and Biosensors aims to enable the researcher or technologist to become acquainted with the latest principles and techniques that keep on enlarging the applications in this fascinating field. Chapter 1 deals with the well-established absorption spectrometry but based on miniaturised integrated optics. Novel luminescence lifetime-based techniques for interrogation of sensor arrays in high-throughput screening are described in Chapter 2, while a modern use of the exotic cataluminescence is the topic of Chapter 3. Going from the uv-vis spectral region into the infrared allows chemical sensing with hollow waveguides (Chapter 4). Chapters 5 and 6 discuss new ways of succeeding in sensor design and fabrication by means of either combinatorial methods or engineered indicator/support couples, respectively. Frontiers in analytical biosensing is the common theme of Chapters 7 to 9, the hot topics of which are the design of DNA optosensors, gold nanoparticle-based assays and reverse symmetry waveguides. The book ends with three chapters concentrating on applications – dealing with pressure-sensitive luminescent paints, optical sensing of enantiomers and the amazing use of digital colour analysis for ions and protein monitoring. All these novel optosensing principles and techniques dramatically increase the analytical power of the current chemical sensors allowing us to predict a promising future for such devices in the years to come.

This monograph is aimed at graduate students in chemistry, physics, biology, engineering, and material science, as well as researchers and technologists in academic and industrial environments who want to keep abreast of the latest trends in chemical sensing and biosensing with photonic devices. The contributing authors are recognised experts in their respective fields so that, in addition to providing an authoritative overview of each selected topic and the advances therein, they have been able to convey to the reader first-hand information and experimental results from their own research and developments. Therefore, the Editors would like to acknowledge and gratefully thank the effort and enthusiasm of all the authors who have provided outstanding, exquisite manuscripts to constitute the present new volume of Springer Series on Optical Chemical Sensors and Biosensors. In Spanish, there is a saying which says it is impossible to make a gate to the countryside, in the same way, it is certainly difficult to define frontiers in the optical sensing field. However, we hope the contents of this book will guide the reader through the novel principles and techniques that are setting the pace in the advances in our continuously evolving field.

September 2005

Guillermo Orellana
Maria C. Moreno-Bondi

Contents

Springer Ser Chem Sens Biosens (2005) 3: 1–44
DOI 10.1007/5346_001

Published online: 28 September 2005

Absorbance-Based Integrated Optical Sensors

Mar Puyol[1] (✉) · Francisco Villuendas[1] · Carlos Domínguez[2] · Víctor Cadarso[2] · Andreu Llobera[3] · Iñigo Salinas[4] · Ignacio Garcés[4] · Julián Alonso[5]

[1]Dpto. Física Aplicada, Universidad de Zaragoza, Saragossa, Spain
mariadelmar.puyol@uab.es

[2]Centro Nacional de Microelectrónica, Bellaterra, 08193 Barcelona, Spain

[3]Institut für Mikrotechnik, Technische Universität Braunschweig, Alte Salzdahlumer Str. 203, 38124 Braunschweig, Germany

[4]Dpto. Ingeniería Electrónica y Comunicaciones, Universidad de Zaragoza, CPS, María de Luna 3, 50015 Saragossa, Spain

[5]Grup de Sensors i Biosensors, Unitat de Química Analítica, Edifici Ciències, Bellaterra, 08193 Barcelona, Spain

Abstract Optochemical sensors have had a huge expansion and in recent years interesting sorts of optical sensor have been developed which make use of the integrated circuit microelectronic technology and the optical technological advances achieved in the telecommunications industry. These devices are based on optical fibers, planar waveguides or the combination of both supports as light-guiding structures and offer an enhanced performance thanks to a miniaturized size, a feasible mass production at low cost, the possibility to measure at large distances and the capability of reaching places hard to access for the in situ monitoring of environmental or medical parameters. In this chapter, we provide an overall view of integrated chemical sensors based on absorbance measurements, describing their main characteristics, advantages and drawbacks. In order to focus on these types of devices, first of all, a brief classification is given, in terms of the interaction mechanisms and the reactive phases or membranes that are involved in the response, and the radiation transmission medium. Next, different waveguide configurations are briefly described because of to the relevancy of these structures as the main constituent of integrated optodes and, finally, some absorbance-based integrated sensors are detailed.

Keywords Absorbance · Antiresonant reflecting optical waveguide · Core-based optodes · Integration · Ion-selective optodes

Abbreviations

A	Absorbance
ATR	Attenuated total reflection
ARROW	Antiresonant reflecting optical waveguide
AU	Absorbance units
C	Concentration (M)
CMOS	Complementary metal oxide semiconductor
CVD	Chemical vapor deposition
d	Membrane thickness
D_m	Diffusion coefficient ($cm^2\ s^{-1}$)
DOS	Bis(2-ethylhexyl)sebacate
DRIE	Deep reactive ion etching
ε	Molar absorptivity ($kg\ mol^{-1} cm^{-1}$)
FIA	Flow injection analysis
IOC	Integrated optic circuit
IWAO	Integrated waveguide absorbance optode
KtpClPB	Potassium tetrakis(4-chlorophenyl)borate
L, l	Path length (cm)
LPCVD	Low-pressure chemical vapor deposition
λ	Wavelength (nm)
n	Refractive index
NIR	Near IR
PECVD	Plasma-enhanced chemical vapor deposition
PIN	p-n photodiode
RIE	Reactive ion etching
RSD	Relative standard deviation
TIR	Total internal reflection
SPR	Surface plasmon resonance
$t_{95\%}$	95% response time (min)
t_r	Response time (min)

1
Introduction

The chemical sensor area has proved to be one of the most dynamic fields in analytical chemistry. During the last few years, an enormous effort has been made in order to design sensors that show appropriate features such as selectivity, low detection limits, reversibility, robustness, portability and easy handling. However, this continuously evolving field is improving to achieve the goals of the conceptual term of sensor and most of the existing ones still present clear limitations.

The collaboration of diverse groups allows conjugating the knowledge of different research fields such as electronics, optics, analytical, inorganic and organic chemistry, biochemistry and materials science, and then adapts sensor technology to industrial, medical or environmental needs. Despite the hindrance of combining such multidisciplinary groups, a great variety of chemical sensors which are designed for very concrete applications and which would not have much utility [1, 2] have been commercialized and show their potentiality and usefulness in the real world.

Optochemical sensors have had a huge expansion during the last few years, and surely the main reason is the wide range of optical phenomena that one can take advantage of for the design of different sensing schemes.

Recently, the greatest improvements have been based on the productive merger of integrated circuit microelectronics and telecommunications technology. This has provided new instrumental platforms and the possibility to apply new optical phenomena, like the evanescent field, in the design of new chemical sensing setups. Miniaturized optical components, such as light sources, photodetectors and light guiding structures (cylindrical and planar waveguides), also offer an enhanced performance and a feasible mass production at low cost, with the possibility to measure at large distances, reaching places hard to access for the in situ monitoring of environmental or medical parameters.

In this chapter, we present an overall view of integrated chemical sensors based on absorbance measurements, explaining their main characteristics, advantages and drawbacks. In order to focus on such devices, first of all, the main sensor components are briefly described and after that a brief explanation is given in terms of the most suitable reactive phases depending on the sensor format to be constructed or the radiation transmission medium employed. Next, because of the relevancy of waveguide structures as the main constituents of integrated optodes, details of some configurations are given and, finally, some absorbance-based integrated sensors are detailed.

2
Design of Optochemical Sensors

The fundamental operation of an optochemical sensor consists of three main steps: the analyte-recognizing element interaction by means of any of the different mechanisms that are schematized in Fig. 1 [3]; the detection and transduction of any physical or chemical variation caused by the recognizing reactions; and the signal processing and the acquisition of results.

Concerning the required instrumentation, usually conventional active and passive optical components are employed; however, their complexity can vary according to the proposed objectives. Two main parts can be distinguished: the *recognition element*, which includes the selective reagents, the intermedi-

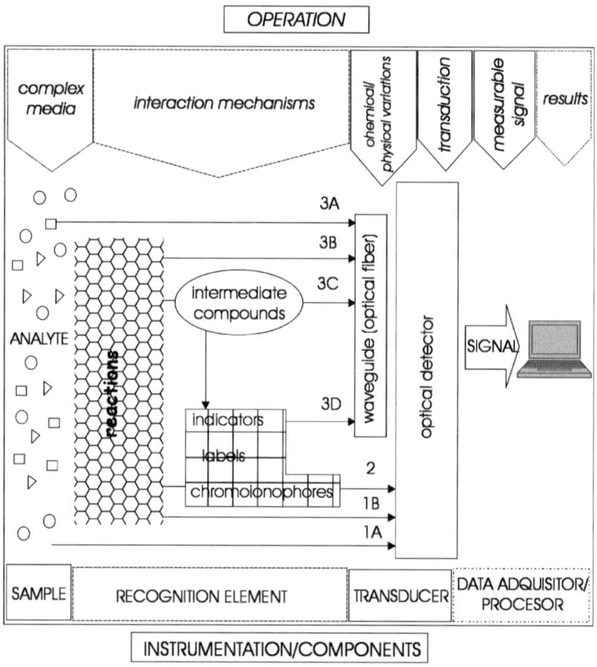

Fig. 1 Chemical interaction mechanisms, basic components of the optical sensor instrumentation and their operation. Mechanisms: direct measurement of chemical compounds that exhibit spectroscopic properties (1*A*) and measurement of light originating from a chemical or a biological reaction in chemiluminescent or bioluminescent phenomena (1*B*); 2 optodes based on the interaction of indicators and labels with light, which are immobilized in a support; and sensors that modify the intrinsic physical or chemical properties of a waveguide (refractive index, phase, etc.) as a result of the presence of the analyte (3*A*), a recognition element (3*B*), an intermediate analyte (3*C*) or an indicator (3*D*)

ate analytes, the indicators or labels, and sometimes a matrix as a support or a membrane, and the *transducer*. The instrumentation in the latter includes the light source, the wavelength selectors, the waveguides and the detector. Finally, amplifiers and processors are used for the signal treatment.

An integrated optical sensor is one including all the different constituents for the signal transduction, amplification and processing in the same substrate. However, the concept of integration is usually employed to designate miniaturized devices, whose main parts (the recognition element and waveguides) lie in the same substrate, and whose waveguides are fabricated with common microelectronic techniques.

From the diversity of the analytical methods and the availability of the instrumentation and possible configurations that can be found, one can infer that the design, the formulation and the integration of the recognition element in the transducer are crucial because quality parameters such as sensitivity, selectivity, lifetime and dynamic range directly depend on them. The initial stages of the development of an integrated optical sensor, based on the direct or the indirect analyte recognition by means of a reactive phase, involve the study and selection of the response mechanism, the choice of the reagents implicated in the recognizing process, the matrix formulation and sometimes the selection of appropriate membrane deposition techniques over the waveguides.

2.2
Reagents and Indicators

When an analyte lacks measurable optical properties by itself, spectroscopically active intermediate molecules are needed. Indicators modify their optical properties when they interact with the analyte and usually are neutral or ionizable dyes.

A great number of indicators are known, but only a few of them are really useful in the optochemical sensor field owing to their properties. As the main characteristics, they must have high molar extinction coefficients and be specific for a certain analyte. Moreover, they must be photochemically stable to be applied in real situations, making viable the development of simple, rugged and low-cost devices. Other operational limitations arise from the use of optical fibers (with low transmission capability in the UV spectral region), from the chosen immobilization conditions and from the low-cost available light sources and detectors (usually LEDs and p-n photodiode, PIN, photodiodes), which are only adequate for a particular spectral range. For instance, indicators with absorption bands located at wavelengths shorter than 500 nm are only appropriate for classical optical instrumentation in absorbance-based systems. Some new blue and violet diode lasers can be found in the market [4, 5]. However, the first ones need an umbral intensity of 500 mA to emit and, moreover, both sources are not optimized for common telecommunication optical fibers. They might be suited if fluorescence is to be measured but they are not optimal for integration in compact systems based on integrated optics.

For the development of new miniaturized and mass-production devices, dyes presenting absorption bands between 550 and 650 nm are needed, because they can be activated with low-cost light sources of high intensity and can work with optical fibers.

A great variety of reagents which fulfill all or part of the requirements mentioned have been employed. Depending on the analytical goal, we face a given recognition principle that is mediated by a certain reagent or indicator. pH indicators (absorbent or fluorescent proton exchangers) have been extensively employed for acidity determinations [6], and when they are combined with ionophores, for indirect measurements of cations and anions [7–13]. On the other hand, for the direct detection of such ions, various types of chromoionophores (metal-chelating agents or ion-pairing formers) have been used [14–18] and different fluorophores have been used to perform indirect measurements based on the quenching effect [6, 19–28].

2.3 Supports and Matrices

Reagents and indicators are immobilized, occluded or dissolved in supports which are formed by cross-linked polymers, plasticized polymers or organic and inorganic activated surfaces. The waveguide itself, the cladding of an optical fiber or any other optical element can be the support. However, it must obey two basic functions: act as a liquid–solid or gas–solid interface and, if radiation crosses through it to allow the signal transmission, be an optically transparent material.

On the other hand, it must stay inert while facing the chemical reactions that are involved in the recognizing process and it must assure an efficient radiation coupling with the light source, if optical fibers are used. In addition, the matrix sometimes plays the role to improve selectivity and regulates the charge transference.

A large variety of different support materials can be used. The preferred ones are often organic polymers even though they sometimes swell significantly as a function of the external medium and are less resistant to abrasion and mechanical shock than the inorganic ones. The main reasons are that indicators and labels are frequently organic compounds and that a great variety of organic supports are available. The selection depends on the affinity towards the reagents, the stability, and the permeability to the analyte, the compatibility towards other materials employed for the sensor construction (e.g., in cases where it is the cladding of an optical fiber) and the compatibility with the sample. The polymeric material has a great influence on the sensor performance.

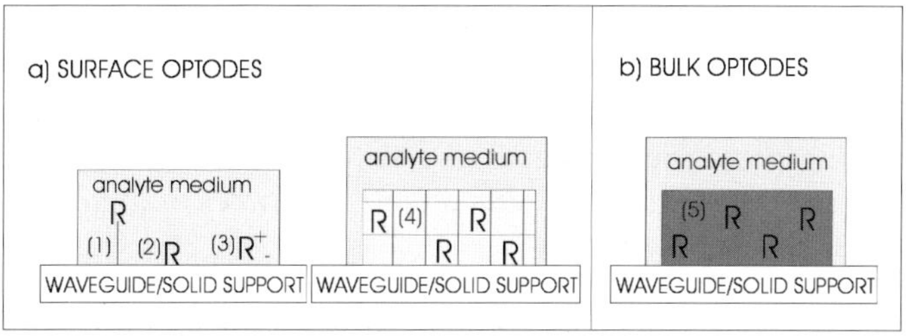

Fig. 2 Schematic representation of optochemical sensors depending on the arrangement of the optically active reagents. **a** Surface optodes: reagents are directly immobilized over a solid support by (1) covalent bonding, (2) adsorption or (3) electrostatic interactions or a waveguide or are trapped in a porous matrix (4). **b** Bulk optodes: reagents are dissolved in a plasticizer (5)

Other inorganic polymers, such as silicones, have excellent optical and mechanical properties for optical membranes. A great number of easily handled commercial silicone prepolymers are available but they have some disadvantages towards other materials. The surface is not easily modifiable to covalently immobilize indicators, they are not suited to combine with the customarily used support structures, owing to hard adhesion, and they are bad solvents for most of the indicators.

Other types of common inorganic supports are silicon gels, quartz/glasses or sol–gels. They are rigid materials and play an important role in the optical transmission when acting as waveguides. In that sense, indicators are usually covalently immobilized in the surface or are trapped in the porous material. Depending on the arrangement of the reagents and indicators in the reactive phase, we can make a distinction between two types of optical sensors which follow different response mechanisms. Reagents are directly immobilized over the support or the waveguide in surface-phenomena-based optical sensors, while in bulk optodes they are dissolved in a polymeric membrane. In Fig. 2 both schemes are depicted and the nature of the interaction concerning the indicators and the matrix or the waveguide is indicated.

2.3.1 Surface Optodes

Their main characteristic is that the optically active reagents are immobilized over the surface of an optical component (waveguide, metallic plate, glass, prism, etc.) or in a porous matrix in a way that they are in direct contact with the analyte in the sample solution. This immobilization can be carried out by covalent bonding [29], by mechanical interactions such as physical adsorption [30], by physical entrapment or by electrostatic interactions [31].

The selective recognizing reactions take place only in the surface that is in contact with the solution. Then, the response time depends on the equilibrium rate in the surface of the reactive phase and the signal relies on the superficial charge density as well as on the ionic strength of the solution. In these conditions, both parameters must be maintained constant.

2.3.2 Bulk Optodes

The reagents are dissolved in an immiscible phase that is held up to an inert polymeric matrix in bulk optodes. The whole constitutes the sensing membrane, and the signal relies on concentration changes within the bulk.

They are often plasticized PVC matrices, which occlude an ionophore as the key selective element, a chromoionophore or a fluoroionophore as the chemical-optical transducer and, sometimes, ionic additives to maintain electroneutrality. Such optodes follow ion-exchange mechanisms between the membrane and the aqueous solution and the analytical response originates from the ratio of the concentration of ions in the solution or from their product (Fig. 3). Moreover, selectivity is ruled by the ion distribution coefficients between both phases and by the formation constants of complexes within the membrane.

The main advantages of using this approach are that response is not affected by the ionic strength of the solution and that the development of specific ion-selective optodes is easily adapted from their analogous well-studied ion-selective electrodes [32–34], because the same selective reagents and membrane constituents are employed. Most of the ion-selective optodes

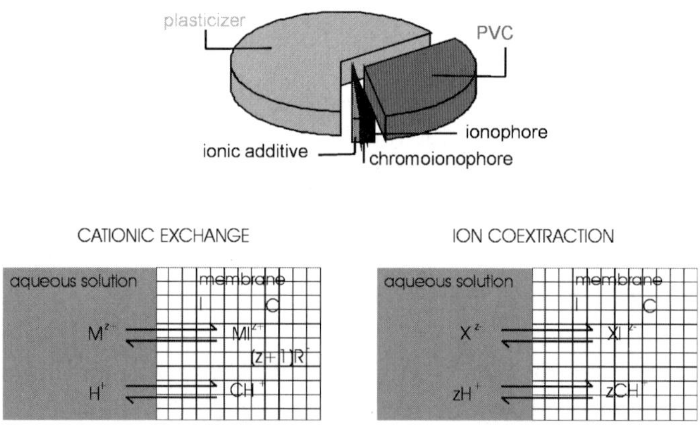

$$\frac{a_M}{(a_H)^z} = K \frac{\alpha^a}{(1-\alpha)^b}$$

$$(a_{H^+})^z a_{X^{z-}} = K \frac{(1-\alpha)^a}{\alpha^b}$$

$$a_{H^+} = \frac{1}{Ka_{MEM}} \frac{(1-\alpha)^a}{\alpha^b}$$

Fig. 3 Basic composition of ion-selective bulk optodes and extraction mechanisms. This type of sensor follows an extraction equilibrium between the aqueous solution and the membrane and the signal is related to the analyte activity in the aqueous solution

described belong to this class of sensors. Table 1 shows some organic compounds employed as near-IR (NIR) chromoionophores in optical sensors based on bulk optodes [35].

2.4 Sensor Configuration

The optical sensor construction is more or less complex depending on its design, the type of formulated membrane, the optical phenomenon that generates the signal and how this signal is measured. Taking into account whether radiation transmission structures are employed or not, we can distinguish two different basic configurations: non-guided radiation devices (conventional configurations) and waveguide-based sensors (Fig. 4).

2.4.1 Non-Guided Radiation Devices

The main characteristic of non-guided radiation devices is that the recognition element or the chemical membrane is not physically integrated in any waveguide structure. Variations of the membrane optical properties lead to variations of the transmission of a non-guided light beam (Fig. 4a).

This class of sensors is the easiest to build but prevents the development of miniaturizable sensors for in situ applications. In general, this conventional configuration represents an easy and rapid way to optimize the response characteristics of any recognition element, before any possible integration in

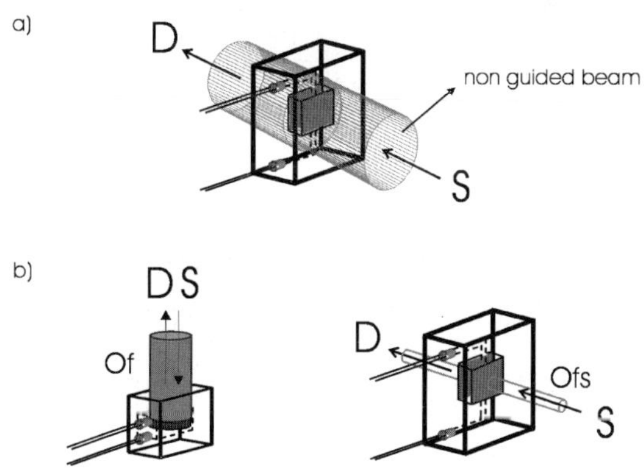

Fig. 4 Different optochemical configurations for flow-cell-based sensors. Notice that flow cells are expendable. *S* source, *D* detector, *Of* optical fiber. **a** Non-guided sensors: the transmission of a non-guided collimated radiation beam is measured. **b** Radiation is guided though optical fibers from the source to the recognition element and from this position to the detector

Table 1 Near-IR chromoionophores applied in bulk optodes. The absorption maxima, the type of matrix employed, the sensor configuration, the optical properties measured, the analyte and year of publication are listed

Type of dye	Name	λ (nm)	Matrix	Configuration optical property	Analyte	Year (Ref.)
Structure 1 polymethinic dyes	Crococyanine	815_{EtOH}	PVC/DOS	Flow cell/A	H^+	1997 [36]
	Squarocyanine	$819_{PVC/DOS}$	PVC/DOS	Flow cell/A	H^+	1997 [36]
		$821_{PVC/oNPOE}$	PVC/o-NPOE			
	Streptopolymethine cyanine	–	PVC/DOS	Flow cell/A	H^+	1996 [11]
		810_{Nafion}	Nafion	Flow cell/A	H^+	1996 [11]
	Carboxyheptamethine Cyanine	780_{Nafion}	Nafion	Flow cell/A	H^+	1996 [11]
	Styrylcyanine	$660_{MeOH:H_2O}$	PVC/DBS	Flow cell/A	H^+, K^+, Na^+, Li^+, Ca^{2+}, Mg^{2+}, Ba^{2+}, $NH_4{}^+$	1999 [17]
		$670_{ISP:H_2O}$	PVC	OF/A	H^+	1995 [37]
		$680, 720_{PVC/TOP}$	PVC/TOP	Flow cell/A	H^+	1994 [9]
	Asymmetric trimethinecyanine	$826_{ISP:H_2O}$	PVC	OF/A	H^+	1995 [37]
		$850_{PVC/TOP}$	PVC/TOP	OF/A	H^+	1994 [9]
	Ketocyanine	$727–770_{PVC/TOP}$	PVC/TOP	Flow cell, IOS/A	H^+	2002 [38]
		$768_{PVC/TOP}$	PVC/TOP	Flow cell, IOS/A	Cd^{2+}	2002 [39]
Structure 2 diphenyltriphenyl methane	Benzophenoxazine	$690_{PVC/DOS}$	PVC/DOS	IOS/R	Ca^{2+}	1995 [40]
	ETH 5294 (Fluka)	–	PVC/DOS	Flow cell/A	Pb^{2+}	1992 [41]
		–	PVC/DOS	Flow cell/A	NH_3	1991 [42]
	Benzophenoxazine	660	PVC/DOS	Flow cell/A, F	NO_2	1997 [43]
	ETH 2439 (Fluka)	–	PVC/DOS	Flow cell/A	Pb^{2+}	1992 [41]
		–	PVC/DOS	Flow cell/A	NH_3	1991 [42]

Table 1 (continued)

Type of dye	Name	λ (nm)	Matrix	Configuration optical property	Analyte	Year (Ref.)
	Benzophenoxazine ETH 5418 (Fluka)	665	PVC/DOS	Flow cell/A	Pb^{2+}	1992 [41]
	Benzophenoxazine ETH 5350 (Fluka)	640	PVC/DOS	Flow cell/A	NH_3	1991 [42]
Structure 3 mixed chromophoric system	Dicyanovinyl alkylated dye ETH 5003–5009 (Fluka)	$576–759_{MeOH}$	PVC/DOS	Flow cell/A	H^+, Ca^{2+}	1998 [44]
	Dicyanovinyl	$755_{PVC/DOS}$ $768_{PVC/oNPOE}$	PVC/DOS PVC/o-NPOE	Flow cell/A	H^+	1997 [36]
Structure 4 metal complexes	Copper complexes	$760_{PVC/DOS}$	PVC/DOS	OF/ev.	$NH_3(g)$	1998 [45]

OF optical fibers, *IOS* integrated optical sensors, *A* absorbance, *R* reflectance, *F* fluorescence, *ev* evanescent wave, *ISP* isopropyl alcohol, *DOS* bis(2-ethylhexyl)sebacate, *o – NPOE ortho*-nitrophenyl octyl ether, *TOP* tris(2-ethylhexyl)phosphate

waveguide structures. That is why a large number of the reported optodes were firstly assayed in such a configuration using different conventional spectroscopic techniques (absorbance, luminescence, reflectance, etc.).

2.4.2 Waveguide-Based Devices and Integrated Optochemical Sensors

Optical fibers and planar waveguides are the main components of waveguide-based sensors, where the recognizing element is directly integrated (Fig. 4b). Waveguides were developed for and by the telecommunications industry and have been widely applied for the procurement of chemical sensors in the field of analytical chemistry because they offer optimum characteristics concerning their dimensions, flexible geometry, immunity to electric interferences and the possibility to transmit light over large distances. The recognition element–waveguide contact can occur by means of a chemical activation of the waveguide surface (covalent immobilization of the reagents) or by physical deposition of homogeneous reactive phases, such as bulk optodes and sol–gel matrices.

Different situations can be distinguished depending on the waveguide configuration. Devices can give discrete information as probes or continuous information, if they are incorporated in flow cells and, from a spatial point of view, measurements can be punctual or distributed all along the guiding structures such as in optical time domain reflectometry [46, 47]. In order to understand their operation, it is essential to learn the waveguide characteristics and optical properties.

2.4.2.1 Light Propagation in Waveguides and Light Coupling

Optical waveguides generally confine light by means of the total internal reflection (TIR) effect, which is accomplished at the interfaces of a high refractive index dielectric medium (core) surrounded by a lower refractive index medium (cladding and substrate) (Fig. 5) [48, 49].

Light propagates inside a waveguide under some restricted conditions. The propagation characteristics of optical fibers are well known and can be found in the literature [50]. In contrast, the analysis of light propagation characteristics in planar waveguides is much more complicated and a theory of propagation is not well established yet. The preferred analysis tool is the simulation of their behavior by means of beam propagation methods [51] and taking into account several approximations. In any case, it is usually necessary to overcome some cycles by trial and error before relatively fair featured waveguides can be fabricated and used to develop integrated sensors on a chip.

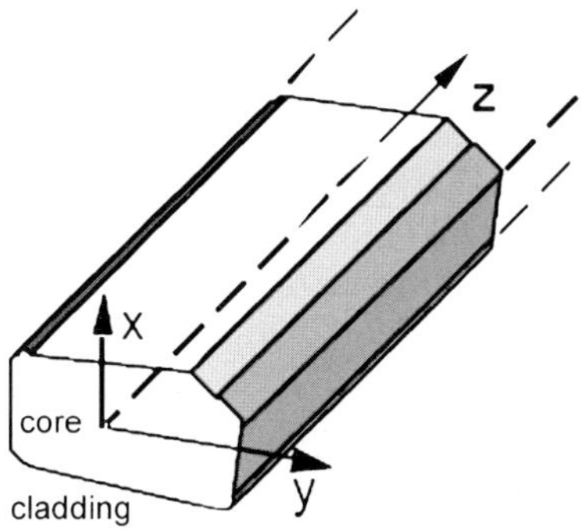

Fig. 5 Elementary parts of a waveguide, propagation and confinement directions

Moreover, to thoroughly study the propagation characteristics of a given waveguide it is important to previously know how light is coupled into it. Light is frequently coupled into integrated waveguides by means of optical fibers, as there are a great number of commercially available ones which have been designed for all the existing optical sources and detectors. Other coupling alternatives are prisms and grating couplers [52], but their application is restricted to planar waveguides and they are not suitable if the goal is to achieve a total integration level or to develop small and compact devices, because bulky optical components such as the light sources and detectors are required.

On the other hand, there are some sensor configurations such as Mach–Zehnder [53–55] interferometers or even evanescent waveguide-based devices [56] that require a single-mode behavior. Such waveguides are difficult to fabricate and require very thin cores to attain low insertion losses with the coupling to standard optical fibers. The only way to prevent such losses is by using more complex coupling configurations as lensed fibers, spot transformers or grating couplers [57].

Optical waveguides can be very different depending on the application they are designed for, and so their light propagation properties will also be very different. For instance, optical fibers are used whenever information is needed to be transported far away from any position, but integrated waveguides are used when designing photonic circuits on a chip. More often, variations in their intrinsic guiding properties are exploited as the measured signal.

2.4.2.2
Waveguide Designs

Waveguide-based devices can take on different configurations depending on the guiding structures employed: cylindrical (optical fibers and capillary waveguides) or planar.

A great number of different optical fibers exist depending on the structure, geometry and fabrication materials. Each of these parameters determines the performance (attenuation, dispersion), physical properties (refractive indexes) and optical properties (propagation range wavelengths). The main materials for fabrication of optical fibers for use in the IR range are SiO_2 glasses and metallic fluoride glasses such as $MF_3 - AlF_3$ and $MF_3 - ZrF$, where M can be Ca, Sr, Ba or Pb. For medical applications plastic fibers of polycarbonate, polystyrene or poly(methyl methacrylate) are employed. They are commonly used in fluorescence and evanescent field applications or for illumination of samples [58, 59]. Multimode glass fibers are also used for fluorescence [60] and single-mode glass fibers for evanescent sensors and surface plasmon resonance (SPR) [61].

Standard single-mode fibers are often 10-μm-thick and multimodal ones 50- and 62.5-μm-thick. The latter are the easiest to apply in the optochemical sensor field, because they allow a proper light transmission at medium and short distances and can be employed with a great number of commercial optical components.

The recognition element can be set out at the distal end of the fiber or sidelong between the core and the cladding.

Capillary waveguides are a special type of guiding structures as they not only transmit light but can also act as flow cells and as the mechanical support for the chemical immobilization of the recognition element [62]. They are multilayered structures, where a liquid or a gas, which is in direct contact with a reactive layer that is deposited in the inner wall or in an extra coating layer, can flow inside. Light can propagate through the capillary cladding if a reflective layer of lower refractive index than the core is used. Then, an evanescent wave optode can be developed. On the other hand, core-based op-

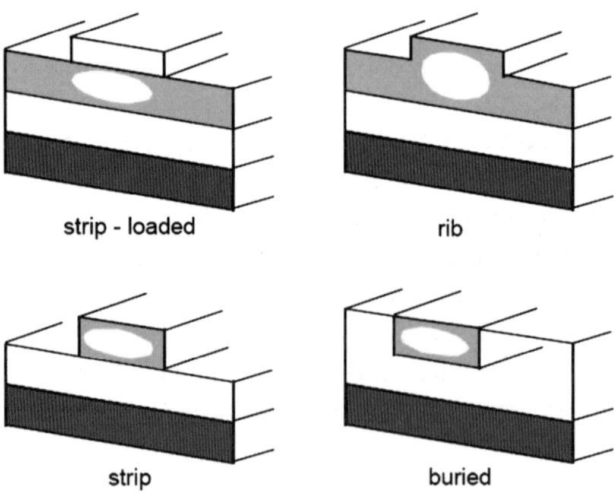

Fig. 6 Types of planar waveguides depending on the processes employed for their fabrication

todes can be developed by using membranes with higher refractive indexes than the surrounding layers, as light propagates though the membrane.

Planar waveguides can be differentiated depending on the fabrication techniques employed: those fabricated by the deposition of layers onto a substrate (such as silica on silicon waveguides) and those fabricated by means of diffusion techniques, by generating an area of a higher refractive index than the rest of the substrate, such as titanium:lithium niobate waveguides (Fig. 6).

The so-called buried waveguides have squarelike or cylindrical geometry depending on the fabrication method. If diffusion or ionic exchange is used, they are often cylindrical, but if they are fabricated by deposition techniques, they can be square. The diffusion of materials in the substrate gives rise to a refractive index gradient in the core and a smooth core–cladding interface. Figure 6 schematically shows the square geometry, whose waveguiding characteristics are similar to that of optical fibers [63].

On the other hand, deposition-based waveguides usually have a squarelike geometry in a strip, a strip-loaded or a rib configuration (Fig. 6). They are fabricated by depositing several layers onto a planar substrate by means of microelectronic processes, covering and etching them until the final geometry is obtained.

2.4.2.3
Waveguides in Integrated Optical Circuits

During the last 50 years, research on microelectronics has overcome the most optimistic expectations regarding achievements as well as new equipment. Integrated optical circuits (IOCs) are still far from microelectronics, but the technology is expected to progress much faster since, instead of developing new equipment, the existing systems can be adapted so as to provide good results in integrated optics.

In order to develop an integrated optical sensor, it has to be taken into account that the main material used in microelectronics is silicon, which absorbs at wavelengths below 1.12 μm. Then, the first step would be the design of waveguides that could operate in the visible region with acceptable absorption losses while keeping their single-mode behavior.

Integrated guiding structures can be based on several optical principles. Among the most interesting, TIR and antiresonant reflecting optical waveguides (ARROW) configurations are highlighted, since they are more versatile, simpler and more robust than the others. The simplest TIR waveguides consist of two layers with different refractive indexes but their main problems arise during the fabrication process and the structure characterization [64].

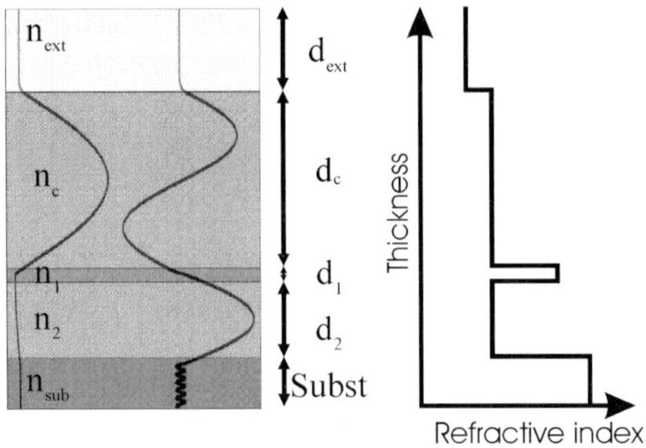

Fig. 7 Fundamental and first mode of an antiresonant reflecting optical waveguide (*ARROW*) waveguide. Layer thickness *d* and refractive indexes *n* are detailed for the external medium (*ext*), the core (*c*), the first cladding (1), the second cladding (2) and the substrate (*sub*)

2.4.2.4
ARROW Structures

ARROW waveguides were firstly developed in 1986 [65]. Their layered structure is presented in Fig 7. As a distinction from TIR waveguides, two different layers are placed just below the core. The refractive index and thickness have to be chosen, for a given working wavelength, so as to provide very high reflections. At the upper air-core interface there is still a TIR. But in the antiresonant pair (d_1 and d_2), a high reflection happens at the antiresonant wavelengths. It could be thought that the fabrication conditions of these structures are extremely strict and that they only work in a very narrow wavelength range. Nevertheless, since the layers are tuned in to antiresonance, the tuning region is broad, providing a high degree of robustness and tolerance to fabrication errors. The major advantage of ARROW structures is that, owing to their principle of operation, the modal properties do not depend on the core thickness, but on the antiresonant pair located just beneath the core. Hence, it is possible to obtain waveguides with a core thickness of the same magnitude as for coupling with optical fibers. Thus, insertion losses are reduced [73].

A brief review of the literature concerning the several materials employed in the fabrication of both TIR and ARROW structures is given in Table 2. The processes employed are completely different, ranging from molecular beam epitaxy to several chemical vapor deposition (CVD) systems, such as low-pressure CVD (LPCVD) or plasma-enhanced CVD (PECVD). As a rule, all suitable materials for ARROWs (and in general for IOCs) should have homogeneous refractive indexes, high mechanical and chemical stability, few

Table 2 Different materials employed in the fabrication of total internal reflection (*TIR*) and antiresonant reflecting optical waveguide (*ARROW*) waveguides

	Substrate	d_c	d_1	d_2	λ (μm)
TIR					
[66]	Si	As_2S_3			1.3
[67]	Si	SiN			1.3
[68]	Si	$Si_{1-x}Ti_xO_2$			1.3
ARROW					
[69]	InP ($n = 3.16$)	InP ($n = 3.16$)	InGaAsP ($n = 3.553$)	InP ($n = 3.16$)	1.55
[70]	Si	SiO_2 ($n = 1.46$)	TiO_2 ($n = 2.3$)	SiO_2 ($n = 1.46$)	0.633
[71]	Si	Si	GeSi	Si	1.33
[72]	Si	SiO_2 ($n = 1.46$)	TiO_2 ($n = 2.3$)	SiO_2 ($n = 1.46$)	0.633
	Ge	NA45 ($n = 1.54$)	Si_3N_4 ($n = 2.0$)	NA45 ($n = 1.54$)	
	GaAs	C7059 ($n = 1.54$)	ZnO ($n = 1.98$)	C7059 ($n = 1.54$)	
	InP		Al_2O_3 ($n = 1.65$)		
			ZrO_2 ($n = 1.92$)		
			Ta_2O_5		

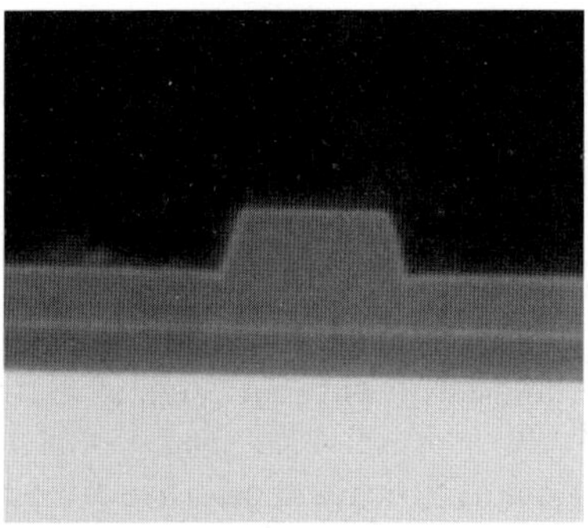

Fig. 8 Photograph of the cross section of an ARROW structure obtained using the technological processes described in the text

defects and impurities and elastic properties compatible with further processing.

The main disadvantage of ARROWs is that the standard structure is polarization-dependent; however, this can be overcome by means of a proper tailoring of the layers. In any case, for sensing applications the advantages of the ARROW configurations outweigh the disadvantages and ARROW configurations are a very good choice for the fabrication of cheap and reliable integrated devices [65].

The structure of an ARROW has the shape of a slab waveguide, i.e., there is only light confinement in the direction orthogonal to the surface of the layers. On the other two axes light is able to propagate freely. This type of waveguide

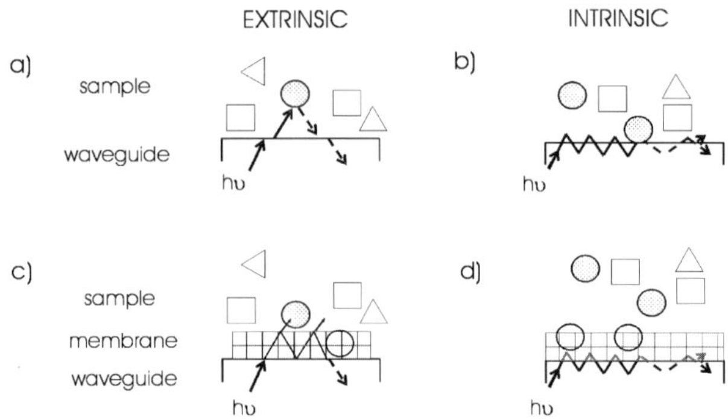

Fig. 9 Four waveguide-based sensors. *Left* Extrinsic sensors **a** in direct analyte measurements or **c** in indirect measurements by means of indicators or other reagents immobilized in membranes. *Right* Intrinsic sensors for measurements of changes in the light guided though the waveguide by **b** the adsorption of the analyte into its surface or by **d** the interaction of the analyte with a recognition phase, which is in direct contact with the waveguide

is rarely used since it does not permit integration of several IOCs in the same substrate. But this drawback is solved by defining a lateral confinement, by way of partially etching the core of the waveguide by forming a rib structure (Fig. 8).

A final protective layer is necessary to insulate devices from the external medium and prevent scratches that could dramatically increase light losses.

2.5 Waveguide-Based Optochemical Sensors

A fundamental classification of optochemical sensors can be done depending on the area where the interaction of light with matter occurs and results in changes in the light guided towards the detector. Light can be allowed to exit the waveguide and be modulated in a separate zone before being directed into either the same or a different waveguide in *extrinsic sensors*. If it continues within the waveguide and is modulated in response to the analyte while it is still being guided, *intrinsic sensors* are developed. Figure 9 depicts four examples of chemical sensors based on waveguides, as well as in cases where the analyte exhibits optical properties such as in cases where immobilized indicators are employed.

2.5.1 Extrinsic Sensors

In terms of the interaction of light with matter, extrinsic sensors can be compared with non-guided radiation sensors because light temporally exits the waveguide to interact directly or indirectly with the analyte or with an active phase before being directed and guided again. This operation principle is derived from the conventional concept of an optochemical sensor. Therefore, waveguides are not really required but offer some advantages: easy miniaturization and the possibility of transmitting information over large distances, using, for instance, optical fibers for luminescence measurements [74, 75], for Raman spectrophotometry [76, 77] or for reflectance [78, 79]. In these situations, the waveguide only acts as the radiation transmission medium.

Other types of sensors are the so-called active waveguides [80] or integrated waveguide absorbance optodes (IWAOs) [81]. They are based on the radiation transmission through the core of a chemically active membrane yielding one part of a light-guiding planar structure. The response mechanism consists of the absorbance/transmittance phenomena of the recognition optode membrane as it interacts with the analyte in the sample and while light is guided through it. Sensors that make use of all the light transmitted though the core/membrane offer some remarkable advantages, such as high sensitivity and an adequate selectivity (regulated by the immobilized reagents), compared with other absorbance sensors derived from waveguides. Hence, they will be discussed in more detailed in due course.

2.5.2
Intrinsic Sensors

The interaction phenomenon of the analyte with radiation is governed or modifies the principle on which light is transmitted through the waveguide. Depending on that interaction, we can differentiate three basic types of intrinsic optical sensors [82]: refractive-index-type sensors, luminescence-based sensors and absorption-type sensors.

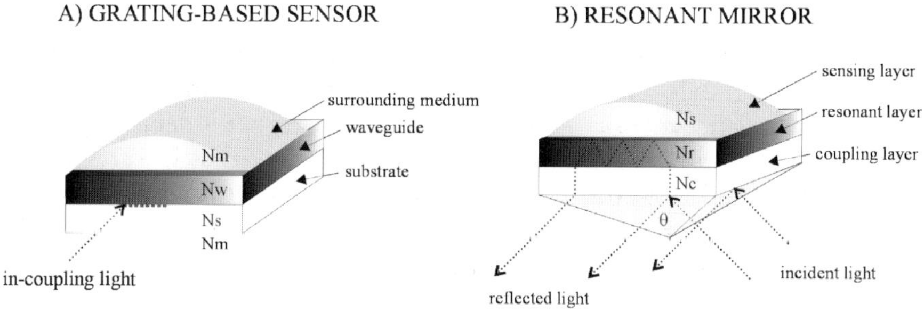

Fig. 10 Different refractive-index-based sensors. **a** A grating coupler, where N_s, N_w and N_m are the refractive indexes of the substrate, the waveguide and the surrounding medium respectively. **b** A resonant mirror, where N_m, N_r and N_c are the refractive indexes of the medium, the resonant layer and the coupling layer

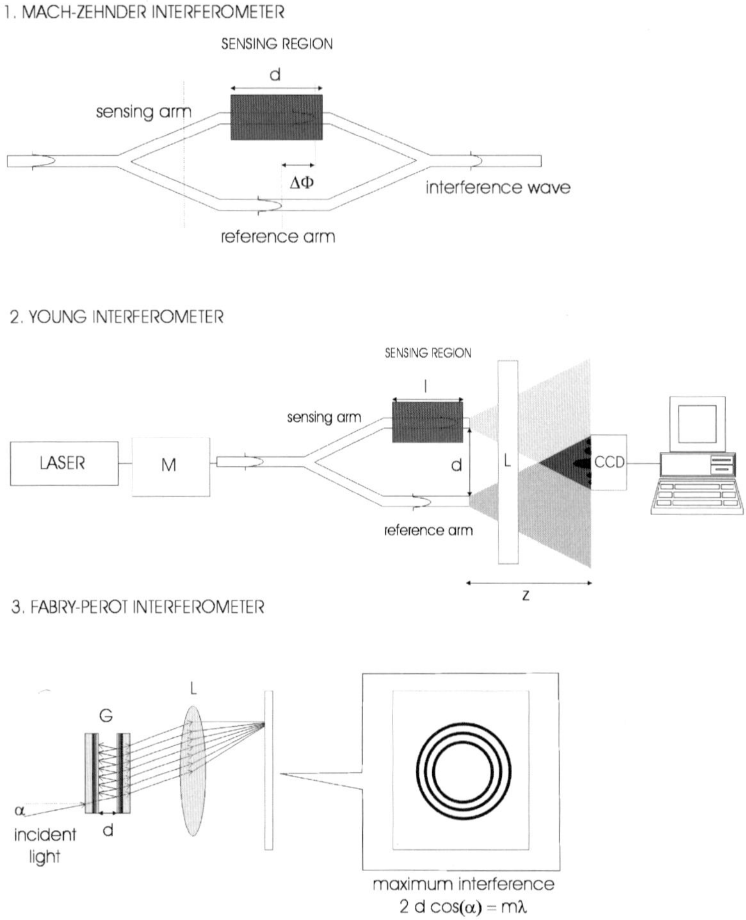

Fig. 11 Geometry of interferometers and basic components. 1 Mach–Zehnder: *d* interaction distance of light and sensing region. 2 Young: *M* microscope objective, *l* length of the sensing window, *d* distance between the parallel channels, *L* lens, *Z* distance between the CCD camera and the end face of the interferometer. 3 Fabry-Perot: *G* parallel mirrors, *d* distance between mirrors or the medium filling the cavity, *L* lens

2.5.2.1 Refractive-Index-Type Sensors

For refractive-index-type sensors, the waveguide is intrinsically the sensing element, and then chemical processes are measured with the variations of the refractive index, *n*, of guided modes. The majority of the various sensor types and published configurations make use of grating couplers (Fig. 10a) [83, 84], resonant mirrors (Fig. 10b) [52, 86–89] or interferometric sensors (Fig. 11) as

Mach–Zehnder [53, 90–94], Young [95–97] and Fabry–Perot [98–100] interferometers.

One of the most widely studied refractive index sensor, especially for the detection of proteins and other bioanalytes, is the SPR-based evanescent-wave-type sensor. In comparison with evanescent-based sensors, the intensity of the field in the recognition layer deposited over the metal is 2 orders of magnitude higher than that deposited over the glass.

These sensors have been used in very concrete applications for gas analysis or analysis of liquid mixtures as well, but following up immunological reactions is an interesting field of application as the interaction of an antigen with an immobilized antibody leads to large changes in the resonance angle and real-time continuous monitoring of affinity reactions can be performed.

Integrated waveguides for SPR have to be manufactured in such a way that the propagation constant matches that of the surface plasmon, which is very difficult to achieve for a given range of refractive indexes, even if an optical fiber is used. Nevertheless some research groups [101, 102] have studied this possibility.

Refractive-index-type sensors present poor specificity in their response and instabilities due to temperature or pressure variations and on the other hand, interferometric sensors are complex, but more sensitive than other integrated chemo-optical evanescent-field-based sensors.

2.5.2.2
Luminescence-Based Sensors

The luminescence techniques can be more selective than the absorption ones as it is possible to distinguish between two different wavelengths (the emission and the excitation one) and because only few molecules are potentially capable of emitting light when relaxing. This fact and the high sensitivity are the reasons why they have been widely applied in the field of optical sensors. If the excitation energy comes from a (bio)chemical reaction, the phenomenon is called (bio)chemiluminescence, meanwhile if it comes from a light source, it is called fluorescence or phosphorescence, depending on whether the excited state where the emission is produced is singlet or triplet.

Different situations can be found: direct determinations, where the luminophore is the analyte itself, or indirect determinations, where the analyte interacts with a luminophore, changing its optical properties, the quenching effect. And different measurement possibilities can be followed: intensity or lifetime of emission.

Even though sensors based on the measurement of fluorescence intensity are the most numerous, the high-intensity light sources employed accelerate the possible photodecomposition of the active molecules. Fluorescence-lifetime-based sensors avoid this problem by the excitation of molecules with light pulses.

Luminescence core-based optodes have been reported as waveguide capillary flow cells in liquid-filled optical fibers [103] or with polymers attached to the inner surface of a glass capillary [104].

Luminescent evanescent wave-based sensors use optical fibers and planar waveguides [105, 106] as light-guiding structures, and they are more complex than the absorbance ones. However, such optodes have been satisfactorily applied to measure fluorescence of indicators or labels for the measurement of gas molecules, proteins or labeled antigen–antibody interactions as well as directly in solution [24, 107] when immobilized in matrices [23, 109].

2.5.2.3
Absorption-Type Sensors

Core-based optodes, such as hollow fibers, and direct and coating-based evanescent wave spectroscopic optodes are included in the classification of absorption-type sensors.

3
Absorbance-Based Integrated Devices

Despite the fact that direct absorbance/transmittance measurements are well established in analytical chemistry owing to the simplicity of the instrumentation and their broad applicability and versatility towards a large number of analytes, most of the reported miniaturized optical devices are based on the measurement of variations of the real part of the refractive index, such as SPR sensors [84, 109–111] or interferometric sensors [94, 112].

Some optical sensors exploit light absorption phenomena and most of them are based on evanescent wave spectroscopy [113, 114]. To avoid problems related to the inherent reduced sensitivity, they require long interaction distances or a great amount of the field out of the waveguide, which in turn makes the response very dependent on variations of the real part of the sample refractive index. Although they have high sensitivity, only a few optical sensors which make use of the guided field in the core of a waveguide have been developed.

The most significant reported absorbance-based miniaturized devices are detailed next.

3.1
Intrinsic Core-Based Optodes

A hollow fiber directly uses the fiber core as the recognition element. This configuration allows much greater interaction of light and the sensing material, so this technique gives sensors with much greater sensitivity. Moreover the optical setup is simple as, once the hollow fiber is made, it can be coupled to commercial optical fibers. Doping the hollow fiber with different sensing materials permits the fabrication of selective and sensitive chemical sensors. Hollow fibers can be fabricated using metal, glass or plastic tubing.

In the previous section, the advantages of using ARROW structures for IOCs, which work in the visible range, were explained. However, for some applications, wavelengths located in the IR range are required. Several attempts have employed low molecular weight chalcogenide glasses (As_2S_3), heavy-metal fluorides, silver halides and sapphire but they have severe drawbacks, such as low laser-damage thresholds, poor chemical and mechanical stability and difficult and expensive fabrication processes [115].

Hollow waveguides have emerged as a very attractive alternative to solid-core IR waveguides because of the inherent advantage of their air core. However, only a few applications in medical and industrial fields have been reported.

They not only enjoy the advantage of high laser power thresholds (transmit wavelengths beyond 10 μm, and ultrashort laser pulses at intensities close to 1014 W cm^{-2}) [116], but they also have low insertion losses, no end reflection, ruggedness and small beam divergence. Their main disadvantages include additional losses on bends and a small numerical aperture. Nevertheless, nowadays they are one of the best alternatives for both chemical and temperature sensing as well as for power delivery in IR laser surgery or in industrial laser systems.

In such waveguides, the core can be simultaneously used for confining the light and the fluid/gas to be measured, thus achieving a high interaction between them. As an example, in optical gas detection, hollow waveguides have already improved the detection limits [117].

Hollow fibers have been tried as liquid-core fibers in spectrophotometry [118] and porous fibers have been well adapted for gas measurements [119, 120]. The latter are made of porous polymers, which can trap indicators or reagents in the matrix with a very high permeability of gases and liquids. For liquid applications, porous silica gel has also been used [121].

Propagation into a hollow waveguide has a high dependence on the geometrical parameters of the waveguide and on the reflection coefficient of the walls. It has to be noted that in order to obtain specular reflections in the waveguide walls the roughness must be as low as possible, reaching values below the working wavelength. Generally, hollow waveguides can be grouped in two categories: if the fiber walls have a refractive index higher than unity (e.g., metallic or dielectric films deposited on the inside of metallic, plastic or gas tubing), they are called leaky hollow waveguides. In contrast, if the material in the inner walls has a refractive index lower than unity (e.g., using a dielectric material that matches this condition at the working wavelength) they are called attenuated total reflection (ATR) hollow fibers.

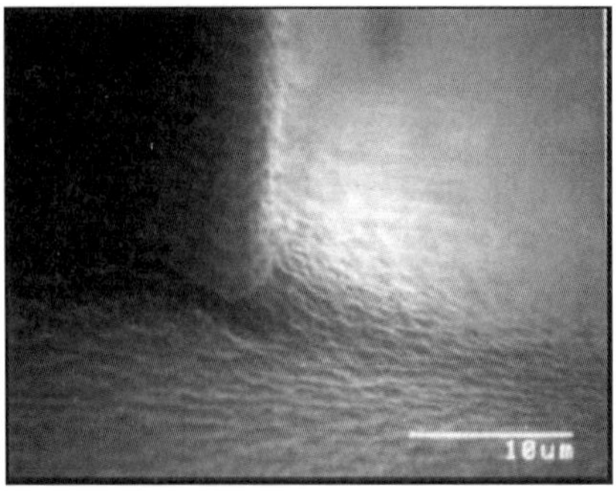

Fig. 12 Scanning electron microscopy (*SEM*) microphotographs of the walls and bottom of a hollow waveguide after the deep reactive ion etching process

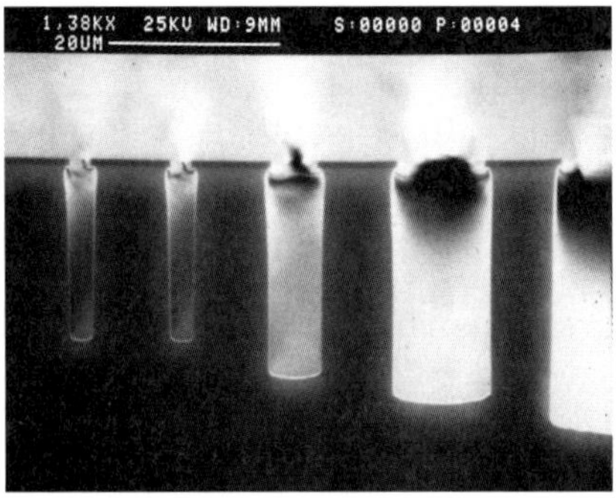

Fig. 13 SEM microphotographs of Si-based hollow waveguides

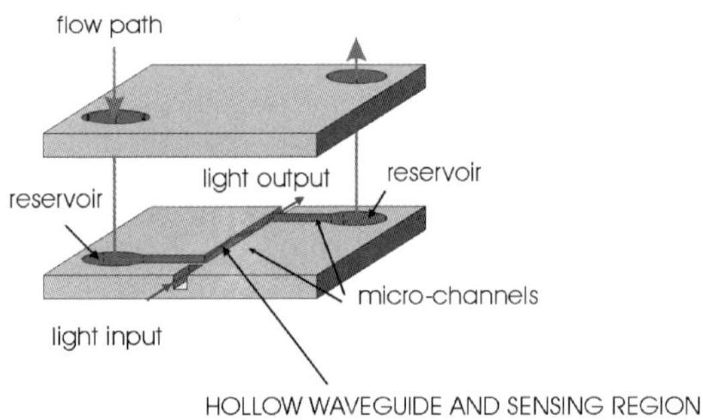

Fig. 14 Standard configuration of hollow-based sensors. Flow is injected from a reservoir to the microchannel that also forms the hollow waveguide. Light interacts with the flow after crossing the input membrane and the signal is read after the output membrane

Hollow planar waveguides have been fabricated by several techniques, including physical vapor deposition and CVD of silver and dielectric layers on metallic substrates. Nevertheless, better results can be obtained by taking advantage of silicon micromachining techniques. Perhaps the most important advantage of silicon hollow waveguides over other hollow structures is the simplicity of the fabrication process. Moreover, hollow planar waveguides do not need a polishing step. Silicon confers a high chemical and mechanical durability to the structure. Then, the combination of the hollow fiber features with the silicon micromachining capability offers an outstanding flexibility for the development of integrated optical devices.

The most important process in the fabrication of silicon hollow fibers is deep reactive ion etching (DRIE). This process is very similar to the standard reactive ion etching (RIE) [122] process but permits us to achieve perfect vertical structures with low roughness. The results shown in Figs. 12 and 13 clearly confirm the required verticality of the walls to assure good confinement can be achieved.

Moreover, since the process is completely complementary metal oxide semiconductor (CMOS) compatible, more complex systems can be developed, with a large variety of components as different as valves, coolers, and photodetectors.

As an example, a proposed integrated sensor based on hollow waveguides is presented in Fig. 14. As can be observed, it consists of a silicon wafer into which the reservoirs (or fluid chambers), together with the fluid channel and the hollow fiber have been defined by a DRIE process. Leakage between these last two structures is prevented by defining a thin chemical membrane between them. The structure is completed by bonding on a second micromachined silicon chip, in such a way that only the input/output reservoirs are exposed to the external medium. Light is injected and crosses the chemically sensitive membrane, which interacts with the flow in the sensing region. After going through the second (output) membrane, light is collected again using an optical fiber or is directly pointed at a photodetector.

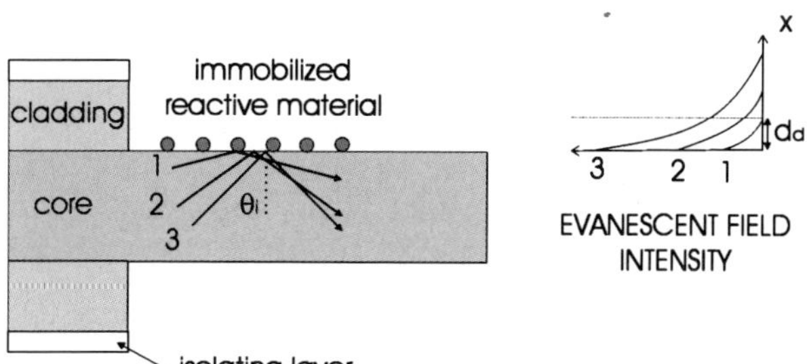

Fig. 15 Light confined in the core interacts with the immobilized reactive material or with the sample though the evanescent field. It decays exponentially in the recognizing medium with the distance from the waveguide-coating interface and its maximum depth (d_d) is obtained at the light incident angle θi. 1, 2 and 3 are different light incident angles

3.2 Intrinsic Direct or Coating-Based Evanescent Wave Spectroscopic Optodes

The evanescent wave phenomenon takes place in the core–cladding interface of a waveguide. Part of the radiation transmitted through the core permeates into the cladding a certain depth (between 50 and 12 000 nm for visible light), which depends on the transmitted wavelength, the incident light angle and the refractive indexes of the core and the cladding media. The evanescent field intensity decays exponentially with the interface distance (Fig. 15). Such sensors are usually obtained by removing a small longitudinal section of the cladding of a standard commercial fiber and replacing it by an absorbing medium. This can be the solution itself or reagents immobilized directly on its surface, in direct evanescent wave spectroscopic optodes, or reagents trapped in a chemically sensitive layer, in coating-based evanescent wave spectroscopic optodes [123].

Absorbance evanescent-based sensors are based on the absorption or dispersion of light outside the core. They rely on light attenuation in the evanescent field following the Beer–Lambert law (ATR sensors), but owing to the low intensity of the field, they offer poor sensitivity. This can be improved because the effective optical path length can be increased, especially when using optical fibers, capillary [62] or planar waveguides [114].

These techniques have usually been applied to the measurement of intrinsic absorptions of the analyte in direct evanescent-wave-based optodes [124], to follow absorption changes of immobilized indicators over the core of a waveguide [125] or as integrated NIR evanescent spectrophotometers [126]. In the latter, a broadband light source from the NIR range is coupled into a planar structure, via optical fibers, and the evanescent wave part of the light field penetrating into the a sensing layer, deposited on the structure, is absorbed at the characteristic frequencies of the molecules trapped on it. The spectral dispersion or absorption of the light coming from the waveguide is then coupled to a conventional NIR spectrometer or a diode array [127].

3.3 Extrinsic Active Waveguides and IWAOs

From the transduction point of view, the optical phenomena that provide sensing in a waveguide configuration may occur in the core layer or outside it. Despite the fact that the former is the more sensitive one because the sensing area is probed by the major part of the energy, as has been stated before, the sensing region is located in the cladding in most optical sensors and is probed by the evanescent field. This kind of sensor requires long interaction distances or a great amount of field out of the waveguide, which in turn makes the response very dependent on variations of the real part of the refractive index.

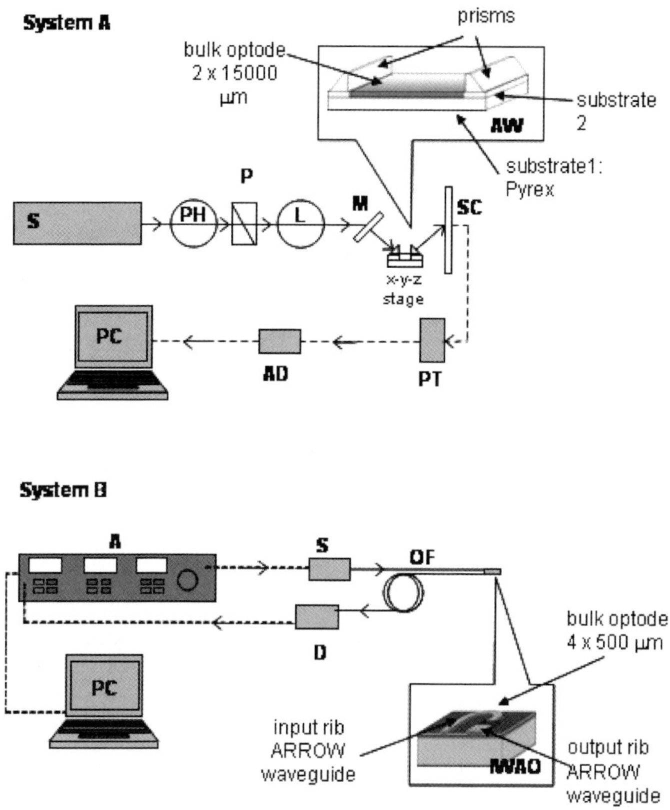

Fig. 16 Optical setup of the extrinsic active waveguides and Integrated waveguide absorbance optodes (*IWAOs*). *System A*: *S* argon laser 488 nm, *PH* pin hole, *P* polarizer, *L* lens, *M* mirror, *SC* screen, *PT* photosensor/photometer, *AD* analog-to- digital converter, *PC* personal computer and *AW* active waveguide composed of a 15 000 µm × 2 µm bulk optode (PVC copolymer), deposited over a 26 mm × 76 mm × 1 mm substrate (Pyrex glass) and *substrate 2* corning 7059 and coupling and decoupling prisms. *System B*: *A* lock-in amplifier, *S* diode laser 778 nm, *OF* optical fibers, *D* p-n photodiode, *PC* personal computer and 30 mm × 20 mm × 0.6 mm IWAO composed of a 20-µm-input ARROW waveguide, a 500 × 4 µm PVC-selective membrane and a 50-µm-output ARROW waveguide

Very few extrinsic core-based sensors have been described. They are focused on polymeric active waveguides (bulk optodes) to develop planar sensors, which possess the advantages of the selective analyte determination and the optical signal propagation in the same structure. This new optical sensor concept consists of light propagation through the absorbing sensing layer, which incorporates a recognition element that leads to changes in the transmitted light. Light has no interaction with the solution matrix, so interference effects are minimized and the active membranes are the same as the ones employed in conventional configurations. Therefore, the same intrinsic characteristics such as selectivity and dynamic range are expected.

It is important to remark that light propagates longitudinally through the waveguides and the membrane and not transversally, so longer path lengths can be employed and higher sensitivities can be achieved without changing the membrane thickness.

Two different approaches with the same philosophy appeared simultaneously in 1997 [80, 128], but their main differences concern the formulation of the active waveguide, the optical setup and its integration, the guiding elements themselves and the technological processes to construct the sensors. Figure 16 represents both measurement systems.

System A is based on the deposition of a specific 2-μm-thick polymeric active membrane over a Pyrex glass substrate. Against the simplicity of the guiding element preparation process, the complex optical setup employed hinders the application possibilities of such an approach [129]. Prisms are employed to couple light from the source to the waveguide and from it to the detector. This coupling technique shows high light losses, requiring the use of powerful emission light sources, such as an argon ion laser, which increases instrumental costs. An added drawback is that such high-intensity light sources worsen the photobleaching of membrane components. Decoupled light is measured with a power meter connected to a computer via an analog-to-digital converter.

System B is based on the deposition of a specific 4-μm-thick polymeric active waveguide over a micromachined silicon circuit of curved ARROW waveguides.

The optical source is a diode laser and fiber-emitter and fiber-detector coupling is accomplished using standard optical fiber connectors. The detector is a PIN photodiode connected to a transimpedance preamplifier and the signal is amplified and filtered using a lock-in amplifier that also tunes the modulation frequency of the laser source. The analytical signal is collected and treated by a PC.

The latter system has been called an IWAO and the results obtained have demonstrated its suitability to be employed as a sensing platform, which confers versatility, robustness and mass-production capabilities besides high sensitivity on conventional bulk optodes. For these reasons it is going to be described in more detail next.

3.4 IWAO Design and Construction

IWAOs have been designed to address some of the weak points concerning absorbance optochemical sensors based on bulk optodes. In order to reach a practical application level, new research is needed to overcome some significant limitations. First, optical transmission signals may be affected by light scattering or matrix absorptions if light crosses the sample solution. On the other hand, bulk optodes usually show long response times to reach equilibrium, which is unfortunately worsened when thicker membranes are used to improve sensitivity. In addition, most of the reagents suffer from photobleaching or leaching.

Besides increasing the dye concentration in the membrane, which is limited by its solubility in the plasticizer, there are two effective alternatives to improve sensitivity: on the one hand, increasing the optical path length and, on the other hand, changing the optically active reagent to another one with higher molar absorptivity. With that in mind, IWAOs are based on radiation transmission through the core of a waveguide.

Such devices consist of microfabricated planar waveguide circuits, based on ARROW structures and on specific chemically active membranes (bulk optodes) deposited in a defined region of the former and yielding one part of the light-guiding planar structure. Concerning the transduction mechanism, it is established by absorbance/transmittance phenomena of the recognition optode membrane as it interacts with a given concentration of the analyte in the sample.

These novel sensors are conceived to optimize four essential characteristics that define a high-quality optochemical sensor: enhanced and sizable sensitivity (depending on the free propagation region length); an adequate selectivity towards a certain analyte (regulated by an ionophore); low response time: and minimization of interference effects caused by the solution matrix. The scheme is presented in Fig. 17.

The basic structure is composed by a rib ARROW waveguide (as the input waveguide), an open region, where the bulk optode is deposited (as the

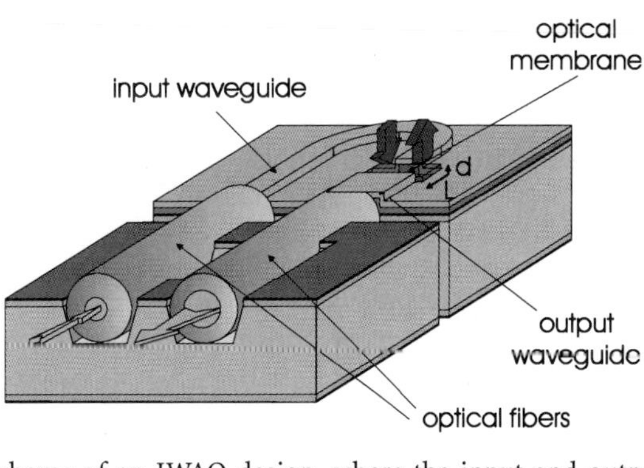

Fig. 17 General scheme of an IWAO design, where the input and output ARROW waveguides and the active membrane and the optical fibers are indicated. Notice that the analyte diffusion direction is transverse to the light transmission direction. *L* membrane length and optical path length, *d* membrane thickness

Scheme	Step properties
	Silicon substrate. One-side polished, N-Type, 4′ diameter, 500 μm thick 2nd cladding: 2μm wet thermal silicon dioxide. n=1.46 1st cladding: 0.38μm LPCVD silicon nitride. n=2.00 Core: 4 μm PECVD silicon oxide. n=1.48
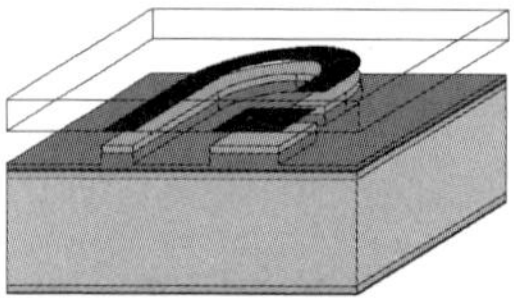	**Definition of the rib structure** Input waveguide: 20 μm Output waveguide: 60 μm Radius: 2, 4, 6 mm Step 2 μm positive photoresist 3.5 μm SiO RIE etching with CHF_3
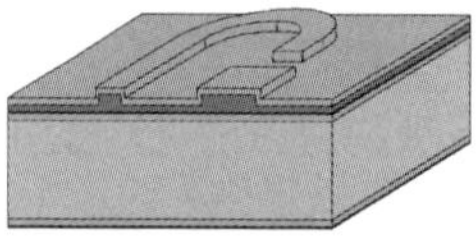	**Protection** 2 μm PECVD silicon oxide. n=1.46
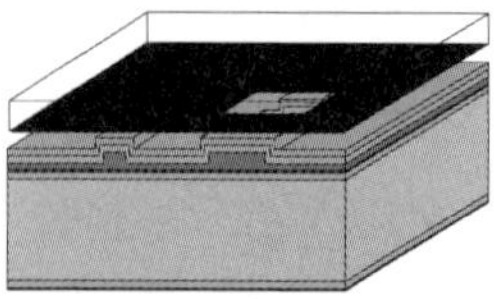	**Protection layer opening** 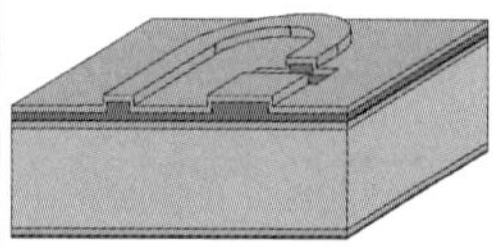 Sensing region: 500 μm Step 2 μm positive photoresist 2 μm SiO_x RIE etching with CHF_3
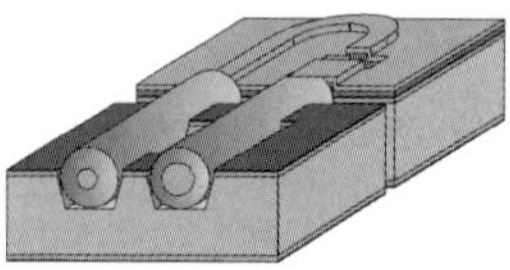	Fiber optics mounting on V- grooves and fiber-waveguide alignment
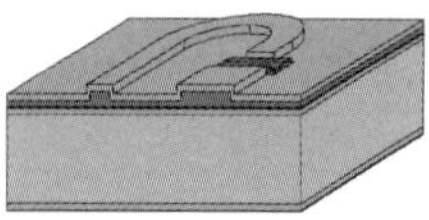	Sensor activation with an ion-selective membrane by spin-coating or a screen printing like technique

Fig. 18 Technological steps used for the fabrication of IWAOs

recognition region), and an output waveguide (as the collection waveguide), all in a continuous arrangement. With this assembly, light from the source is propagated inside the input waveguide to the optical membrane. For certain thickness and refractive indexes of the membrane, light remains guided while simultaneously interacting with the analyte that has diffused in the membrane from the solution and changes its optical properties. Then, the output waveguide collects light and directs it to the detector by means of optical fibers.

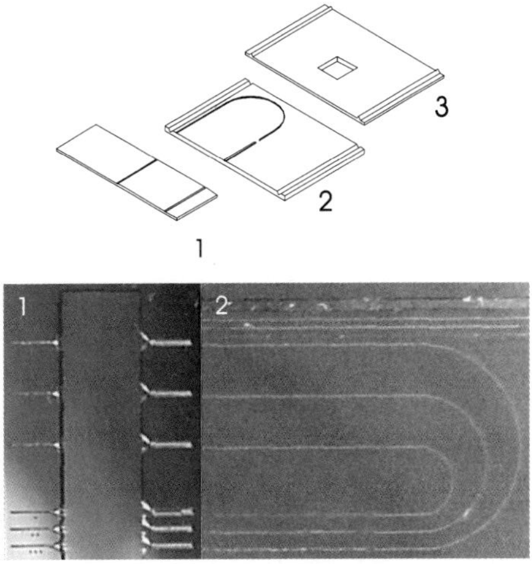

Fig. 19 The sensor can be separated into three main parts. 1 A V-grooved auxiliary support, where the optical fibers are fixed, 2 the exchangeable transducer based on curved ARROW waveguides, where the membrane is deposited and 3 a connection platform, which allows the alignment of the transducer waveguides to the optical fibers, which are fixed in the auxiliary support by the aid of two microfabricated channels at both sides of the surface

ARROW waveguides are constructed using CMOS compatible processes [130, 131] over a silicon wafer. The technological steps that allow the integrated absorption sensors to be obtained are schematized in Fig. 18. A first mask defines the waveguide width, while a second one allows the protective external cladding to be opened to obtain the sensing region. The SiO_2 substrate–cladding layer is made by thermal oxidation of the silicon wafer. The Si_3N_4 ARROW layer is deposited using LPCVD [132] at 800 °C and the upper cladding and core layers correspond to different stoichiometry silicon oxide layers deposited using PECVD [132] at 300 °C. The rib structure and the opened region are fabricated by the RIE processes [133] used in integrated optics.

A first generation of optochemical absorption sensors based on rectilinear waveguides was successfully fabricated and characterized [81, 134]. The main technological drawback for those designs is that input and output waveguides and optical fibers were placed at both sides of the device, making difficult the application of the sensors in field, without the aid of flow cells and complicating the possibilities to obtain devices that can be easily reactivated. A second generation of IWAOs is proposed in order to overcome the double-polishing step required for the fiber–waveguide alignment and to provide optical sensors of disposable or easily reactivated transducers, to make them suitable for in situ monitoring [39]. However, the required polishing procedure to align the sensor with the V-grooved platform and the required high precision of the dimensions, in order that transducers can be exchangeable, are still the main problems. To solve them, input and output waveguides are placed at the same side of the transducer. The input waveguide is bent 180° before the sensing region. After that, power is collected by a wider straight waveguide.

To easily align waveguides with optical fibers a three-pieced device is constructed. The main idea is to define a setup that allows changing and replacing sensors without the fiber–waveguide alignment step. The appearance of these structures is shown in Fig. 19. Notice that the sensing transducer has three different open regions together with bent waveguides used for testing purposes. Therefore, it would be possible to perform three measurements of the same parameter or to develop a multisensor if some changes in the optical setup are made.

3.4.1
IWAOs Features

As mentioned at the beginning of the previous section, some instrumental and methodological problems with optical sensors exist when performing absorbance measurements analogous to those made with conventional spectroscopic techniques. However, IWAOs present some technological advantages in comparison with other optical sensors and have some suitable analytical features to be exploited.

3.4.2
Technological Advantages

In relation to other waveguides, ARROW structures offer some advantages by themselves for the development of chemical sensors.

- As the modal propagation of light in the transversal direction boils down to a 4-μm-thick waveguide, it is easily coupled to a single-mode fiber and fewer losses are obtained.
- The operation fundamentals of these waveguides present high tolerance to variations in the fabrication parameters. This means that they keep on performing as single-mode waveguides even with little changes of the core refractive index.
- They are technologically easier to construct than conventional planar waveguides and the fabrication processes are more economic and rugged. Conventional waveguides need thicker substrates (10 times thicker) than ARROW structures to reach equivalent losses in this layer. Moreover, they are less suitable to develop optical sensors based on the core propagation, because if the membrane refractive index was the same, a thickness of about 100 nm would be needed to achieve single-mode behavior in the transversal direction to the layers. ARROWs have a 4-μm-thick core,

which is an appropriate optical membrane thickness to achieve short response times using bulk optodes.

- Cutting the waveguide to define an area where the optical membrane can be deposited and act as a waveguide at the same time is only possible with ARROWs. Conventional waveguides require a membrane refractive index higher than the refractive indexes of the surrounding layers to assure the light transmission in the transversal direction to layers and avoid radiation losses in the substrate. This is a critical requirement because the real part of the refractive index changes with the imaginary part (absorbance). Moreover, in a conventional single-mode waveguide, the core and cladding refractive indexes must be very similar, and any variation in membrane refractive index to values lower than the refractive index of the substrate would lead to high radiation losses.

Concerning the coupling technique, IWAOs are based on the direct alignment of fibers and waveguides using V-grooved channels. This is the so-called end-fire coupling. V-grooved channels are constructed with standard techniques of micromechanization over silicon substrates [135]; therefore, problems such as using complex optical setups to couple light into the membrane or employing difficult fabrication processes as in the case of grating couplers are avoided [136].

3.4.3 Sensitivity and Response Time

Regarding the analytical features of IWAOs, the main one is that sensitivity can be improved without simultaneously increasing the response times to achieve the steady-state signal. This configuration allows an analyte diffusion direction transverse to the light transmission, so the response time is independent of the optical path length.

As is well known, the response of a bulk optode follows the Beer–Lambert law, which linearly relates the absorbance to the concentration of an absorbent species (C), its molar absorptivity (ε) and the optical path length (d):

$$A = \varepsilon d C . \quad (1)$$

Therefore, the possible strategies to increase sensitivity are to change the active optical species for another one with higher molar absorptivity, to increase its concentration or to increase the optical path length.

The first two alternatives are impracticable as very few molecules are suitable for use as chromoionophores and the ones that could be employed present a limited solubility in the very lipophilic membrane matrix. In conclusion, the feasible option would be to increase the optical path length.

The membrane thickness usually defines the optical path length of conventional optical sensors based on absorbance measurements. If in order to improve sensitivity, the optical path length is increased, the response time is impinged on as well. The relation of membrane thickness and response time can be theoretically explained if one assumes that the response rate is controlled by the analyte diffusion within the membrane.

Taking into account that D_{m} (centimeters squared per second) is the medium diffusion coefficient from all the mobile species in a membrane of thickness d (centimeters), the solution of the diffusion equation gives the following mathematical expression [137]:

$$A(t) = A(\infty) - \{A(\infty) - A(0)\} \times \frac{8}{\pi^2} \sum_{m=0}^{\infty} \frac{1}{(2m+1)^2} \exp\left(-\frac{(2m+1)^2 \pi^2 D_{\mathrm{m}} t}{4d^2}\right), \qquad (2)$$

where t is time, $A(t)$ is the total amount of species that diffuse in time t, m is an integration counter, D_{m} is the diffusion coefficient and d is the membrane thickness.

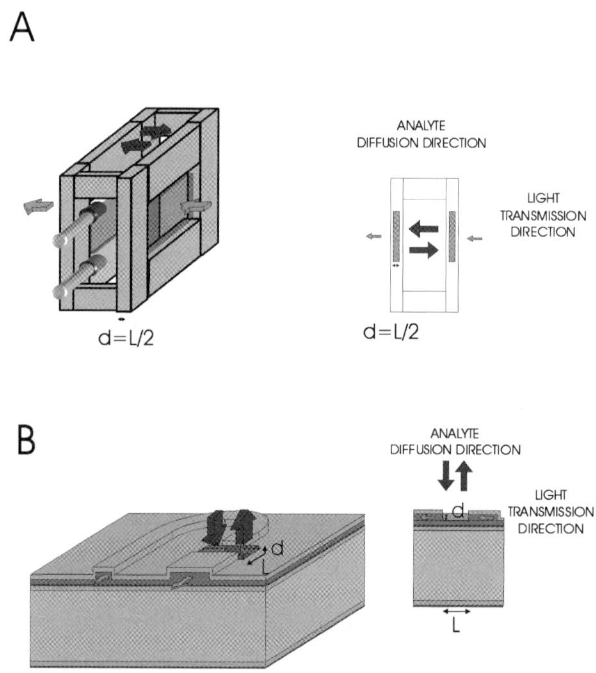

Fig. 20 **a** Conventional configuration, where the optical path length is related to the membrane thickness by $L = 2d$. **b** IWAO, where the optical path length (L) is independent of the membrane thickness (d)

One can infer from Eq. 1 that the time to achieve 95% of the steady-state signal is

$$t_{95\%} = 1.13\frac{d^2}{D_m} . \quad (3)$$

From Eqs. 1 and 3, one can conclude that an increase in the membrane thickness leads to a proportional rise of sensitivity and a response time increase, related to the square of the membrane thickness, if a conventional configuration is used.

On the other hand, in IWAOs the optical path length is not defined by the membrane thickness but by the length of the open region of the waveguiding circuit. Figure 20 shows both configurations. In IWAOs the optical path length, that is the open region length (L), can be incremented in order to improve sensitivity, while the membrane thickness (d) remains unaltered, and then the response times are not affected.

A comparison of the theoretical results obtained with these equations and taking into account $D_m = 1 \times 10^{-8}\ cm^2\ s^{-1}$, $C = 1 \times 10^{-3}\ mol\ kg^{-1}$ and $\varepsilon = 1 \times 10^5\ kg\ mol^{-1} cm^{-1}$ gives an example of the results that one can expect using IWAOs in regard to sensitivity and response time. If the optical path length is increased in both configurations (d for the conventional and L for the IWAO) in order to improve sensitivity, the response time (t_r) is expected to be increased only in the conventional one. It directly depends on the membrane thickness (d). In the case of using an IWAO, only the path length (L), which is defined by the length of the open area, is changed, so the response time is seen to remain unaltered, as the membrane thickness remains the same. The theoretical results predict the possibility of increasing the sensitivity by 100 times, using an open area of 500-μm length and maintaining reduced response times, in comparison with a conventional optical sensor with the same membrane thickness.

Moreover, theoretical studies based on the radiation transmission in ARROW structures demonstrate that light remains confined in the membrane and does not exit, so the interaction of light with the solution matrix is avoided. In this way, any possible interference caused by residual absorbance, turbidity or refractive index changes of the solution matrix is prevented.

3.4.4 Applications

Different optical membranes have been applied to IWAOs to validate their properties. They consist of the well-studied bulk optode membranes. As stated in the "Introduction," they contain a selective ionophore, a lipophilic ionic salt to maintain the electroneutrality and a second ionophore with spectroscopic characteristics, denominated chromoionophore. Such optodes rely on concentration changes within the bulk of the sensing membrane and fol-

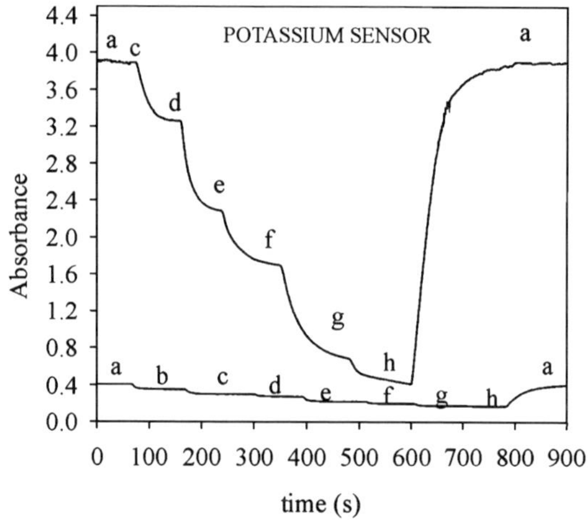

Fig. 21 Comparison of two calibration experiments obtained in the same conditions with a conventional configuration and a potassium IWAO at $\lambda = 670$ nm. *a* Acetic/acetate buffer 0.05pH = 5.55, and KCl solutions, *b* 1×10^{-5} M, *c* 1×10^{-4} M, *d* 1×10^{-3} M, *e* 1×10^{-2} M, *f* 1×10^{-1} M, *g* 5×10^{-1} M, *h* 1 M

low the ion-exchange mechanism between the membrane and the aqueous solution.

They sense the analyte in an aqueous solution by extracting it into the membrane to form a positively charged complex with the ionophore used. To preserve electroneutrality in the membrane, the protonated chromoionophore (CH^+) sheds its proton to the aqueous phase and the absorbance measured decreases. More detailed mechanistic descriptions can be found in the literature [137–141].

Simple IWAO sensing schemes have been designed following the optical response towards different analyte activities depending on the ionophore employed and keeping constant the solution pH.

3.4.4.1
Potassium-Selective IWAO

The membrane used to activate this potassium-selective IWAO [134] consists of a potassium bulk optode based on 0.5 wt % chromoionophore ETH 5294, 1.0 wt % ionophore valinomycin, 0.5 wt % ionic additive potassium tetrakis(4-chlorophenyl)borate (KtpClPB), 31.0 wt % polymer PVC, 67.5 wt % organic solvent and plasticizer bis(2-ethylhexyl)sebacate (DOS) [142]. This commercially available optode not only acts as an example of the development of an enhanced ion-selective IWAO, but also serves to validate the previously remarked features, because results can be compared with the ones obtained with membranes of the same composition and thickness in a conventional configuration [143]. Taking into account that the conventional device employed for comparison purposes, which contains two membranes and has a path length corresponding to twice the thickness of each membrane, the results demonstrate the fact that the sensitivity of the IWAO is much higher although the conventional device has a larger total membrane thickness. The sensitivity expressed as the slope of the calibration curves is -0.86 AU dec^{-1} using the IWAO and -0.03 AU dec^{-1} using the conventional instrumentation. Whereas the total absorbance change in a conventional configuration is 0.05 AU, the total absorbance change achieved using an IWAO is 3.49 AU (Fig. 21). The dynamic range considered is from 1×10^{-4} to 1 M KCl concentration range using an IWAO, similar to that obtained with a conventional configuration (1×10^{-5} -1×10^{-1} M). In short, the absorbance change and sensitivity achieved using a conventional configuration is multiplied by using an IWAO, while the response times for both devices are comparable.

3.4.4.2 pH IWAO

The response characteristics of an ion-selective optode rely on the formation constant of the ionophore–analyte complex and the acidity constant of the chromoionophore in the mediim. Therefore, for an ionophore of a specific target analyte, it would be useful to select a chromoionophore that gives the best response characteristics.

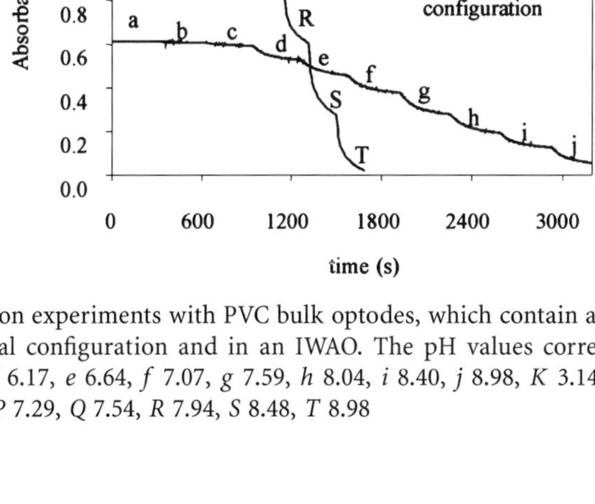

Fig. 22 Calibration experiments with PVC bulk optodes, which contain a ketocyanine dye in a conventional configuration and in an IWAO. The pH values correspond to *a* 3.09, *b* 4.04, *c* 5.10, *d* 6.17, *e* 6.64, *f* 7.07, *g* 7.59, *h* 8.04, *i* 8.40, *j* 8.98, *K* 3.14, *L* 5.09, *M* 5.99, *N* 6.44, *O* 6.97, *P* 7.29, *Q* 7.54, *R* 7.94, *S* 8.48, *T* 8.98

For this reason, the most studied application has been the use of optically active pH indicators as H^+-selective ionophores to develop pH-selective optical sensors [38]. First, the optical and analytical characteristics in PVC membranes of a series of novel neutral pH indicators, a class of ketocyanine dyes [144], were studied in a conventional configuration. The absorption maxima of the dyes are located in the NIR region, which is very appropriate for applications using telecommunications components and the membranes show good performance during calibrations in the pH range 3–8.5.

The studies prove the suitability of ketocyanine dyes as chromoionophores in plasticized PVC membranes and exploit their features by using them as recognition reagents of IWAO devices to provide miniaturized sensors with an enhanced performance.

Two representative calibration experiments obtained using both configurations are shown in Fig. 22. As can be observed, the great advantage offered by the IWAO is the possibility of attaining high sensitivity without increasing the dye ratio or the membrane thickness. The slopes of the experimental calibration curves are at least doubled. Sensors show good repeatability of the measurement, because the values are obtained with an acceptable relative standard deviation (RSD) (pH 6.07: 1.50 AU, RSD = 0.8%; pH 7.08: 1.35 AU, RSD = 0.4%).

To verify that the sensitivity increase is not related to the use of thicker membranes, the response times are calculated for both configurations and then compared. The results show that using the IWAO, the average of the response times calculated for every step change (usually 0.2 AU) are of the same order as or lower than the ones obtained using the conventional configuration.

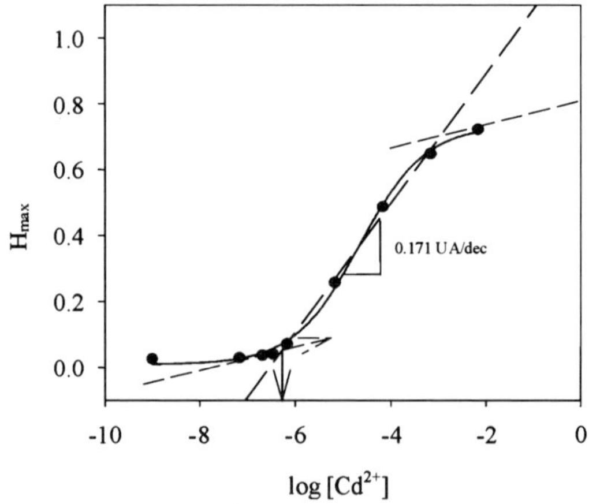

Fig. 23 Calibration experiment for the determination of cadmium. Detection limit and sensitivity are graphically evaluated

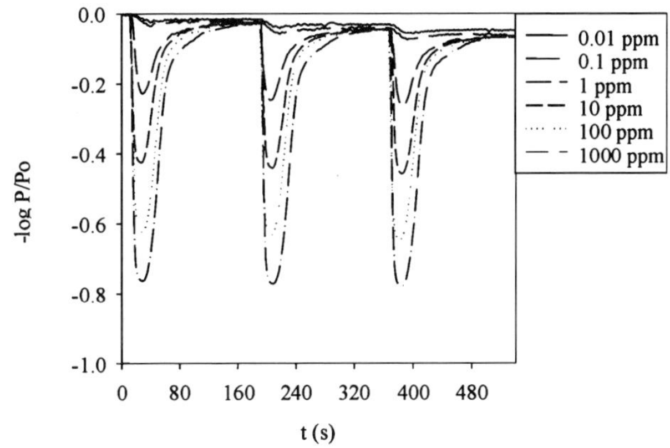

Fig. 24 Recorder output of the flow injection analysis system for triplicate 300 μL´sample solutions of $Cd(NO_3)_2$ at a flow rate of 2.5 mL min^{-1}

The IWAO proves to be a very promising alternative to the conventional optical sensors, as well as those based on optical fibers such as the conventional flow cell configuration. Fully reversible, reproducible, fast and sensitive bulk optodes are obtained.

The combination of ketocyanine dyes with commercial ionophores makes the design of ion-selective optodes feasible, and the use of the IWAO permits increased sensitivity using simple and low-cost equipment so that analysis of a target analyte can be performed in different application fields.

3.4.4.3 Cadmium-Selective IWAO

The first prototype of a technologically improved IWAO was developed and tested with a membrane based on a new H^+-selective ketocyanine dye and a commercial cadmium ionophore [39]. Its incorporation in an IWAO allows a highly sensitive and portable optical system to be obtained for an situ chemical analysis as well. The authors propose a flow injection analysis (FIA) system for the determination of cadmium in water samples using a cadmium-selective IWAO, as an alternative method to the ones generally used in analytical control laboratories. It permits enhanced sensitive signals in short response times by taking advantage of the very thin membranes deposited over the circuit.

The membrane consists of 0.3 wt % of a ketocyanine dye 5ee, 0.3 wt % of KtpClPB as an ionic additive, 12.8 wt % of ETH 5435 as an ionophore, 28.1 wt % of PVC and 58.5 wt % of DOS.

As a compromise between sampling rate and analytical sensitivity, an injection volume of 300 μL and a flow rate of 2.1 mL min^{-1} are chosen. With this optimized system a sampling throughput of 55 h^{-1} is attained for a 0.01–1000-ppm cadmium range. The buffer chosen is a 0.1 M acetic/acetate solution at pH 5 to obtain the lowest detection limit (1.6×10^{-7} M cadmium concentration, approximately 20 ppb) and the highest sensitive response (0.171 AU dec^{-1}) in a reasonable measuring range for environmental applications (Fig. 23).

The obtained FIA peaks are relative to the base line signal, determined by the pH of the conditioning system. In comparison to conventional transmission configurations, slopes of 0.063 AU dec^{-1} are obtained using the same membrane composition. Figure 24 shows the typical signal obtained with a FIA system.

In general, the selectivity over alkali and alkaline-earth metals and transition metals is good with the exception of lead and copper ions. Selectivity over lead is sufficient for environmental sensing and good enough taking into account that usually this ionophore (ETH 5435) is reported as a lead ionophore [145, 146]. However, copper ion irreversibly interferes in the response.

In conclusion, the proposed technologically improved IWAO made evident its appropriateness as a versatile platform for the development of ion-selective optical sensors, depending on the membrane composition, which also permits disposable transducers to be obtained.

Actually, investigations of IWAO devices are focused on improving ion-selective membranes as well as giving ruggedness and portability to the optical system for the future transference of technology. Even though advantage is taken of the technology employed , research is centered on developing optical sensors based on other optical phenomena such as fluorescence.

References

1. Capitán-Vallvey LF, Fernández Ramos MD, Al-Natsheh M (2003) Sens Actuators B 88:217
2. Schult K, Katerkamp A, Trau D, Grawe F, Cammann K, Meusel M (1999) Anal Chem 71:5430
3. Boisdé G, Harmer A (1995) Chemical and biochemical sensing with optical fibers and waveguides (eds) Artech House, London
4. http://www.photonic-products.com/products/sanyo_violet_laser_diodes/sanyo_violetblue.htm
5. http://www.roithner-laser.com/violet_LD.html

6. Orellana G, García-Fresnadillo D (2004) In: Narayanaswamy R, Wolfbeis OS (eds) Springer series on chemical sensors and biosensors, vol 1, chap 13. Springer, Berlin Heidelberg New York
7. Patonay G, Casay GA, Lipowska M, Strekowski L (1993) Talanta 40:935
8. Bakker E, Lerchi M, Rosatzin T, Rusterholz B, Simon W (1993) Anal Chim Acta 278:211
9. Lindauer H, Czerney P, Mohr GJ, Grummt UW (1994) Dyes Pigments 26:229
10. Kuznetsov VV, Yakunina IV (1995) Mendeleev Commun 52
11. Citterio D, Rásonyi S, Spichiger UE (1996) Fresenius J Anal Chem 354:836
12. Lobnik A, Oehme I, Murkovic I, Wolfbeis OS (1998) Anal Chim Acta 367:159
13. Delmarre D, Méallet-Renault R, Bied-Charreton C, Pasternack RF (1999) Anal Chim Acta 401:125
14. Wang E, Zhu L, Ma L, Patel H (1997) Anal Chim Acta 357:85
15. Wallington S-A, Labayen T, Poppe A, Sommerdijk NAJM, Wright JD (1997) Sens Actuators B 38–39:48
16. Tarazi L, George A, Patonay G, Strekowski L (1998) Talanta 46:1413
17. O'Neill S, Conway S, Twellmeyer J, Egan O, Nolan K, Diamond D (1999) Anal Chim Acta 398:1
18. Balogh IS, Maga IM, Hargitai-Tóth A, Andruch V (2000) Talanta 53:543
19. Geddes CD, Lakowicz JR, Rosenfeld MY (2002) J Fluorescence 12:119
20. Huang HM, Wang KM, Xiao D, Yang RH, Yang XH (2001) Anal Chim Acta 439:55
21. Chan WH, Yang RH, Wang KM (2001) Anal Chim Acta 444:261
22. Narayanaswamy R, Brook TE (1997) Sens Actuators B 39:195
23. MacCraith BD, McDonagh CM, O'Keeffe G, Keyes ET, Vos J (1993) Analyst 118:385
24. Choquette SJ, Locascio-Brown L, Durst RA (1992) Anal Chem 64:55
25. Braun S, Shtelzr S, Rappoport S, Avnir D, Ottolenghi M (1992) J Non-Cryst Solids 147–148:739
26. Wu S, Ellerby LM, Cohan JS, Dunn B, El-Sayed MA, Valentine JS, Zink JI (1993) Chem Mater 5:115
27. Wang R, Narang U, Prasad PN, Bright FV (1993) Anal Chem 65:2671
28. Koncki R, Mohr GJ, Wolfbeis OS (1995) Biosens Bioelectron 10:653
29. Wang Y, Robitt DR (1994) Anal Chim Acta 298:105
30. Kawabata Y, Kamichika T, Imasaka T, Ishibashi N (1990) Anal Chem 62:2054
31. Zhujun Z, Seitz R (1984) Anal Chim Acta 160:47
32. Leiner MJP, Wolfbeis OS (1991) In: Wolfbeis OS (ed) Fiber optic chemical sensors and biosensors, vol 1. CRC, Graz, p 364
33. Oehme I, Prokes B, Murkovic I, Werner T, Klimant I, Wolfbeis OS (1994) Fresenius J Anal Chem 350:563
34. Suzuki K, Ohzora H, Tohda K, Miyazaki K, Watanabe K, Inoue H, Shirai T (1990) Anal Chim Acta 237:155
35. Encinas, C (2003) PhD thesis, Universidad Autónoma de Barcelona
36. Citterio D, Jenny L, Rásonyi S, Spichiger UE (1997) Sens Actuators B 38–39:202
37. Lehmann H, Schwotzer G, Czerney P, Mohr GJ (1995) Sens Actuators B 29:392
38. Puyol M, Miltsov S, Salinas I, Alonso J (2002) Anal Chem 74:570
39. Puyol M, Salinas I, Garcés I, Villuendas F, Llobera A, Dominguez C, Alonso J (2002) Anal Chem 74:3354
40. Freiner D, Kunz RE, Citterio D, Spichiger UE, Gale MT (1995) Sens Actuators B 29:277
41. Lerchi M, Bakker E, Rusterholz B, Simon W (1992) Anal Chem 64:1534
42. Preininger C, Mohr GJ (1997) Anal Chim Acta 342:207

43. Demuth C, Spichiger UE (1997) Anal Chim Acta 355:259
44. Citterio D, Jenny L, Spichiger UE (1998) Anal Chem 70:3452
45. Malins C, Landl M, Simon P, MacCraith BD (1998) Sens Actuators B 51:359
46. Kvasnik F, McGrath AD (1989) Proc SPIE Int Soc Opt Eng 172:75
47. http://www.iti.northwestern.edu/tdr/tdr2001/proceedings/
48. Ramo S, Whinnery JR, Van Duzer T (1994) Fields and waves in communication electronics. Wiley, New York
49. Sudbo AS (1992) J Lightwave Technol 10:418
50. Snyder AW, Love DJ (1983) Optical waveguide theory. Chapman & Hall, New York
51. Okamoto W (2000) Fundamentals of optical waveguides. Academic, San Diego
52. Lukosz W, Tiefenthaler W (1988) Sens Actuators B 15:273
53. Prieto F, Llobera A, Jiménez D, Domínguez C, Calle A, Lechuga LM (2000) J Lightwave Technol 18:966
54. Heideman RG, Kooyman RPH, Greve J (1993) Sens Actuators B 10:209
55. Luff BJ, Wilkinson JS, Piehler J, Hollenbach U, Ingenhoff J, Fabricius N (1998) J Lightwave Technol 16:583
56. Veldhuis GJ, Parriaux O, Hoekstra HJWM, Lambeck PV (2000) J Lightwave Technol 18:677
57. Vinchant JF, Pagnod-Rossiaux P, Le Bris J, Goutelle A, Bissessur H, Renaud M (1994) IEEE Phot Tech Lett 6:1347
58. Bayle JJ, Mateo J (1996) 5th international conference on plastic optical fibers and applications, Paris, France
59. Philip-Chandy R, Scully PJ, Eldridge P, Kadim HJ, Gérard Grapin M, Jonca MG, D'Ambrosio MG, Colin F (2000) IEEE J Quantum Elect 6:764
60. Ueberfeld J, Parthasarathy N, Zbinden H, Gisin N, Buffle J (2002) Anal Chem 74:664
61. Flannery D, James SW, Tatam RP, Ashwell GJ (1997) Opt Lett 22:567
62. Weigl BH, Wolfbeis OS (1994) Anal Chem 66:3323
63. Kapila D, Plawsky JL (1995) Chem Eng Sci 50:2589
64. Hunsperger RG (1995) In: Hunsperger RG (ed) Integrated optics: theory and technology. Springer, Berlin Heidelberg New York, p 32
65. Duguay MA, Kokubun Y, Koch TL (1986) Appl Phys Lett 49:13
66. Baba T, Kokubun Y (1992) IEEE J Quantum Elect 28:1689
67. Baba T, Kokubun Y (1989) IEEE Phot Tech Lett 1:232
68. Suhara T, Handa Y (1982) Appl Phys Lett 40:120
69. Stuitus W, Streifer W (1977) Appl Opt 16:3218
70. Takato N, Yasu M (1986) Electron Lett 22:321
71. Kubica J (1992) J Lightwave Technol 16:767
72. Mao ZM, Huang WP (1993) J Lightwave Technol 11:1183
73. Garcés I, Villuendas F, Vallés J, Domínguez C, Moreno M (1996) J Lightwave Technol 14:798
74. Kawabata Y, Tahara R, Kamichika T, Imasaka T, Ishibashi N (1990) Anal Chem 62:1528
75. Ertaş N, Akkaya EU, Ataman OY (2000) Talanta 51:693
76. McCreery R, Fleischman M, Hendra P (1983) Anal Chem 55:146
77. Dao NQ, Jouan M (1993) Sens Actuators B 11:147
78. Andres RT, Sevilla F (1991) Anal Chim Acta 251:165
79. Kirkbright GF, Narayanaswamy R, Welti NA (1984) Analyst 109:15

80. Hisamoto H, Kim K, Manabe Y, Sasaki K, Minamitani H, Suzuki K (1997) Anal Chim Acta 342:31
81. Garcés N, Villuendas F, Salinas I, Puyol M, Alonso J, Domínguez C, Llobera A (1999) Sens Actuators B 60:191
82. Domínguez C, Rodríguez JA, Lechuga LM (2003) In: Alegret S (ed) Comprehensive analytical chemistry, vol XXXIX. Elsevier, Amsterdam, p 541
83. Ramsden JJ (1998) Biosens Bioelectron 13:593
84. Dübendorfer J, Kunz RE, Jobst G, Moser I, Urban G (1998) Sens Actuators B 50:210
85. Brandenburg A, Gombert A (1993) Sens Actuators B 17:35
86. Skládal P, Deng A, Kolář V (1999) Anal Chim Acta 399:29
87. Fisher MI, Tjärnhage T (2000) Biosens Bioelectron 15:463
88. Lenney JP, Goddard NJ, Morey JC, Snook, RD, Fielden PR (1997) Sens Actuators B 38–39:212
89. Hulme J, Malins C, Singh K, Fielden PR, Goddard NJ (2002) Analyst 127:1233
90. Fabricius N, Gauglitz G, Ingenhoff J (1992) Sens Actuators 7:672
91. Lechuga LM, Lenferink ATM, Kooyman RPH, Greve J (1995) Sens Actuators B 24–25:762
92. Schipper EF, Bergevoet AJH, Kooyman RPH, Greve J (1997) Anal Chim Acta 341:171
93. Kempen LU, Kunz RE (1997) Sens Actuators B 38–39:295
94. Qi Z, Matsuda N, Itoh K, Murabayashi M, Lavers CR (2002) Sens Actuators B 81:254
95. Brandenburg A (1997) Sens Actuators B 38–39:266
96. Brynda E, Houska M, Brandenburg A, Wikersta A (2002) Biosens Bioelectron 17:665
97. Ymeti A, Kanger JS, Wijn R, Lambeck PV, Greve J (2002) Sens Actuators B 83:1
98. Woodruff SD, Yeung ES (1982) Anal Chem 54:1174
99. Correia JH, Bartek M, Wolffenbuttel RF (1999) Sens Actuators A 76:191
100. Bartek M, Correia JH, Wolffenbuttel RF (1999) J Micromech Microeng 9:162
101. Harris RD, Wilkinson JS (1995) Sens Actuators B 29:261
102. Ctyroký J, Homola J, Lambeck PV, Musa S, Hoekstra HJWM, Harris RD, Wilkinson JS, Usievich B, Lyndin NM (1999) Sens Actuators B 54:66
103. Fujiwara K, Simeonsson JB, Smith BW, Wineforder JD (1988) Anal Chem 60:1065
104. Kieslinger D, Draxler S, Trznadel K, Lippitsch ME (1997) Sens Actuators B 38–39:300
105. Duveneck GL, Pawlak M, Neuschäfer D, Bär E, Budach W, Pieles U, Ehrat M (1997) Sens Actuators B 38–39:88
106. Srivastava R, Bao C, Gómez-Reino C (1996) Sens Actuators A 51:165
107. Plowman TE, Durstchi JD, Wang HK, Christensen DA, Herron JN, Reichert WM (1999) Anal Chem 71:4344
108. Mac Craith BD, Ruddy V, Potter C, O'Kelly B, McGilp JF (1991) Electron Lett 27:1247
109. Wiki M, Gao H, Juvet M, Kunz RE (2001) Biosens Bioelectron 16:37
110. Wiki M, Kunz RE, Voirin G, Tiefenthaler K, Bernard A (1998) Biosens Bioelectron 13:1181
111. Challener WA, Edwards JD, McGowan RW, Skorjanec J, Yang Z (2000) Sens Actuators B 71:42
112. Rowe CA, Tender LM, Feldstein MJ, Golden JP, Scruggs SB, MacCraith BD, Cras JJ, Liegler FS (1999) Anal Chem 71:3846
113. Tóth K, Nagy G, Lan BTT, Jeney J, Choquette SJ (1997) Anal Chim Acta 353:1
114. Saavedra S, Yang L (1995) Anal Chem 67:1307
115. Saggese S, Harrington JA, Sigel G (1991) Opt Lett 16:27
116. Davies J, Mendoça J (2000) Phys Rev E 62:7168
117. Saito Y, Kanaya T, Nomura A, Kano T (1993) Opt Lett 18:2150
118. Wei W, Qushe H, Tao W, Minzhao F, Yuanmin L, Gouxia R (1992) Anal Chem 64:22

119. Shahriari MR, Zhou Q, Sigel GH (1988) Opt Lett 13:407
120. Zhou Q, Sigel GH (1989) Proc SPIE Int Soc Opt Eng 1172:157
121. Shahriari MR, Ding JY (1994) In: Klein LC (ed) Sol-gel optics: processing and applications, chap 13. Kluwer, Boston
122. Wolf S, Tauber RN (1986) Silicon processing, vol 1. Lattice, Sunset Beach, CA
123. Zaatar Y, Zaouk D, Bechara J, Khoury A, Llinaress C, Charles JP (2000) Mater Sci Eng B 74:296
124. DeGrandpre MD, Burgess LW, White PL, Goldman DS (1990) Anal Chem 62:2012
125. Lee JE, Saavedra SS (1994) Anal Chim Acta 285:265
126. Bürk J, Zimmermann B, Mayer J, Ache HJ (1996) Fresenius J Anal Chem 354:284
127. Ge Z, Brown CW, Sun L, Yang SC (1993) Anal Chem 65:2335
128. Garcés I, Villuendas F, Subías J, Alonso J, del Valle M, Domínguez C (1997) Opt Lett 23:225
129. Kim K, Minamitani H, Hisamoto H, Suzuki K, Kang S (1997) Anal Chim Acta 343:199
130. Moreno M, Domínguez C, Muñoz F, Calderer J, Morante JR (1997) Sens Actuators A 62:524
131. Moreno M, Garcés I, Muñoz J, Domínguez C, Calderer J, Villuendas F, Pelayo J (1995) Advances in science and technology: advanced materials in optics, electrooptics and communication technologies, vol 11. Vinzencini, Faenza, p 465
132. Moreno M, Muñoz J, Garrido B, Samitier J, Calderer J, Domínguez C (1995) Advances in science and technology: advances in inorganic films and coatings, vol 5. Vinzencini, Faenza, p 149
133. Domínguez C, Muñoz J, González R, Tudanca M (1993) Sens Actuators A 37–38:779
134. Puyol M, del Valle M, Garcés N, Villuendas F, Domínguez C, Alonso J (1999) Anal Chem 71:5037
135. Kovacs GTA, Petersen K, Albin M (1996) Anal Chem News Features 407A
136. Kang L, Dessy RE (1990) Anal Chem 21:377
137. Bakker E, Bühlmann P, Pretsch E (1997) Chem Rev 97:3083
138. Morf WE, Seiler K, Lehmann B, Behringer C, Tau S, Hartman K, Sorensen PR, Simon W (1989) Ion selective electrodes, vol 5. Pergamon, New York, p 115
139. Seiler K, Simon W (1992) Sens Actuators B 6:295
140. Spichiger U, Simon W, Bakker E, Lerchi M, Bühlmann P, Haug JP, Kuratli M, Ozawa S, West S (1993) Sens Actuators B 11:1
141. Bühlmann P, Pretsch E, Bakker E (1998) Chem Rev 98:1593
142. Spichiger UE, Seiler K, Wang K, Suter F, Morf WE, Simon W (1991) Proc SPIE 1510:118
143. Wang K, Seiler K, Morf WE, Spichiger UE, Simon W, Linder E, Pungor E (1990) Anal Sci 6:715
144. Miltsov S, Encinas C, Alonso J (2001) Tetrahedron Lett 42:6129
145. Hisamoto H, Satoh S, Satoh K, Tsubuku M, Siswanta D, Shichi Y, Koike Y, Suzuki K (1999) Anal Chim Acta 396:131
146. Sotomayor PT, Raimundo IM Jr, Neto GO, de Oliveira WA (1998) Sens Actuators B 51:382

Springer Ser Chem Sens Biosens (2005) 3: 45–92
DOI 10.1007/5346_002

Published online: 28 September 2005

Luminescence Lifetime-Based Imaging of Sensor Arrays for High-Throughput Screening Applications

Michael Schäferling

Institute of Analytical Chemistry, Chemo- and Biosensors, University of Regensburg, 93040 Regensburg, Germany
Michael.schaeferling@chemie.uni-regensburg.de

Abstract This review highlights optical imaging technologies for the fluorescent read out of sensor arrays. Chemosensor arrays for the determination of pH, oxygen partial pressure or metal ions found particular applications in biomedical and environmental analysis. On the other hand, the monitoring of biomolecular interactions, e.g. of DNA sequences or proteins, is an important tool in pharmaceutical research and medical diagnosis. Microwell plate-based assays provided the possibility to analyze a large number of samples in parallel in a very short time. The development of microarray technologies was a step forward in miniaturization of high-throughput (or multiplexed) assay formats. The analysis of both microwell plate and microarray-based assays are subject of this survey, focussing on fluorescence lifetime imaging methods.

1 Introduction

Luminescence lifetime imaging is nowadays an important and widespread tool in life sciences. It supports the study of complex chemical or biological samples, either at the macroscopic or the microscopic level. Optical imaging methods generally have an essential impact on analytical biochemistry and clinical diagnostics. The combination of fluorescence analysis with imaging paves the way for such diverse applications as mapping of biological samples, e.g., tissue or cell cultures for different significant parameters, or monitoring biological reactions and interactions. Imaging can also be used in high-throughput screening (HTS) modes, e.g., by insertion into microwell plate techniques or microarray formats. The implementation of fluorescence microscopes permits insight into cellular processes. (Immuno)histochemical staining [1–4], fluoroimmunoassays [5] or the monitoring of antisense experiments [6–8] are prominent examples of diagnostic applications in this field. The two- or three-dimensional spatial distribution of a fluorescent marker or sensor probe can be directly visualized by using (confocal) microscopic methods, which makes them ubiquitous in medical applications.

Finally, the use of CCD cameras as detectors in combination with short-pulsed light sources enables not only spatially resolved imaging, but also a time-resolved fluorescence detection. Time resolution offers two different new alternatives:

1. Online monitoring of cellular processes in real time. Current research activities in this field deal with the visualization of fluorescence-labeled viruses to detect the infection pathway of a single virus in a living cell [9], the diffusional behavior of single protein molecules in living cells [10, 11], the restriction of single DNA molecules coupled to a microbead [12], and the analysis of signal transduction processes using fluorescent fusion proteins [13], just to name a few. Single-molecule spectroscopy and single molecule imaging are the most challenging issues in this context.
2. Detection of changes in fluorescence lifetime as analytical information. The decay time and decay profile of an excited fluorophore can be strongly influenced by external physical or chemical stimuli, mainly quenching agents, and by its microenvironment. Usually, optical sensor platforms monitor the intensity of the emitted light of the applied fluorescent probes or labels during their excitation. The lifetime approach has some convenient advantages over the steady-state or intensity-based methods: sensitivity to unsteadiness of the light source intensity and to fluctuations in the light field are strongly reduced, and also to inhomogeneities of the dye concentration and to bleaching effects [14]. Furthermore, interferences from background signals with fast decay times can be eliminated; these may be caused by the autofluorescence of biological samples or by the

polymer substrates used as supports for the sensor platform. This noise suppression can be achieved if luminescent dyes with expanded decay times in the microsecond range are used [15]. Fluorescence lifetime imaging microscopy (FLIM) can be performed either in the time domain or the frequency domain [16]. Imaging of the distribution of the oxygen concentration in biological samples, e.g., for tumor detection [17], or of pH gradients [18] are first practical outcomes of luminescence lifetime-based methods.

This overview is mainly related to the visualization and quantitative analysis of sensor arrays for HTS applications using time-domain lifetime imaging technologies.

2 Lifetime-Imaging of Sensor Arrays Using Luminescent Probes

The first section will highlight the utility of fluorescent dyes with decay times in the range of microseconds as probe molecules for different chemical analytes. Starting from the basic principles of fluorescence lifetime imaging microscopy (FLIM) methods, some practical approaches will be demonstrated on how these molecular sensors can be used in microwell plate formats for a parallel analysis of a high number of samples, using straightforward time-resolved imaging techniques. For these purposes, the fluorescent dyes can be applied either in homogenous solution or immobilized on solid polymer substrates.

2.1 Basic Methods of Fluorescence Lifetime Imaging

Fluorescence spectroscopy is a useful method for analyzing the properties of excited states of atoms and organic molecules. In life sciences, many applications are related to a quantitative determination of biomolecules with a very high precision and sensitivity, both in vivo and in vitro. The low detection limits of quantitative fluorimetry are unsurpassed up to now. Many essential biomolecules (such as some proteins and other biopolymers, NADH, flavins, porphyrins) exhibit an intrinisic fluorescence after excitation with light in the UV or visible wavelength band [19]. In those cases the emission spectra can be used as a fingerprint of the compound and therefore for its identification, while the intensity of the emitted light at a specific wavelength affords a quantitative determination. On the other hand, many biopolymers, supramolecular structures, or cell membranes may be labeled or stained with fluorescent dyes [20, 21]. Furthermore, a growing number of probe molecules for the monitoring of biologically relevant parameters (e.g., pH distribution, oxygen partial pressure) are available [15]. These fields of applications turn fluorescence spectroscopy into a versatile tool in biological research and a promising diagnostic method in medicine.

Basic principles and applications of time-resolved fluorescence spectroscopy have been outlined in a very illustrative way by Valeur [16]. Although "punctiform" spectroscopy is still the best way to get a detailed knowledge of all the important parameters that characterize fluorescence emission (exact spectral properties, decay time behavior, polarization), imaging is always preferred whenever the localization of the distribution of any biomolecule of interest is required or a great number of samples have to be analyzed [22].

The easiest way to perform FLIM consists in triggering a pulsed light source synchronized with a gated detector (e.g., CCD camera) with the help of a digital pulse generator. These time-domain measurements very quickly reveal basic parameters like luminescence lifetimes of the selected dyes and their spatially resolved distribution in a sample [23]. Frequency-domain FLIM needs a continuous-wave (CW) laser as light source, usually coupled to an acousto-optical modulator [24], but the development of directly modulated light sources is making fast progress. The phase shifts and modulation ratios are measured relative to a fluorescence standard and the lifetime can be rapidly calculated from these parameters.

Lifetime imaging of sensor arrays is usually carried out in the time domain. This fact requires a closer look at methods for time-gated data acquisition. One early approach to a time-gated image microscope is illustrated in Fig. 1 [25].

Digital gated image intensifier modules (ICCDs) can now operate at the picosecond level. The principle of lifetime imaging by a gated image intensifier is shown in Fig. 2. Assuming a single exponential decay of the emitted fluorescence with an amplitude α after a short excitation pulse, the time-gated

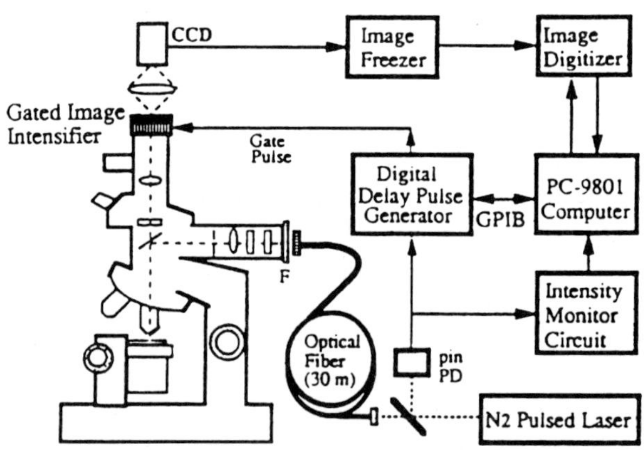

Fig. 1 Fluorescence microscope for lifetime imaging according to Wang [25]

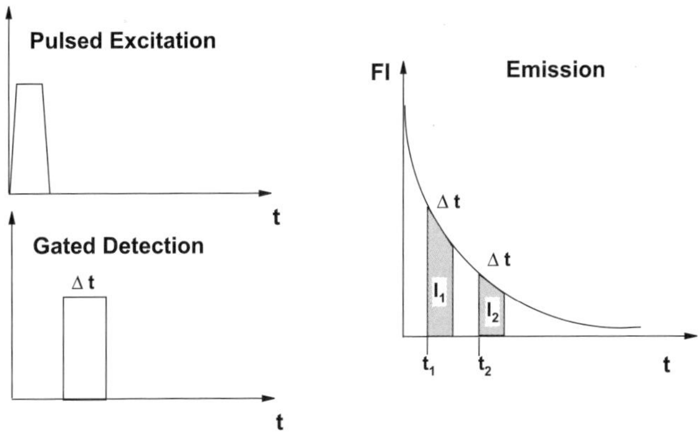

Fig. 2 Data acquisition for time-domain FLIM. *FI* fluorescence intensity, I_1 gated image no 1, I_2 gated image no 2. *Left* Excitation pulse of the light source and synchronized time-gated detection with a CCD camera. *Right* Lifetime determination by two subsequent time-gates according to Eq. 2

intensity I is given by:

$$I = \int_{t}^{t+\Delta t} \alpha \exp(-t/\tau)\mathrm{d}t \tag{1}$$

Fluorescence can be detected at various delay times with adjustable time gates Δt in consecutive acquisition cycles (multigate detection). At best, lifetime τ can be calculated from four experimental data (for the case of monoexponential decay):

$$\tau = \frac{t_2 - t_1}{\ln(I_1/I_2)} \tag{2}$$

CCD cameras equipped with a multichannel-plate image intensifier can be operated at internal frequencies of up to 12.5 MHz with a time-resolution (minimal gate width) of 3 ns. This is of course insufficient for very rapidly decaying probes. Faster CCD detectors can be obtained with a microchannel plate photomultiplier (MCP), working with gating rates of around 100 MHz and a time-resolution of 20 ps, which matches the decay profiles of the commonly used organic fluorescent markers. These camera systems are comparatively high priced and demand highly sophisticated accessories for short pulsed excitation, synchronization, and data acquisition. Usually, high gating frequencies are accompanied by a high background noise. A favorable alternative is offered by CCD systems with an electronic (or mechanical) shutter, which achieves internal frequencies of typically 100 kHz and a minimal gating time down to 100 ns. Such cameras can only be applied for luminescent dyes with decay times in the microsecond range. These include luminescent complexes of metals of the platinum group (ruthenium, palladium, platinum) or lanthanides (primarily europium and terbium) with organic chelating ligands. A great number of such metal complexes have been developed in the last few years, with lifetimes of up to several milliseconds, which can serve as molecular probes and as labels for biomolecules (Fig. 3).

From a purely analytical perspective, determination of the exact lifetime of the probe is not necessary. Ratiometric imaging methods exhibit all the advantages of lifetime imaging, like background suppression and internal referencing, and are straightforward in processing. The output of these techniques is the simple ratio of the intensities of two successive time gates. In principle, there are two different ways to perform ratiometric imaging. If the first image is taken during the excitation period (light source on), and the second image is acquired in the emission period (light source off), the acquisition process can be referred to as *phase delay imaging* (PDI) [26, 27]. In the second approach, adapted from the RLD method (*rapid lifetime determination*) [28–30] and here termed *rapid lifetime imaging* (RLI), both images are taken in the decay period (see Fig. 4).

In practice, the images of the two different gates are taken separately in subsequent acquisition cycles. The evaluation software integrates the two sets of pictures, followed by a subtraction of the corresponding background dark pictures (detected with the same time gates and frequency but without illumination). Figure 5 illustrates the whole process for the PDR imaging where four sets of images are obtained: Image 1 (I_1), image 2 (I_2), dark image 1

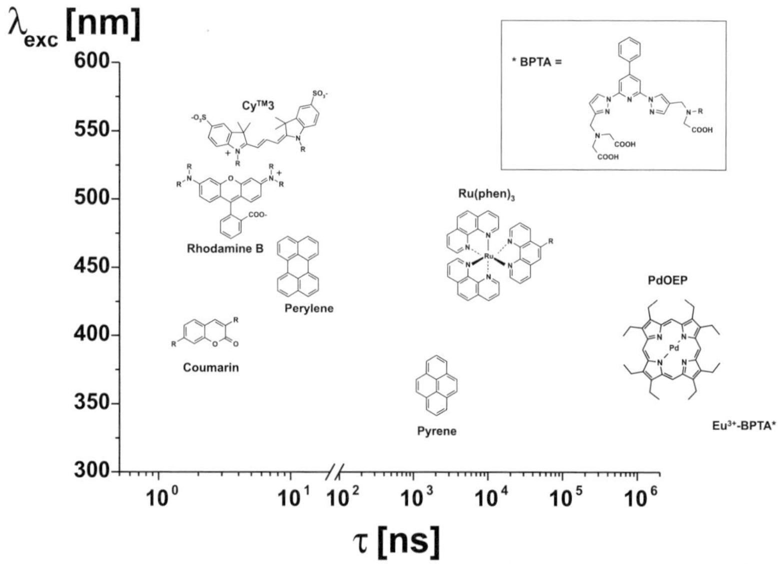

Fig. 3 Luminescence lifetimes τ and excitation wavelength λ_{exc} of typical organic dyes and metal chelates

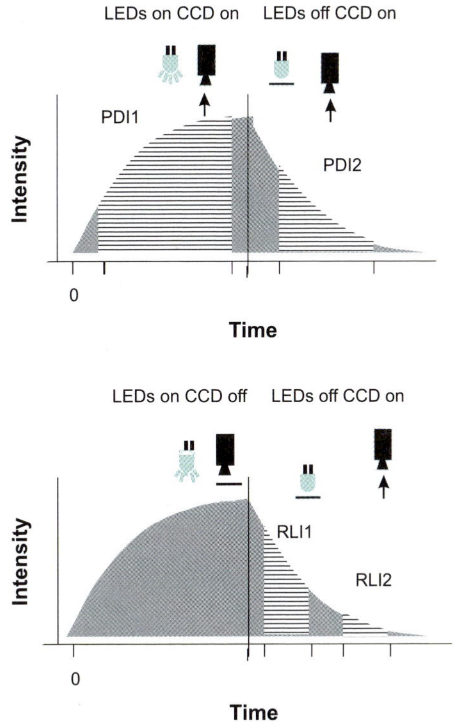

Fig. 4 Ratiometric lifetime imaging methods: PDI and RLI

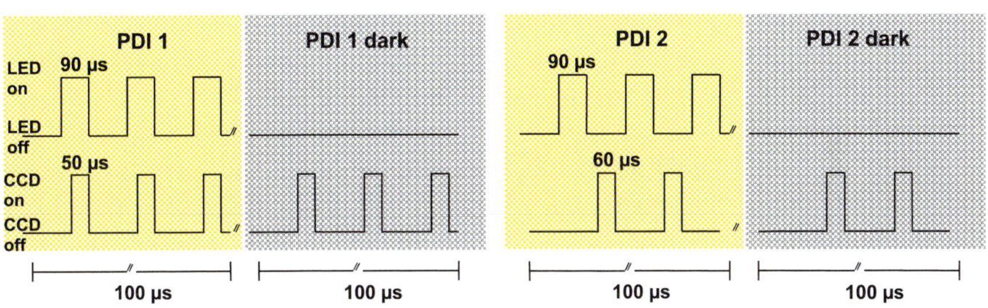

Fig. 5 Data acquisition cycles for ratiometric PDI using luminescent dyes (e.g., europium complexes) with decay times in the µs range

(D_1) and dark image 2 (D_2). The resulting ratio R is calculated according to:

$$R = \frac{\sum I_1 - \sum D_1}{\sum I_2 - \sum D_2} \tag{3}$$

of images 1 and 2 are both set in the decay period, when the light source is off. Both imaging modes have their specific advantages, which will be discussed later. By executing these processes, thousands of single screens can be added and averaged within split seconds. This affords a high sensitivity and a reduced influence of imprecisions in the lengths of single pulses. With simple imaging systems, one microwell plate or several sensor microarrays can be read within 1 s.

2.2 Imaging of Oxygen Partial Pressure

Oxygen supply and pH value are important factors for the cell function. Changes in oxygen partial pressure (pO_2) and the pH distribution in extracellular fluids can monitor status and dysfunctions of the cell metabolism (e.g., after exposure to toxins or drugs [31]) and play an important role in the development of radio- and chemotherapeutic treatments of tumors [32]. Determination of the oxygen demand by fluorescence imaging is also an useful method for controlling tissue engineering [33] and for tumor detection [17]. Fluorescence imaging is an established method for the detection of tumors or virally infected cells. In vivo imaging of lysosomal and mammalian protease activities by means of resonance energy transfer (RET) technologies is a promising example [34, 35]. Photosensitizers, developed for the treatment of skin tumors by photodynamic therapy (PDT) can also be used for diagnostic applications if they accumulate in cancerous tissues.

An outstanding example is the exogenous administration of 5-aminolevulinic acid (ALA), a naturally occurring heme precursor, which induces the biosynthesis of the photosensitizer protoporphyrin IX (PpIX) in cells [36]. This process occurs nearly selectively in neoplastic [37] or malignant cells [38]. Thus, it can be used for diagnostic tumor detection [39] and in photodynamic therapy, as PpIX is photoactive and strongly fluorescent [40]. It can be excited with a 405-nm blue laser diode and has an emission maximum at around 635 nm. The fluorescence lifetime is 18 ns and therefore much longer than that of natural tissue, which usually does not exceed 6 ns [41]. These properties were used by Cubeddu et al. for the detection of basal cell carcinomas by time-gated fluorescence imaging of the PpIX distribution [22].

Since the $pO_2$is significantly reduced in malignant tissue [42], the application of oxygen-sensitive probes is another successful approach for skin tumor imaging. The sensitivity of fluorescent probes relies on quenching effects by molecular oxygen. A broad range of oxygen sensitive probes is now available, covering excitation wavelengths from the near infrared to the UV region, with luminescence lifetimes up to the microsecond region in case of transition metal complexes. These comparatively long lifetimes are perfectly suited for applications in ratiometric time-resolved imaging methods. Some examples of oxygen-sensitive fluorescent probes and their physical properties are collected in Table 1. A large number of phenanthroline or bipyridyl derivatives are synthetically available as ligands and can alter the luminescence proper-

Table 1 Luminescence properties of oxygen-sensitive probes applicable to time-resolved imaging

Probe	QY [a] Φ_0	Absorption λ_{max} [nm]	Emission λ_{max} [nm]	Lifetime [a] τ_0 [μs] (Solvent)	Ref.
Pyrene	0.65	335 [b]	376, 396, 474 [c]	0.45 (Cyclohexane)	[16, 45, 46]
$Ru(bipy)_3$	0.062	452	630	0.62 (Water)	[47, 48]
$Ru(phen)_3$	0.080	447	590	0.92 (Water)	[47–49]
$Ru(dpp)_3$	0.366	462	618	5.9 (EtOH/MeOH) [d]	[47–49]
PtOEP	0.5	380, 534	647	95 (Toluene)	[50]
PdOEP	0.5	393, 512, 546	663	770 (Toluene)	[51]
PdTPFFP	0.001	406, 519, 552	562, 608, 660	1500 (3-MP)	[52, 53]
$ReL(CO_3)(py)$	0.354	240, 320	550	17.6 (CH_2Cl_2)	[54, 55]

$Ru(bipy)_3$ Ru(II)-tris-2,2′bipyridiyne; *$Ru(phen)_3$* Ru(II)-tris-1,10-phenanthroline; *Ru(dpp)* Ru(II)-tris(4,7-diphenyl)-1,10-phenanthroline; *PtOEP* platinum(II)-octaethylporphyrin; *PdOEP* palladium(II)-octaethylporphyrin; *PdTPFFP* palladium(II)-tetra(pentafluorophenyl)porphyrin; *L* 2,4,7,9-tetraphenyl-1,10-phenanthroline; *py* pyridine; *MP* 3-methylpentane

[a] In deaerated solution

[b] A series of further strong absorption maxima exist in the region between 320 and 240 nm

[c] Excimer

[d] v/v 4 : 1

ties of the transition metal complexes [43]. Metal complexes with mixed ligands are also accessible. Many of these compounds are commercialized. Osmium complexes show a behavior similar to that of their ruthenium analogs, but are highly toxic. Pyrene has been derivatized with fatty acids (e.g., pyrene butyric acid, PBA), which results in decay times around 200 ns [44]. Among the metalloporphyrins, tetra(pentafluorophenyl)-porphyrin complexes have found great attraction because of their high photostabilities.

There is a certain solvent effect on the quantum yields (QY) and lifetimes τ of these probes. Even for the same solvent, deviant values can be found in the literature, e.g., for pyrene in cyclohexane ($\tau = 382$ ns) [56] or the ruthenium complexes in water [57]. Besides their long lifetimes, these transition metal complexes show another advantage with respect to time-resolved imaging, namely large Stokes' shifts of around 200 nm, which simplifies the suppression of background signals emanating from the light source by low-end optical filters. Ruthenium probes can be added to living cells for FLIM for intracellular oxygen sensing [58]. Numerous approaches have been described for immobilizing luminescent metal complexes on polymer supports, e.g., on siloxanes [27, 51], silicon rubbers [33, 47, 49], cellulose, poly(vinyl chloride), polystyrene [59–62], poly(methyl methacrylate) [48, 51, 63], poly(acrylo nitrile) [60], fluoropolymers [51, 63, 64], or copolymeric hydrogels on the basis of poly(acrylic acid) derivatives [15]. The dyed polymers can be cast as planar fluorescent sensors, which can monitor the spatial distribution of pO_2 in tissue samples exposed to the sensor surface, or as oxygen flux sensors [27]. If the luminophore is distributed uniformly in a homogeneous polymer matrix, the quenching of its luminescence is found to fit a simple Stern–Volmer equation of the type:

$$I_0/I = 1 + (K_{SV} + K_{eq})[O_2] + K_{SV}K_{eq}[O_2]^2 \tag{4}$$

where $[O_2]$ is the oxygen partial pressure, K_{SV} the Stern–Volmer constant (collisional quenching) and K_{eq} the bimolecular quenching constant (static quenching). However, this is only an idealized correlation and divergences are usually observed if the probes are immobilized on polymer supports.

The overall oxygen sensitivity exhibited by an optical sensor is basically predefined by the Stern–Volmer constant K_{SV}. The sensitivity of the final optical oxygen sensor increases with K_{SV} [65]. Generally, high K_{SV} values are provided by the Pd- and Pt-porphyrin complexes, by $Ru(dpp)_3$, and by pyrene. Fluorescence quenching by oxygen not only affects the fluorescence intensity of the dye, but also has an influence on its lifetime τ (Fig. 6):

$$\tau_0/\tau = 1 + K_{SV}[O] \tag{5}$$

While the natural lifetime τ_0 of a fluorophor is usually not much affected by immobilization in a polymer matrix, the encapsulating medium can cause a significant change in the sensitivity towards oxygen [62]. It is apparent that a high permeability and diffusion coefficient for oxygen in the encapsulating polymer improves the oxygen sensitivity of the film. It was found that silicon-based polymers as well as fluoropolymers have a higher oxygen permeability and solubility than organic polymers like polystyrene or poly(methyl methacrylate) PMMA [49]. In an intriguing example, a sensor foil containing PtOEP was mounted on the back of a hamster, with help of a dorsal skinfold chamber. An amelanotic melanoma was detected by measuring the tissue oxygenation with FLIM [17].

These luminescent sensor films can easily be inserted into microarray formats. Two general ways for the fabrication of ready-made sensor arrays are feasible:

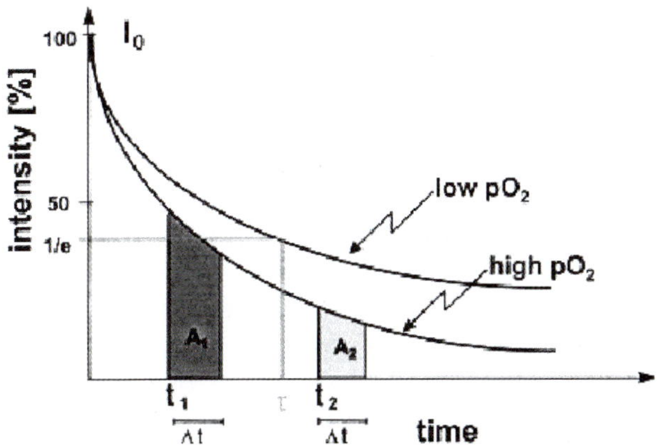

Fig. 6 Schematic representation of a time-resolved measurement of pO_2. Oxygen quenches the luminescence of the sensor probe and decreases its decay time τ [33]

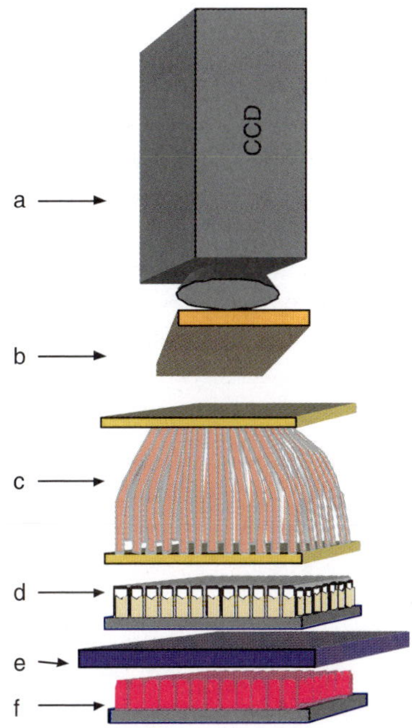

Fig. 7 Set-up for time-resolved imaging of 96-microwell plate formats. **a** Fast gateable CCD-camera, **b** Optical emission filter, **c** Light-guiding adapter, consisting of 96 optical fibers, **d** 96-microwell plate (black with transparent bottom), **e** Optical excitation filter, **f** Pulseable LED array with 96 light emitting diodes

1. The polymer solution is spotted on a solid support or dispensed in the single wells of a microwell plate. After drying, the fluorescent dyes can be immobilized on these matrices. Direct spotting and drying of a mixed polymer/luminophore "cocktail" can be an alternative.

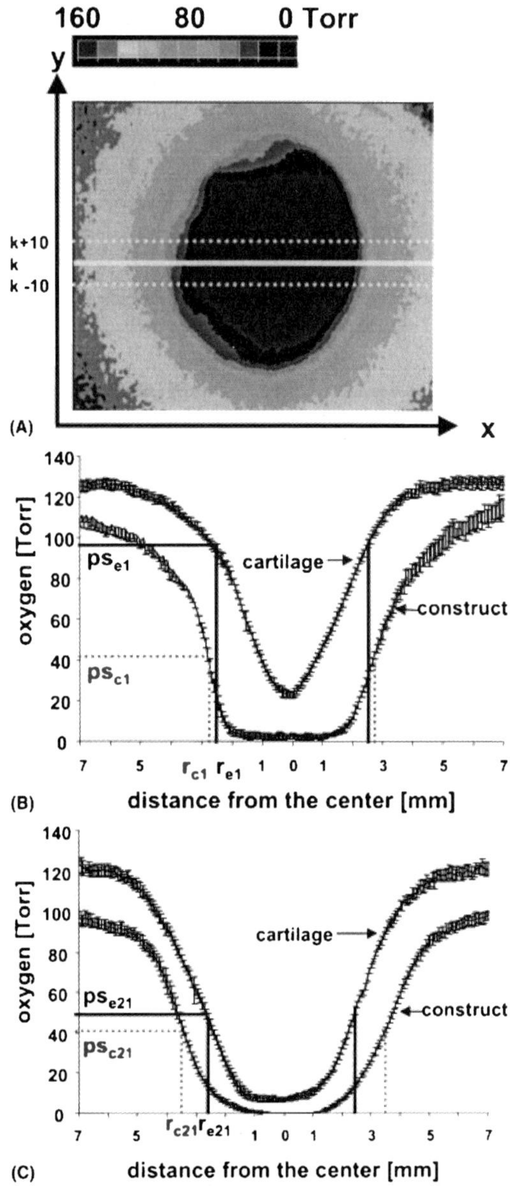

Fig. 8 **a** Different gray values of the 2D surface plot of the horizontal cross-section of a tissue represent different pO_2 values. Means and standard deviations were calculated from 20 pixel rows through the center of the samples. Means and standard deviations of a representative cartilage explant and cell–polymer construct were plotted versus their tissue size on day 1 (**b**) and day 21 (**c**) of cultivation. Data represent the average ± SD of 20 oxygen partial pressure values. Sample radius of cartilage explants (r_e) and cell–polymer constructs (r_c) and the pO_2 at the sample surface of explants (ps_e) and constructs (ps_c) were marked [33]

2. A solid support is coated with the sensor membrane, and after drying a plastic matrix can be glued on the sensor foil which leaves blank single sensor spots. In the easiest case, this matrix can be a 96-microwell plate-format, where the back panel of the plate has been removed.

An oxygen sensor array using these principles was developed by Liebsch et al. [60]. A modification of the applied imaging set up is illustrated in Fig. 7. It consists of a CCD camera with a fast mechanical shutter, an array of 96 pulsed LEDs, a computer-steered pulse generator, a constant power supply, a light-guiding adapter with 96 optical fibers, and the corresponding excitation and emission filters.

A PtOEP-doped polystyrol layer was spread on a transparent polyester support with a thickness of 125 μm. After evaporation of the solvent, 3–5 μm thick sensor films were obtained. Circular spots were punched out of the sensor foil and fixed at the bottom of microtiter plate wells. These ready-for-use sensor arrays can be applied for a parallel determination of pO_2 in aqueous solution in multiplexed samples. The read-out of the sensor arrays was performed by means of ratiometric imaging via the RLI method shown in Fig. 4.

In the same way, a 6-well culture plate has been prepared in order to monitor gradients in the oxygen supply of engineered tissue samples over a 21 day period. Figure 8 shows the horizontal cross-section of a tissue and the resulting charts of different samples after different time periods. Light-scattering particles (TiO_2) had been added to the polymer mixture to increase the excitation efficiency within the sensitive layer [33].

These luminescent polymeric materials can also be used for the coating of fiber-optic oxygen sensors, with applications in clinical, environmental, and process analysis [49]. The single microsensors can be arranged in a multisensor array system for the determination of oxygen distributions within a sample. These systems are usually detected by phase-modulation techniques [66]. The monitoring of the consumption or generation of oxygen also plays an important role in oxidase-based bioassays.

2.3 Imaging of pH

Applications of fiber-optic pH sensors in environmental analysis, biomedical research, medical monitoring, and industrial process control have been reviewed by Lin [67]. A multitude of luminescent systems for pH monitoring are commercially available, mostly under special trademarks. Pyrene [68–70], coumarin, bromothymol blue [71] and fluorescein [72–74] derivatives are typical examples that have been used in research in the past two decades. Carboxyfluorescein derivatives have been directly applied to skin tissue samples for the lifetime imaging of pH gradients in the extracellular matrix of the epidermis [75]. Two-photon excitation microscopy became an established method for this kind of application, because of the small penetration depth of the excitation light at the required wavelengths for the common pH probes [76, 77]. There are numerous approaches for using these fluorescent probes inside microbial biofilms [78] or inside living cells [79]. An alternative approach is the addition of near-infrared dyes [80] for determining the pH or other parameters in biological samples, because longer wavelengths exhibit a better penetration depth into tissue samples.

These can be immobilized by either electrostatic adsorption on ion-exchangers [81] or sulfonated polystyrene [82], by hydrophobic interactions on organic polymers [83], by covalent binding to cellulose [68, 84] and chlorosulfonated polystyrene [71], or by entrapment in hydrogels [69, 85]. Sol-gel based methods were established particularly for the preparation of fiber optic sensors [86–88].

Ping et al. have fabricated an integrated microsensor array on a silicon wafer for pH imaging [89]. Six different pH-sensitive colorimetric dyes (methyl violet 6B, phenolic red, alizarin complexone, 5-carboxy-fluorescein, alizarin red and methylthymol blue) were used to cover the whole pH range. The dyes were adsorbed on microbeads and placed in etched microwells on the silicon wafer. The indicator array was also used as a cation sensor chip (see Sect. 2.4).

A time-resolved fluorescent pH imaging method was first developed by Liebsch et al. [18]. The sensor foil was prepared by immobilizing pH-sensitive microbeads consisting of a poly(acrylonitrile)-based core that contains the inert luminophore $Ru(dpp)_3$ and a hydrogel shell with covalently bound carboxyfluorescein in a polyurethane layer. The hydrogel cocktail was spread on a polyester support and dried. The ruthenium complex in this case served as reference dye and fluorescein as the pH-sensitive dye. With the *t*-DLR method (time-domain dual lifetime referencing), whose basic principles are illustrated in Fig. 9, pH gradients could be imaged. For an ideal performance of the *t*-DLR scheme in optical sensors, the following conditions must be met:

1. The inert reference luminophore and indicator fluorophore need to have highly different decay times
2. The decay time and the quantum yield of the reference dye is not affected by the sample
3. The indicator changes its fluorescence intensity as a function of the analyte concentration
4. The excitation spectra of reference and indicator overlap; thus, simultaneous excitation at a single wavelength becomes possible
5. The luminescence of reference and indicator are detectable at a common wavelength
6. Both luminophores are in close spatial proximity
7. The ratio of the two dyes remain constant

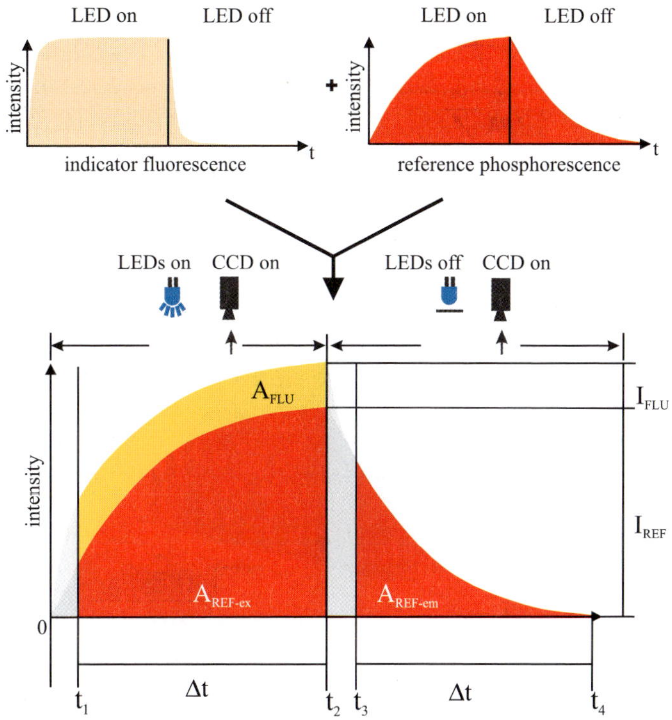

Fig. 9 Schematic representation of the time-domain DLR (*t*-DLR) scheme. The short-lived indicator fluorescence and the long-lived phosphorescence of the inert reference beads are simultaneously excited and measured in two time gates. The first (A_{ex}) is in the excitation period where the light source is on and the signal obtained is composed of short-lived fluorescence and long-lived luminescence. The second gate (A_{em}) is opened in the emission period where the intensity is exclusively composed of the reference luminescence [18]

Similarly to the PDR method (Sect. 2.1), *t*-DLR imaging is based on the acquisition of two images, one taken in the excitation period when the light source is on, and one in the decay period when the light source is off. The luminescence in the decay period is measured after a certain time delay in order to eliminate interferences. The lifetime of the reference dye should be longer than that of the indicator by at least a factor of 100. The resulting image A_{ex} represents the sum of short-lived indicator and long-lived reference luminescence, while image A_{em} originates exclusively from the long-lived reference dye (Fig. 10). By ratioing both images A_{ex} and A_{em}, all interferences can be referenced out.

Single spots can be punched out of the sensor foil and placed in the wells of a microtiter plate. Figure 11 shows the results of a ratiometric *t*-DLR screen for sensor spots exposed to buffered solutions with pH values from 5.0 to 8.5, displayed as a 3D plot and the corresponding calibration curve [18]. In

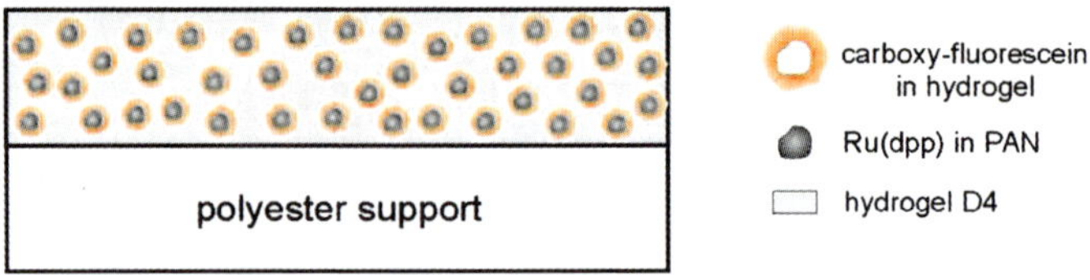

Fig. 10 Cross-section of the pH sensor membrane. The polyester film serves as an inert and optically transparent support [18]

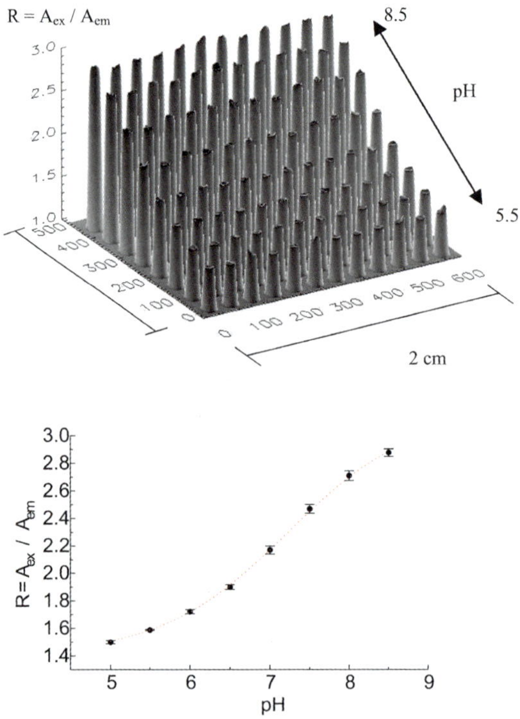

Fig. 11 *t*-DLR-based pH imaging of several independent samples simultaneously. **a** Surface plot of a sensor array contained on the bottom of the wells of a microtiter plate (8 × 12 wells). The wells were filled with buffer of defined pH to obtain 12 independent data points for each pH. The acquisition time was 800 ms. **b** Calibration plot R versus pH of the *t*-DLR pH sensor. The error bars represent the standard deviation of the 12 wells [18]

a different approach, the pH-dependent efficiency of the non-radiative energy transfer between the excited state of $Ru(dpp)_3$ as donor and a pH-sensitive dye (bromothymol blue) as acceptor was measured. Both were immobilized in polyurethane. Protonation and deprotonation of the acceptor modulate the spectral overlap and the efficiency of energy transfer. As a result, the luminescence intensity and the decay time of the ruthenium complex changes with pH [60]. Again, sensor spots were fixed at the bottom of single wells of a microwell plate and the images were obtained by means of the DLR method. The use of ruthenium complexes with their long lifetimes as reference dye, or as a donor for the energy transfer to a short-lived sensitive dye, facilitates the execution of time-resolved imaging methods which enable a fast read-out of sensor arrays.

Ruthenium complexes with mixed bipyridyl ligands, immobilized inside a Nafion film, may also be used as pH-sensitive sensor layers [90]. A completely different approach for a ratiometric imaging of pH sensor foils was developed for diagenetic studies of marine sediments, using the dual fluorescence excitation ratio of the pH-sensitive fluorophore 8-hydroxypyrene-1,3,6-trisulfonic acid (HPTS) [91]. Commonly used dual fluorophors with different absorption and emission maxima in the protonated and basic form for ratiometric measurements are the naphthofluorescein and seminaphthofluorescein derivates (SNARF and SNAFL) [92]. It should be noted that ammonia or carbon dioxide can also be detected by some of these pH-sensitive materials [55, 93].

2.4 Imaging of Metal Ion and Anion Sensor Arrays

A demanding task for the determination of environmental contaminants is the proper choice of existing optical sensors and the assembly to a single array. An ideal sensor array should be based on a single measurement technology, e.g., fluorescence detection, and a uniform analytical protocol. Preferably, the applied fluorescent indicators should be characterized by similar excitation and emission wavelengths. The integration into microwell plate technologies in combination with imaging paves the way for parallel analysis of aqueous solutions for a large number of parameters within a short time.

Certain pH-sensitive dyes, assembled in microbeads into wells of a silicon wafer (Ping et al. [89], for details see last section) can also be utilized as metal indicators, e.g., for water quality monitoring. Table 2 lists the metal indicators and their associated analytes that were applied for a colorimetric RGB (red-green-blue) imaging.

An extensive calibration by pattern recognition methods and special algorithms for the data processing are required for a qualitative and quantitative

Table 2 Colorimetric metal indicators and associated analytes [89]

Indicator	Analytes	pH-sensitive range
Alizarin complexone	Ce^{3+}, La^{3+}, Zn^{3+}, Pb^{3+}, Co^{2+}	2–12
o-Cresolphthalein complexone	Ba^{2+}, Ca^{2+}, Mg^{2+}, Mo^{6+}	10–12
Methylthymol blue	Al^{3+}, Fe^{3+}, Cu^{2+}, Mg^{2+}	2–12
Calconcarboxylic acid	Ca^{2+}, U^{6+}	~ 10
Xylenol orange	Al^{3+}, Fe^{3+}, Zn^{2+}, Ca^{2+}	6–8
Chromazurine S	Al^{3+}, Fe^{3+}, Cu^{2+}	2–12

analysis of complex samples. In a similar approach, PEG-PS resin beads that are derivatived with the corresponding indicator molecules were positioned in microwells formed in Si/SiN wafers. The indicator components were fluorescein (pH), *o*-cresolphthalein complexone (pH, Ca^{2+}), alizarine complexone (Ce^{2+}, Ca^{2+}, pH), and derivatized galactose (sugars). A CCD camera reads out the RGB (red, green and blue light intensities) color patterns by steady-state imaging [94]. Occasionally, this concept of an array consisting of unspecific sensors combined with a read-out by pattern recognition is described as "artificial tongue".

The combination of different fluorescent metal indicators with inert luminescent reference beads consisting of poly(acrylonitrile) containing $Ru(dpp)_3$ leads to a sensor array in a microwell plate format, suited for ratiometric time-resolved imaging [95]. The data can be acquired with the help of the *t*-DLR method (for details see Sect. 2.3). A cross-reactive sensor array was arranged for the determination of mixtures of calcium(II), copper(II), nickel(II), cadmium(II), and zinc(II) ions by nine different commercially available fluorescent indicators (Table 3). For a successful application, it is mandatory that all luminophores can be excited at the same wavelength range between 400 and 500 nm, and that the excitation and emission spectra of all indicators overlap with those of the reference dye encapsulated in the nanobeads.

The sensing strategy employs a set of unspecific sensors forming a cross-reactive array. Selectivity is achieved by analyzing the response patterns by algorithms like support vector machines. The interaction of the metal ions and the indicators results in fluorescent signals that generate a characteristic pattern for every sample mixture. Figure 12 shows a time-resolved gray-scale and pseudo-color picture reflecting the intrinsically referenced luminescence intensities.

Table 3 Effect of different ions on the fluorescence of indicator dyes [95]

Indicator	Ca^{2+}	Cu^{2+}	Ni^{2+}	Zn^{2+}	Cd^{2+}
FluoZin1			+	+	+
BTC-5N		–	–	+	
Calcein		–	–		
Lucifer Yellow		–	–		
Phen Green		–			–
Newport Green			+	+	+
Oregon Green		–			+
BAPTA-5N	+				
Fluo-5N	+	–		+	

+ Indicates quenching, – indicates fluorescence enhancement
Indicator concentrations 1–5 $\mu mol\,L^{-1}$, ion concentration 100 $\mu mol\,L^{-1}$

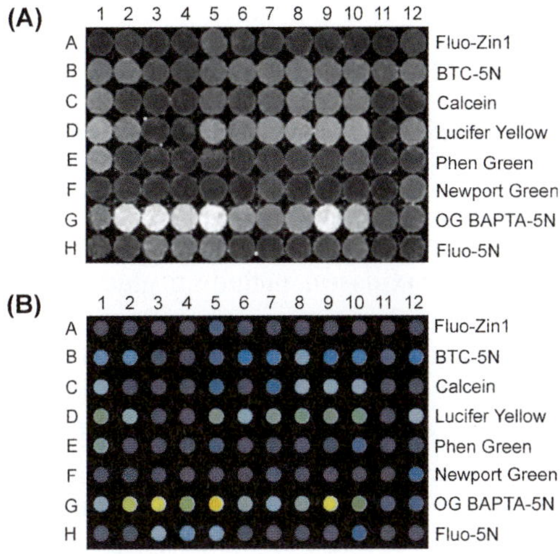

Column No.	Ca^{2+} [μM]	Cu^{2+} [μM]	Ni^{2+} [μM]	Zn^{2+} [μM]	Cd^{2+} [μM]
1	0	0	0	0	0
2	1	1	1	1	1
3	10	10	10	10	10
4	100	100	100	100	100
5	1	0	0	0	0
6	0	1	0	0	0
7	0	0	1	0	0
8	0	0	0	1	0
9	0	0	0	0	1
10	10	0	0	0	0
11	0	10	0	0	0
12	0	0	10	0	0

Fig. 12 **a** Gray-scale pictures of a 96-well microtiter plate (containing 12 cross-reactive arrays) after addition of the ion mixtures given beneath. Each column represents one array consisting of eight elements (*spots*). Rows were filled with solutions of the indicators given on the right side. The uniformity of the distribution of the gray tones indicates that local intensity variations have been successfully referenced out by the *t*-DLR method. Acquisition of the image takes 800 ms. **b** Pseudo-color picture of the response pattern of a 96-well microtiter plate containing 12 cross-reactive arrays on exposure of the ion mixtures given beneath. One row represents one array consisting of eight elements (*spots*). The spots represent the average value of an extracted circular area of the spots in picture **a** [95]

A similar sensor array was developed with fluorescent indicators for Ca^{2+} (Oregon Green Bapta-5N), Na^{+} (Sodium Green), Mg^{2+} (Magnesium Green), SO_4^{2-} (Oregon Green Bapta-5N + $Ba(NO_3)_2$), Cl^{-} (Lucigenin), and Hg^{2+} (Phen Green FL). The basic data sets resulted from *t*-DLR experiments after the exposure of different ion mixtures with the indicators physisorbed in six adjacent wells of a microtiter plate. The sensor was developed to deliver simple on/off patterns of complex ion mixtures [96].

2.5 Imaging of Hydrogen Peroxide

Hydrogen peroxide is one of the so-called reactive oxygen species that can cause substantial damage to biological material. Therefore, methods for determination or continuous sensing of H_2O_2 are of high interest. Probably the largest demand in assays and sensors for H_2O_2 results from the fact that it is produced by almost all oxidases, including glucose oxidase. It is therefore not surprising to see that numerous assays have been developed for the detection of H_2O_2. A large number of optical tests exist as well. Both chemiluminescent and electrochemiluminescent methods have very low limits of detection [97, 98], but work only at pH values above 8.0. In addition, these assays are usually irreversible. Other optical assays are based on the oxidative power of H_2O_2, which can convert a colorless substrate into a colored product, usually in the presence of the enzyme peroxidase [99]. Recently, a new

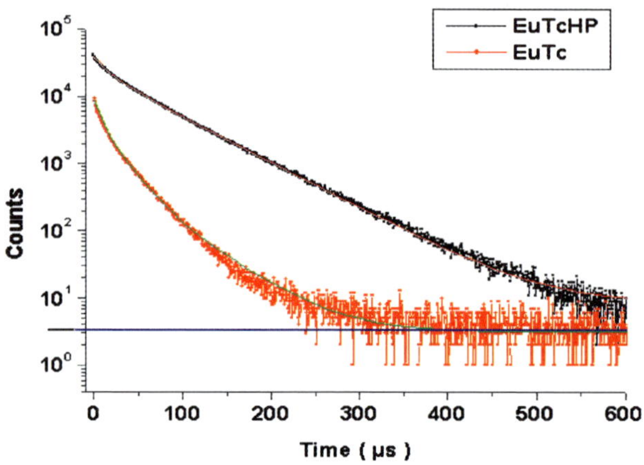

Fig. 13 Luminescence decay time profiles of [Eu(Tc)] and [Eu(Tc)(H_2O_2)] in MOPS buffer

Table 4 Fluorescence decay components of [Eu(Tc)] and [Eu(Tc)(H_2O_2)] in MOPS[a] buffer

	[Eu(Tc)] τ [μs]	[Eu(Tc)] Amplitude [%]	[Eu(Tc)(H_2O_2)] τ [μs]	[Eu(Tc)(H_2O_2)] Amplitude [%]
τ_1	6.8	39.5	10	17.3
τ_2	24	54.7	34	18.2
τ_3	53	5.9	61	64.5
Average[b]	30		60	

[a] (*N*-Morpholino)propanesulfonic acid buffer 10 mM, pH 6.9
[b] Component-weighted average

scheme for imaging of hydrogen peroxide has been developed. It is based on the luminescent probe europium(III)-tetracycline [Eu(Tc)] [100], works best at physiological pH, and does not require the addition of any further reagent or catalyst. This metal complex (with a stoichiometric ratio Eu/Tc of 3 : 1) undergoes a 15-fold increase in luminescence intensity at an emission wavelength of 616 nm after excitation with a 405-nm blue diode laser on exposure to hydrogen peroxide. Simultaneously, large changes in the luminescence decay time can be observed [101]. It is assumed that the [Eu(Tc)] complex can bind H_2O_2, which thereby replaces water molecules ligated to the eu-

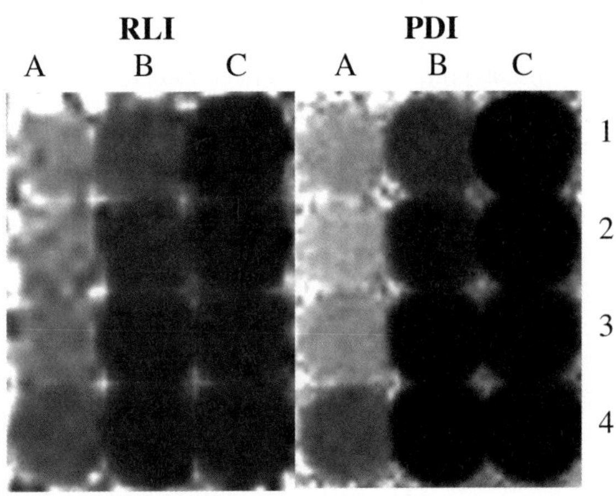

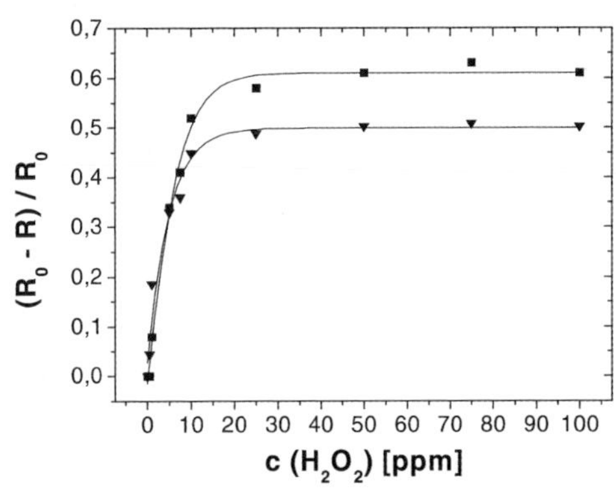

Fig. 14 Ratiometric fluorescence images (*right* RLD, *left* PDR) of sensor spots in a microwell plate exposed to different hydrogen peroxide concentrations: *A*1 0, *A*2 0.5, *A*3 1, *A*4 5, *B*1 7.5, *B*2 10, *B*3 25, *B*4 50, *C*1 75, *C*2 100, *C*3 250, *C*4 500 ppm H_2O_2 (*top*). Calibration plot $(R_0 - R)/R_0$ versus the concentration of hydrogen peroxide in MOPS buffer (pH 6.9). R_0 is the initial ratiometric fluorescence intensity of the sensor membrane, and R the intensity in presence of increasing concentrations of H_2O_2. *Squares* PDR image, *triangles* RLD image (*bottom*)

ropium cation, to form the strongly fluorescent [Eu(Tc)(H_2O_2)] complex. The fluorescence decay profile is shown in Fig. 13. It follows a three-component exponential decay model with the distribution shown in Table 4. The average lifetime of [Eu(Tc)] is therefore 30 μs, while [Eu(Tc)(H_2O_2)] has an average lifetime of 60 μs.

In order to obtain a ready-for-use sensor array, the probe was immobilized in a block copolymer matrix (polyacrylonitrile-*co*-polyacrylamide; Hypan), which is completely penetrated by water if exposed to it [102]. Prior to immobilization, the sensor membrane was cast onto an optically transparent ethyleneglycol-terephthalate polyester support (Mylar). The resulting sensor foil was glued on a black 96-microwell format matrix. The sensor arrays were analyzed by means of time-resolved RLI and PDI methods (see Sect. 2.1) with an optical set up as illustrated in Fig. 6 at an excitation wavelength of 405 nm. The ratiometric images resulted in similar calibration plots for both methods (Fig. 14). The limit of detection and the dynamic range of this sensor foil are comparable to those observed with [Eu(Tc)] in solution [103].

To obtain a high sensitivity and a low limit of detection, the time gates of the ratiometric methods, especially for RLI, must be fitted properly. Generally, the recorded time gates should be set in those ranges of the decay profile where the greatest differences occur on exposure to the analyte. If this is optimized, the slope of the linear range in a Stern–Volmer type calibration plot (Fig. 14) becomes as high as possible. Both the RLI and the PDI method have their specific advantages. RLI usually provides a higher sensitivity for low analyte concentrations. Also, low-end optical filters are adequate, especially if used for probes with large Stokes' shifts, as for example [Eu(Tc)], because both images are recorded when the excitation source is off. PDI, on the other hand, is preferred if only very weak signals are emitted by the fluorophore, as the first gate is applied during the illumination period.

2.6 Imaging of Citrate

Citrate is an important intermediate in the Krebs (citrate) cycle, which is the central metabolic hub of the cell for harvesting chemical energy and for building many biomolecules. It is a chelating agent that assists in the elimination of heavy metal ions by being taken up and biotransformed by bacteria. It is used as an anticoagulant to prevent blood clotting and as an additive in the food and pharmaceutical industry. Its urinary excretion rate is related to certain diseases. Citrate and the other intermediates in the Krebs cycle do not have significant physical and chemical properties suitable for direct determination, so they need to be transformed by enzymes or recognized by synthetic receptors [104–106]. The weakly fluorescent [Eu(Tc)] complex can reversibly associate with citrate to form the strongly fluorescent europium-tetracycline-citrate complex [Eu(Tc)($cit)_2$]. The stoichiometry

Eu(Tc)(cit) is 1 : 1 : 2 in this case. The fluorescence intensity of the 615-nm emission line of [Eu(Tc)(cit)$_2$] is 22 times stronger than that of [Eu(Tc)]. Citrate, as a polydentate ligand, can chelate the Eu^{3+} ion via the oxygen atoms of its carboxy and hydroxy groups. It is assumed that citrate displaces water molecules which ligate to the 8- and/or 9-coordination sites of the Eu^{3+} ion and quench its fluorescence. Table 5 summarizes the luminescence properties of the [Eu(Tc)] complex (1 : 1 stoichiometry) and the corresponding chelate complexes with the intermediates of the citrate cycle. Citrate concentrations can be imaged with this luminescent probe by means of the RLI method (Fig. 15) [107].

The other intermediates in the Krebs cycle (isocitrate, α-ketoglutarate (KG), succinate, fumarate, L-malate and oxaloacetate) also cause large differences in fluorescence intensity on addition to [Eu(Tc)], which can be imaged with the help of a microwell plate (Fig. 16a). As KG and succinate cannot effectively coordinate with Eu^{3+}, no significant fluorescence enhancement was expected, nor indeed observed. By choosing different lag times for time-gated detections, different intermediates can be "seen" in different time windows.

Table 5 Weighted average luminescence decay times and quantum yields of EuTc-L[a] complexes [107]

Species	EuTc-	EuTc-cit	EuTc-isocit	EuTc-ketoglut	EuTc-succin	EuTc-fumar	EuTc-malate	EuTc-oxal
τ_{av} [μs]	44	83	66	37	38	63	77	56
QY[b]	0.004	0.032	0.007	0.004	0.004	0.005	0.008	0.014

cit citrate, *isocit* isocitrate, *ketoglut* α-ketoglutarate, *succin* succinate, *fumar* fumarate, *malate* L-malate, *oxal* oxaloacetate

[a] L is the respective ligand out of the citrate cycle. EuTc: 50 μmol L^{-1} in MOPS buffer pH 6.9. L: same concentrations for all ligands, 150 μmol L^{-1}

[b] Using ruthenium(II)-tris(2,2′-dipyridyl) dichloride hexahydrate as a reference [108]

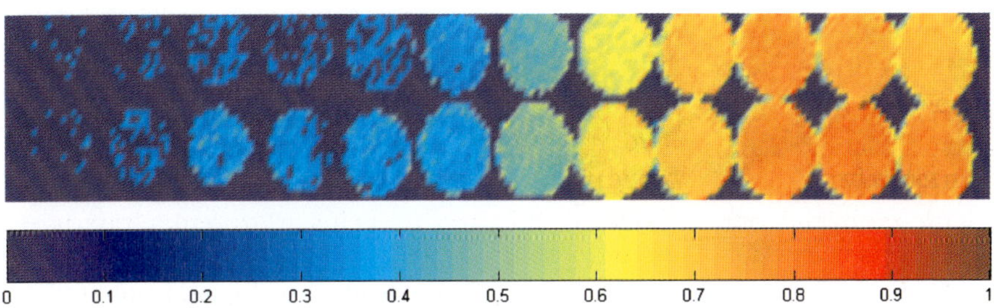

Fig. 15 Fluorescence imaging of citrate in a microtiter plate via the EuTc probe by rapid lifetime determination (in false colors). The concentration of EuTc is 50 μmol L^{-1} throughout; citrate concentrations (*from left to right*) are 0, 0.16, 0.4, 1.0, 1.6, 4.0, 10.0, 16.0, 20., 40.0, 60.0 and 80.0 μmol L^{-1}

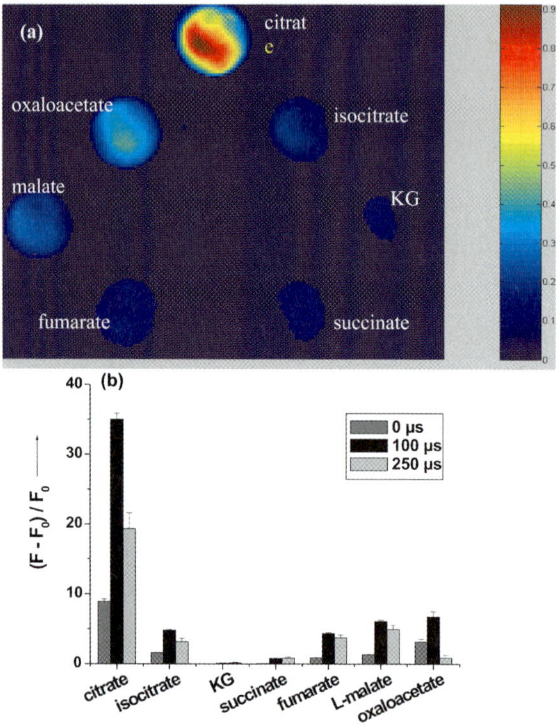

Fig. 16 **a** Fluorescence images of intermediates of the Krebs cycle. **b** Relative fluorescence intensities of the respective EuTc-L complexes at different lag times. The concentration of EuTc is 50 μmol L^{-1}, the concentration of the ligands citrate, isocitrate, ketoglutarate, succinate, fumarate, L-malate, and oxaloacetate is 150 μmol L^{-1} throughout. F_0 and F are the fluorescence intensities of EuTc and the EuTc-L complex, respectively

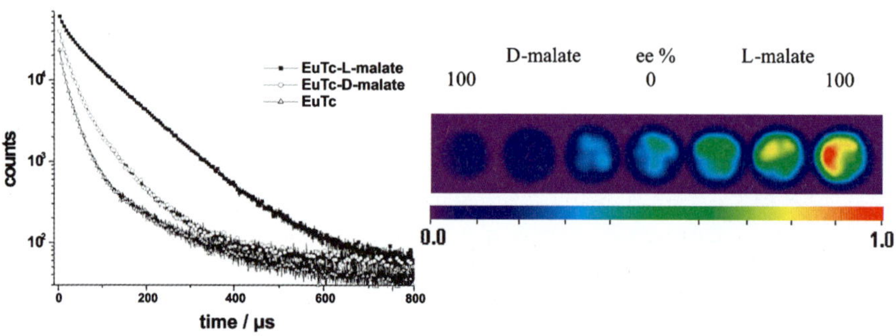

Fig. 17 Decay profiles of [Eu(Tc)] and the complexes with D- and L-malate *(left)*. Time-resolved imaging of enantiomer malates with [Eu(Tc)] (50 μM), [L-malate] + [D-malate] = 500 μM, time gate 140–220 μs *(right)* [108]

Gating obviously can be used to fine-tune between selectivity and sensitivity. On increasing the gating time from 0 to 100 μs, the normalized intensity [defined as $(F - F_0)/F_0$] of all species is increased (Fig. 16b) except for KG. On increasing the gating time to 250 μs, oxaloacetate is widely suppressed and citrate is reduced by 40%, while isocitrate, fumarate, and malate remain much less affected. Obviously, L-malate and oxaloacetate, and citrate and isocitrate can be nicely discerned via different delay times.

It should be noted that this scheme does not require the presence of enzymes (or even multi-enzyme systems). Furthermore, chiral discrimination can also be accomplished by fluorescence imaging with help of the [Eu(Tc)] probe. As shown in Fig. 17, the enantiomeric excess (ee) of L- or D-malates can be screened by means of time-resolved imaging with the help of microwell plates [108].

2.7
Imaging of Biomolecules Using the Europium(III) Tetracycline Probe: Determination of Glucose, Enzyme Activities, and ELISAs

While a seemingly uncounted number of assays has been reported for glucose, it is only in the past 15 years that methods for continuous monitoring of glucose have become available, mainly with the aim of developing sensors that are needed in context of the design of an artificial pancreas. It is interesting to note that practically all methods for determination of glucose are based on either sampling at a discrete sampling site (such as the artery), or on continuous sensing at a single site. Many indirect methods for determination of glucose have been developed based on the electrochemical or optical measurement of one of the species produced in the glucose oxidase (GO*x*)-assisted oxidation of glucose, followed by hydrolysis:

$$\text{glucose} + O_2 + H_2O \xrightarrow{(GOx)} \text{D-gluconolactone} + H_2O_2 \quad (6)$$

$$\text{D-gluconolactone} + H_2O \rightarrow \text{D-gluconate} + H^+ \quad (7)$$

For example, fluorescent sensor membranes have been described that are based on the measurement of the consumption of oxygen during GO*x*-catalyzed oxidation of glucose [109, 110]. This approach, while working well in clinical assays, suffers from the variable oxygenation of the sample, thus requiring the oxygen level to be measured in a second assay. Since glucose and dextran bind reversibly to concanavalin A (Con A), a smart sensing approach has been developed by Schultz et al. that relies on the competitive binding of fluorescence-labeled dextran and glucose for the binding site of Con A, which is immobilized on sepharose [111]. A similar approach has been described by Lakowicz et al. [112], who also applied anthracene derivatives for a lifetime-based sensing of glucose [113]. The ideal species to be detected in GO*x*-based assays is the hydrogen peroxide (H_2O_2) produced according to Eq. 6. H_2O_2 is not present in most samples and thus does not form an unknown background. The [Eu(Tc)] probe is applicable at near-neutral pH values and – unlike other probes for H_2O_2 – does not require the presence of a peroxidase. Formation of H_2O_2 in the presence of glucose and GO*x* leads to a strong increase in the luminescence intensity of [Eu(Tc)] at an emission wavelength of 616 nm after excitation with a 405-nm LED [101].

Europium tetracycline and GO*x* can be coadsorbed on hydrophilic sensor membranes as described in Sect. 2.4, adding up to ready-for-use glucose-sensitive luminescent foils. These can be implemented in microwell plate formats for a parallel screening of different samples [114]. Figure 18 shows eight adjacent wells of the plate spotted with differently concentrated glucose solutions in MOPS buffer, and the resulting calibration plot of the sensor.

The [Eu(Tc)] assay for H_2O_2 can also be used for the determination of enzyme activities. This comprises all enzymes that produce (oxidases) or consume (catalase, peroxidases) H_2O_2. For example, the activity of glucose oxidase can be directly imaged after the addition of [Eu(Tc)] as indicator and glucose as substrate. Steady-state and time-resolved imaging schemes can be applied for the read-out of a microwell plate-based assay. The best results were obtained by the RLI method (Fig. 19) [115].

If an [Eu(Tc)(H_2O_2)] complex is used as the indicator system, the activity of catalase can be monitored by the decomposition of this strongly fluorescent complex to form the weakly fluorescent [Eu(Tc)]. The influence of catalase inhibitors can be directly evaluated with this assay [116]. The same scheme can be applied to the determination of peroxidase (PO*x*) activity, and its in-

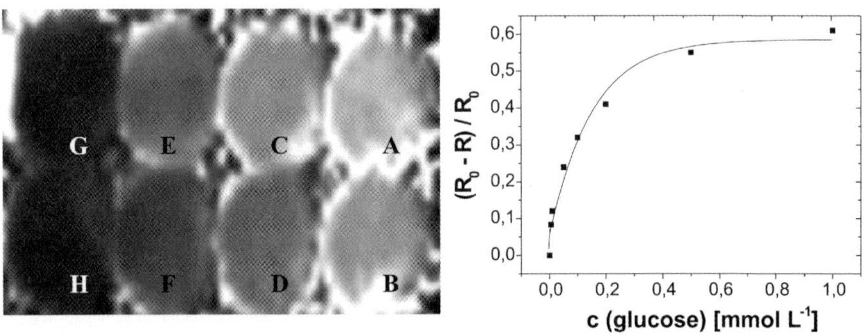

Fig. 18 Gray-scale ratiometric fluorescence image of sensor membrane spots in a microwell plate exposed to different glucose concentrations in MOPS buffer. *A* 0, *B* 0.005, *C* 0.01, *D* 0.05, *E* 0.1, *F* 0.25, *G* 0.5, *H* 1 mmol L^{-1} glucose (*left*). Calibration plot $[(R_0 - R)/R_0]$ versus the concentration of glucose in MOPS buffer at pH 6.9. R_0 is the initial ratiometric fluorescence intensity of the sensor membrane, and *R* the intensity in presence of increasing concentrations of glucose (*right*)

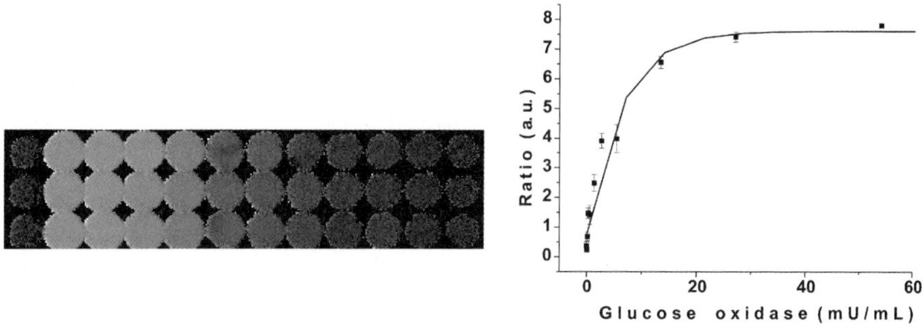

Fig. 19 Rapid lifetime determination imaging of the activity of glucose oxidase. Gray-scale image of the activity of glucose oxidase (*left*) and resulting calibration curve (*right*). Experiments were performed in triplicate (*rows*). The wells in the images contained, from 1 to 12, glucose oxidase activities of 0 (*blank*), 135, 54.1, 27.1, 13.5, 5.4, 2.7, 1.35, 0.54, 0.27, 0.14, and 0.05 mU mL^{-1} respectively, 100 μL of a 0.2 mmol L^{-1} EuTc solution, and 15 μL of a 277.2 mmol L^{-1} glucose solution. The total volume was made up to 200 μL with MOPS buffer

hibitors, by adding phenol as the second enzyme substrate [108]. All these assays work at neutral pH.

The use of europium chelates, with their unusually long fluorescence decay times, as labels for proteins and antibodies has provided techniques that are referred to as time-resolved fluoroimmunoassays (TRFIA). Fluorophores as labels for biomolecules will be the topic of Sect. 3. Nevertheless, TRFIAs always have to compete with ELISA (enzyme-linked immunosorbent assays) techniques, which are characterized by their great versatility and sensitivity through an enzyme-driven signal amplification. Numerous studies have been published over the past two decades which compare both analytical methods, e.g., with respect to the detection of influenza viruses or HIV-1 specific IgA antibodies [117, 118]. Lanthanide luminescence detection is another new development, and Tb(III) complexes have been applied, for instance, as indicators for peroxidase-catalyzed dimerization products in ELISAs [119].

The combination of time-resolved fluorescent detection and ELISA techniques leads to new fluorimetric assays which can be referred to as time-resolved fluorimetric ELISA (TRF-ELISA). Based on the fact that $[Eu(Tc)(H_2O_2)]$ undergoes a strong decrease in fluorescence intensity on enzymatic consumption of hydrogen peroxide, this assay can be used for the direct visualization and quantification of multiplexed enzyme-linked antibody samples in array formats by means of time-resolved fluorescence imaging. Commercially available and widely-used PO*x*-conjugates can be applied as antibody probes. In a sandwich type ELISA experiment (Fig. 20), a 96-microwell plate was implemented with rabbit anti-bovine IgG as the immobilized capture probe, bovine IgG as the target antigen, and rabbit anti-bovine IgG/peroxidase conjugate as the detector molecules. After adding phenol as a substrate, the presence of enzyme-labeled antibodies is indi-

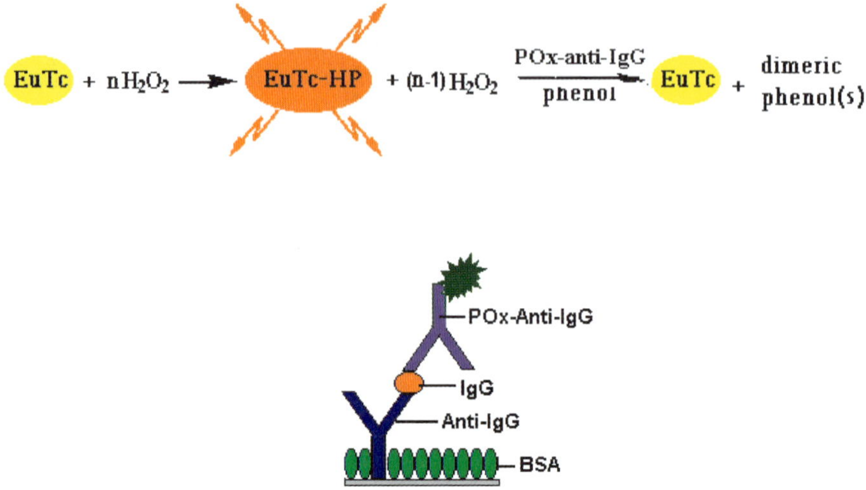

Fig. 20 Mechanism of a sandwich ELISA for the detection of bovine IgG using peroxidase (POx)-labeled anti-IgG and the europium(III) tetracycline system as fluorescent reagent (*top*). Schematic of the microwell surface showing the assembly of the sandwich assay (*bottom*)

cated by a decrease of the fluorescence intensity of the europium complex. A time-resolved gated fluorescence image of a microwell plate, containing triple spotted solutions with different IgG concentrations, is shown in Fig. 21.

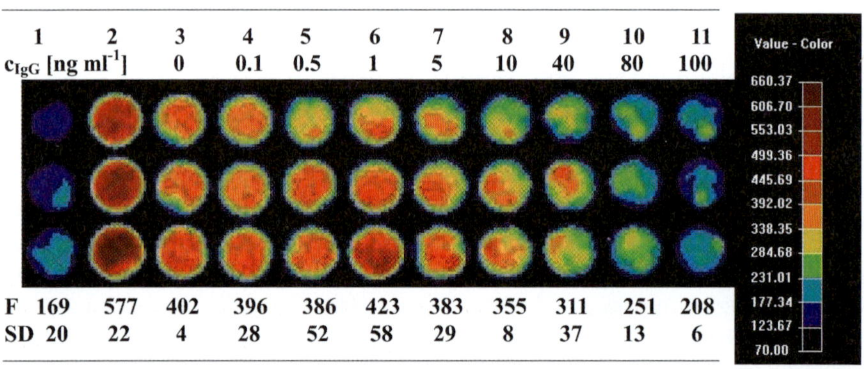

Fig. 21 Time-resolved fluorimetric image of a microwell plate containing the sandwich assay and the fluorescent reagent solution after the addition of different IgG concentrations from 0 to 100 ng/mL (spot rows 3–11). Reference spot rows 1 [Eu(Tc)] and 2 [Eu(Tc)]/H_2O_2/phenol illustrate the maximum possible dynamic signal range of the assay. *F* is the intensity value calculated from an area in the centers of the spots and averaged over three spots of equal concentrations. *SD* is the standard deviation in this spot regions

The dynamic range of the TRF-ELISA covers IgG concentrations from 5 to 100 ng/mL [108]. Detection limits of more than one magnitude lower (0.1 ng/mL) can be obtained with commercially available time-resolving microplate readers.

The method may also be used the other way round with GO*x* as the enzyme label, glucose as substrate, and [Eu(Tc)] as the fluorescent indicator. In that type of assay, the presence of enzyme-labeled antibodies is indicated by an increase in the observed fluorescence intensity. It should be noted that commercially available fluorescence microplate readers, preferable equipped with time resolution, are also suited for the screening of all the microwell plate-based assays presented in this chapter. Nevertheless, the imaging process is much faster, accomplished in the order of one second, and enables ratiometric measurements.

3
Imaging of DNA and Protein Arrays with the Help of Luminescent Labels

While the last section focused on luminescent probes and sensors for the determination of different analytes, this part will deal with the application of luminescent dyes as labels or markers for biomolecules. Typical applications include the staining of electrophoresis gels and blotting membranes [89], the staining of cell membranes or intracellular compartments [120], fluoroimmunoassays [121], and the screening of DNA or protein microarrays. The latter will be the main focus of this section. Starting from a rough outline of the current technologies of DNA and protein arrays, possible alternatives for the read-out of such biochips will be highlighted by applying time-resolved imaging methods.

3.1
Current Status of DNA and Protein Array Technology

Numerous surfaces with varying chemical functionalities on a wide range of substrates have been constructed to immobilize complex biomolecules such as DNA, proteins, antibodies, lipids, and oligosaccharides. The interaction of these reagents immobilized on a surface with various targets can be monitored in a high throughput manner using microarrays. Most of the current focus is on investigating DNA interactions to probe gene expressions (e.g., differential gene expression profiling) and single nucleotide polymorphisms (SNPs) in individual genes using DNA microarrays. These methods find applications in genome decoding ("sequencing by hybridization"), functional genomics, pharmagenomic research, clinical diagnosis, identification of genetically modified food, and genotyping.

Although DNA microarrays are interesting tools for deciphering gene expression behavior, this technique has not been able to reliably explain the protein expression levels within the cell. While the use of DNA arrays will continue to increase in functional genomics research, there is an increasing trend to focus on the proteins that act as the "work horses" of the cell. Proteins hold the key to various cellular mechanisms and act as biomarkers in the diagnosis of diseases. Protein and antibody arrays are useful tools for protein expression analysis, proteome research, medical diagnosis (immunoassays), and the discovery of binding domains of target molecules. A large number of review articles and books have appeared in the last years, addressing all aspects of DNA [122–127] or protein arrays [128–132] from fabrication, surface modification, and probe spotting to applications, screening methods, and data evaluation.

A common definition for the term "microarray" is given as: a spatially ordered, miniaturized arrangement of a multitude of immobilized reagents [133]. In the literature the pairs reagent/analyte, receptor/ligand, or probe/target are used alternatively to describe a screening experiment for biomolecular interactions. The reagent (or receptor, probe) is immobilized on a surface, while the analyte (or ligand, target) is usually labeled by a fluorescent dye or can be detected via a sandwich type of assay. Generally, microarrays have a planar surface, so they are distinct from microtiter or nanotiter plates with their isolated wells.

3.2 Imaging Techniques for Microarrays

Direct labeling of a biomolecule involves the introduction of a covalently linked fluorophore in the nucleic acid sequence or in the amino acid sequence of a protein or antibody. Fluorescein, rhodamine derivatives, the Alexa, and BODIPY dyes (Molecular Probes [92]) as well as the cyanine dyes (Amersham Biosciences [134]) are widely used labels. These probe families show different absorption and emission wavelengths and span the whole visible spectrum (e.g., Alexa Fluor dyes show UV excitation at 350 nm to far red excitation at 633 nm). Furthermore, for differential expression analysis, probe families with similar chemical structures but different spectroscopic properties are desirable, for example the cyanine dyes Cy3 and Cy5 (excitation at 548 and 646 nm, respectively). The design of fluorescent labels is still an active area of research, and various new dyes have been reported that differ in terms of decay times, wavelength, conjugatibility, and quantum yields before and after conjugation [135]. New ruthenium markers have been reported as well [136].

To enable a covalent labeling of target molecules, the dyes are functionalized with reactive groups. These are either phosphoramidites for the labeling of synthetic oligonucleotides (solid-phase synthesis), or uridine-5′-triphosphates for enzymatic coupling in PCR (random primed labeling technique [137, 138]) or 3′-terminal DNA marking with terminal transferase [139]. On the other hand fluorophores with amino-reactive (*N*-hydroxysuccinimide ester, isothiocyanate) or thiol-reactive (iodoacetamide, maleimide, disulfide) groups are available for protein labeling. They can be coupled selectively to lysine or cysteine units in the polypeptide backbone, respectively.

Currently, fluorescent read-out methods for microarrays are exclusively based on steady state intensity measurements. Two alternative types of microarray readers are on the market: laser scanner or imaging systems. Typically, scanners are equipped with one to three lasers with different excitation wavelengths, a movable x/y stage, and a photomultiplier tube (PMT) as detector (Fig. 22). Array imaging systems represent further developments of gel or blot documentation stations and make use of a CCD camera as detector and an array of light-emitting diodes (LEDs) in combination with fiber optic waveguides or a white light source provided with an appropriate set of optical filters for excitation (Fig. 23). Scanners with various lasers can screen dual-fluorophore samples either sequentially or simultaneously, which is an important feature for differential analysis, but makes this kind of instrument rather expensive. CCD-based imagers can cover a large area, so that several slides can be imaged simultaneously in a short time. The great advantage of a scanner system is that lasers can focus more energy to excite fluorophores, and thus they collect more light in less time. Therefore, they generally have a higher sensitivity and shorter signal integration times than CCD-based systems. However, this higher output is not essential since fluorophores can reach a saturation point at which further excitation leads to photobleaching or a non-linear emission response [140, 141]. Imagers can easily be equipped

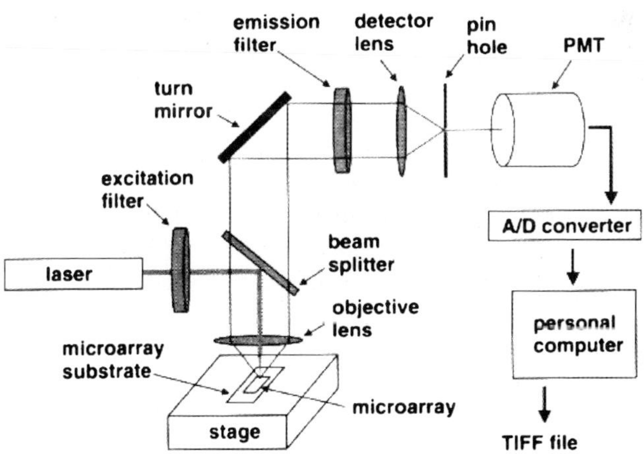

Fig. 22 Basic set up of a laser scanner system for the read-out microarrays according to Schena [127]

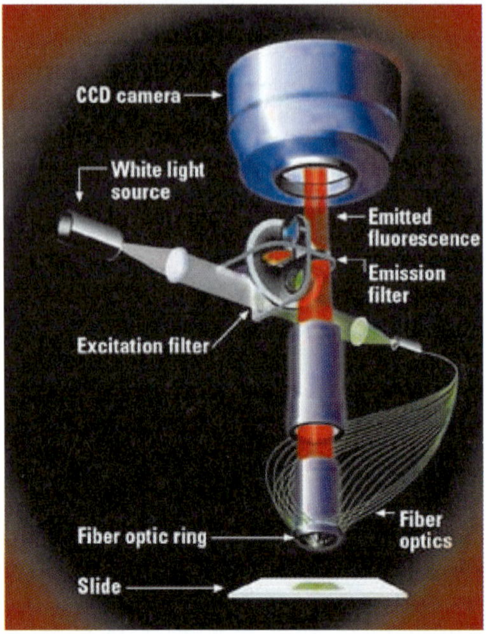

Fig. 23 Scheme showing the components of a CCD camera-based imaging system for the read-out microarrays (Applied Precision, Issaquah, Wash. [142])

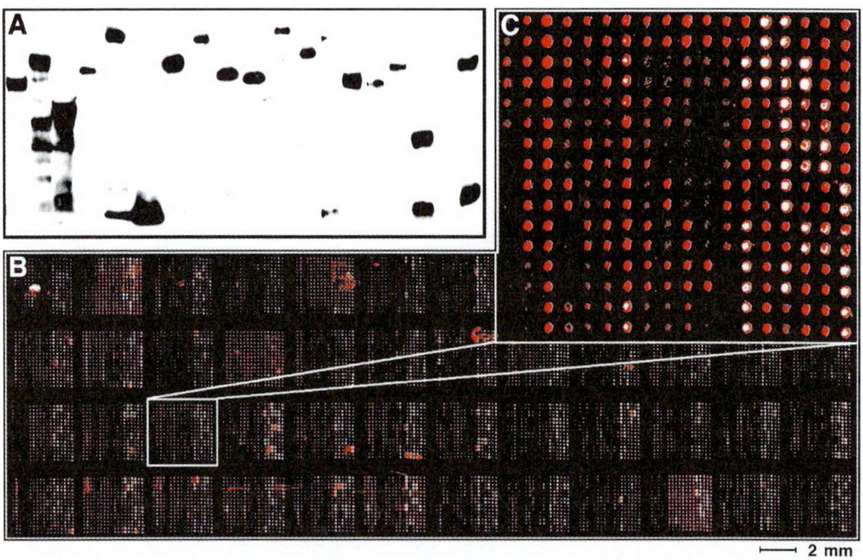

Fig. 24 GST yeast protein analysis. **a** 60 samples were examined by immunoblot analysis using anti-GST; 19 representative samples are shown. **b** 6566 protein samples representing 5800 unique proteins were spotted in duplicate onto a single nickel-coated microscope slide which was then probed with anti-GST. **c** Enlarged image of one of the 48 blocks [143]

with an assortment of different excitation wavelengths, which makes them versatile and, compared to scanners, relatively low-cost instruments.

An outstanding example for the screening of protein interactions is shown in Fig. 24. In this study by Snyder et al. [143], 6556 GST yeast proteins were spotted onto a glass slide. In this particular case, the fusion proteins attach via their His-tags to a nickel complex-coated surface and therefore the binding sites orient away from the surface. Low cost diagnostic tools for the imaging of proteins arrayed onto a membrane, consisting of a LED, a CCD camera, and a gelatine filter are also within the scope of testing [144].

3.3
Luminescence Lifetime-Based Imaging of DNA and Protein Microarrays

While fluorescence lifetime measurements have already been applied to improve DNA sequencing [145, 146], only very few approaches have been published up to now concerning FLIM techniques for the read-out of fluorophore-labeled DNA or protein chips. Cubeddu et al. [22, 147] have developed an experimental set-up with a fast light intensifier coupled CCD camera (minimum gate 300 ps) and a diode laser, synchronously pumped by an argon laser or by a second-harmonic mode-locked Nd : YLF laser as the excitation source for time-resolved imaging of DNA arrays. Two types of experiments were conducted in a 10×10 spot matrix (Fig. 25): mutation DNA arrays with targets modified by a single marker (Cy3), and cDNA arrays for expression profiling using target mixtures bearing two different dyes. In the latter case, a bi-exponential fit had to be used to discriminate the contribution of the two markers in order to calculate their relative concentration. The authors found a higher contrast and signal-to-noise ratio for the time-domain approach than for continuous wave measurements. However, the imaging set-up was composed of rather expensive and complex units such as an intensified CCD camera and mode-locked lasers.

Nearly simultaneously, time-resolved imaging of DNA microarrays in the subnanosecond range was described by Waddell et al. [148]. They used near-infrared dyes like aluminium naphthalocyanine tetrasulfonate as labels and aminated PMMA substrates for the immobilization of the oligonucleotide probes. The detection of hybridization events was carried out with a modification of a laser scanner set-up with a single photon avalanche diode, a pumped laser diode as excitation source, and a x/y stage controlled by a stepper motor. Also in this case the results obtained by lifetime-based imaging showed a higher sensitivity than by intensity-based imaging.

A promising way to avoid the high demands on the instrumentation is the use of labels with longer decay times, preferential in the microsecond range. Lanthanide complexes like europium or terbium chelates are extensively used for time-resolved fluoroimmunoassays (TRFIA). They enable the suppression of fast decaying fluorescent background signals (= noise) by collecting the

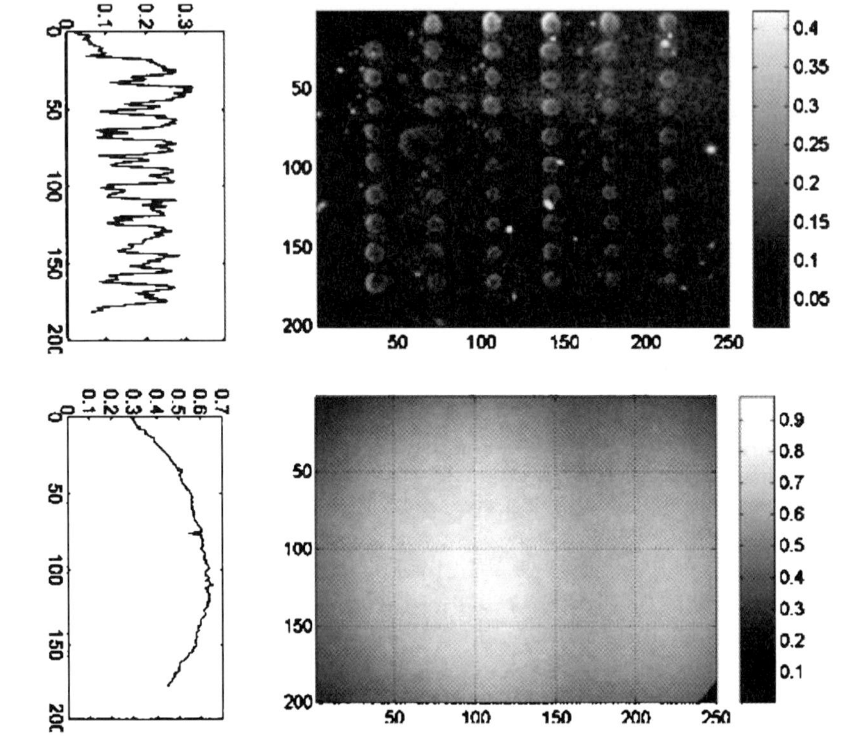

Fig. 25 Amplitude of Cy3 (*top*) and background (*bottom*) fluorescence obtained from a bilinear fit carried out on a set of time-delayed images of a DNA microarray. Line profiles across the first column are reported on the left [148]

emitted fluorescence signal within time gates in the microsecond range, and find applications not only in immunological diagnosis, but also in the monitoring of treatments for cancerous or viral diseases [149–153]. The lanthanide chelate complexes can be functionalized with amino- [154] or thiol-reactive groups [155] for the attachment to proteins. Naturally, lanthanide chelates can also be used for the labeling of DNA [156]. Surprisingly, these lanthanide probes have not been implemented for a time-resolved imaging of microwell plate- or microarray-based assays up to now. Related to that, monofunctional *p*-isothiocyanatophenyl derivatives of platinum(II) coproporphyrin-I have been evaluated as phosphorescent labeling reagents for DNA hybridization assays [157], and protein [158] or antibody detection [159] with time-resolved microwell plate readers.

Ruthenium complexes have been applied successfully to the luminescent detection of proteins on blotting membranes like nitrocellulose [160]. The bipyridyl and phenanthroline complexes modified with aminoreactive NHS-ester or isothiocyanate groups are commercially available [161]. An even higher sensitivity and lower detection limit can be obtained by encapsulating the fluorescent dyes into nanobeads [15]. Common particle materials include polystyrene, poly(acryl amide), poly(acrylo nitrile), or silica gel, in which different flourophors like ruthenium [162, 163] and europium [164] complexes or platimun porphyrins [165] can be incorporated. With this approach, a large number of luminophors can be attached to one biomolecule, which are additionally shielded from interferences like oxygen quenching. The surface of these nanometer-sized particles can be carboxylated and further activated for the conjugation to proteins or amino-terminated DNA sequences. Conjugates with biotin or (stept)avidin are commercially available [161].

In a preliminary study, different concentrations of streptavidin-coated nanoparticles filled with platinum porphyrin dyes (40 nm in diameter) were captured by a biotinylated alkylsilane monolayer surface on a 96-spot glass microarray. The spots are surrounded by a hydrophobic teflon poly(tetrafluoroethylene). The biotin serves as a model for a protein capture probe on the surface. The degree of surface biotinylation was optimized regarding the binding capacity towards streptavidin. Figure 26 shows an image of this protein microarray model obtained by the set-up shown in Fig. 7 [166]. The sensitivity of this system is sufficient for the detection of biomolecular interactions on a monolayer level (2D surface) [167]. The labeling with single dye molecules instead of nanobeads requires, however, excitation sources with higher intensities and makes higher demands on the optical components. The results revealed that alkylsilane-coated glass slides provide a much lower background (autofluorescence) than polymeric substrates. This becomes essential when dealing with very low signal intensities, as is the case for protein (or DNA) monolayers.

Whereas the advantages of time-resolved fluorescence imaging like higher spot homogeneity or the elimination of background signals have been out-

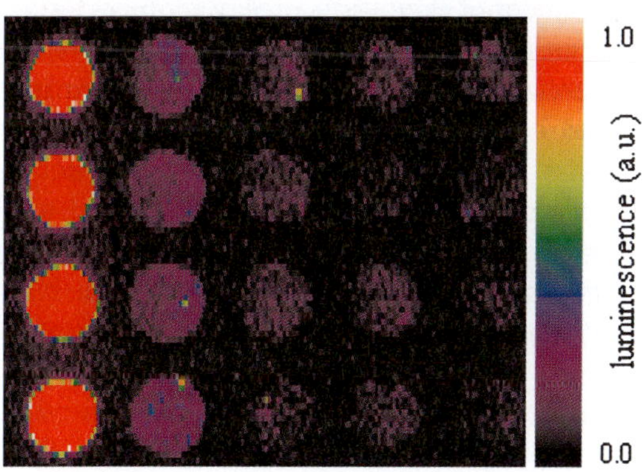

Fig. 26 Pseudo-color time-resolved image of polystyrene nanobeads containing Pt porphyrin and conjugated to streptavidin immobilized to a biotinylated microarray surface (black teflon coated 96-well glass slide, spot diameter 1 mm, Erie Scientific) in different concentrations (25, 15, 10, 5, 0 ng streptavidin per well) [167]

lined, the potential of ratiometric imaging as a screening method for biochips has not been addressed so far. The possible options are obvious. For applications in differential expression profiling, either on the cDNA or proteome level, the two samples can be marked with proper fluorescent dyes having different decay times. If these two dyes can be excited by one common wavelength, and if the difference between the lifetimes is high enough (> factor 10), the ratio of the expression profiles from two samples can be directly imaged within one scan by applying acquisition methods similar to *t*-DLR (Fig. 9). This approach can reduce the time needed for screening and data evaluation and the costs for microarray read-out, because only one excitation source is needed instead of two subsequent scans with different excitation wavelengths, as in the case of steady state measurements.

In the last few years a new generation of fluorescent labels had been developed. These are either so-called quantum dots consisting of semiconductor materials [168] or lanthanide-doped inorganic nanocrystals [169]. The biocompatible surface functionalization of these nanocrystalline materials is now under way [170–172]. Conjugates of quantum dots, whose emission wavelengths can be tuned with the size (3–10 nm) of the nanocrystal (e.g., CdSe) [173] with antibodies or streptavidin are commercially available [174], so they can be used in different bioassays and for bioimaging. They are applicable as luminescence markers for DNA microarrays, e.g., for SNP detection [175]. Their luminescence lifetimes are in the range of 20 to 50 ns, so quantum dots seem to be interesting candidates as labels for time-resolved imaging.

3.4 Resonance Energy Transfer Assays and Single Photon Counting Technology

The non-radiative transfer of excitation energy requires a spectral overlap between a donor and an acceptor molecule and can be used for the detection of biomolecular interactions, e.g., DNA hybridization, protein/protein interactions, antigen/antibody binding, or ligand/receptor recognition. It occurs if the emission spectrum of the donor partially coincides with the absorption spectrum of the acceptor. Such transitions are said to be in resonance, so the term *fluorescence resonance energy transfer* (FRET) is often used in the literature. Strictly speaking, this expression is incorrect as it is not the fluorescence that is transferred but the electronic energy of the donor. Therefore the terms *resonance energy transfer* (RET) or *excitation energy transfer* (EET) are recommended [16]. Many luminescent donor/acceptor pairs are well-established [176]. Biomolecular recognition brings the donor/acceptor couple in close spacial proximity (Fig. 27), and the resulting energy transfer can be detected via the fluorescence quenching of the donor or the specific emission of the acceptor dye if the donor is excited. This Förster resonance energy transfer takes place over a distance of around 1–10 nm between donor

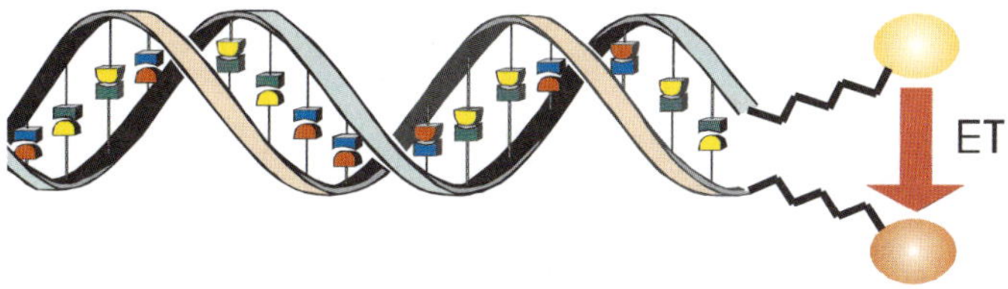

Fig. 27 Labeling of two complementary DNA strands results in a close proximity of the two labels in the duplex, thus resulting in efficient energy transfer [15]

and acceptor and results from long range dipole–dipole interactions. RET technologies have found numerous applications in cellular imaging and diagnosis [177] and the analysis of conformational distributions and changes in biopolymers [45, 178, 179].

A related approach is the molecular beacon. In this case, the immobilized DNA probe has complementary ends and forms a stem-and-loop structure. Both fluorophore and quencher are linked to the two ends of the stem. This non-fluorescent structure becomes fluorescent if the loop hybridizes with its target and the conformational reorganization into a rigid double helix separates the quencher from the fluorophore [180]. RET as well as the molecular beacon method can be combined with the polymerase chain reaction (real-time PCR) for the detection of SNPs [181] or the monitoring of DNA polymerase functions [182]. A related approach results in a method for in vitro as well as in vivo imaging of viral protease activities or the screening for protease inhibitors. A near-infrared fluorescent probe is connected to a quencher via a cleavable peptide spacer. The peptide serves as specific ligand for protease. After cleavage, the fluorescence of the near-infrared dye is activated and can be imaged [183, 184]. A protease assay based on RET was further performed with cleavable peptides fused to GFP (a green fluorescent protein) and DsRed (a red fluorescent protein) [185].

Time-resolved RET is capable of very sensitive detection of DNA hybridization. With a lanthanide chelate as the donor and an organic fluorophore like tetramethylrhodamin as the acceptor, time-resolved measurements can indicate the hybridization by strong changes in the intensity decay of the donor [186]. The development of new dyes for time-resolved RET with improved properties still is a major task [187, 188]. But, so far, the detection of biomolecular interactions by time-resolved RET has not entered real applications in the DNA or protein array market.

Zeiss have developed a high-throughput reader system (the Plate::Vision multimode reader) for microwell plates that is capable of fast time-resolved fluorescence (Fast-TRF) imaging with nanosecond resolution [189] (Fig. 28), which has found applications in pharmaceutical screening. This microplate imager is suitable for time-resolved RET assays. Kinase assays, for instance, can be performed with antibodies labeled with a $Ru(batho)_2bipy$ complex or

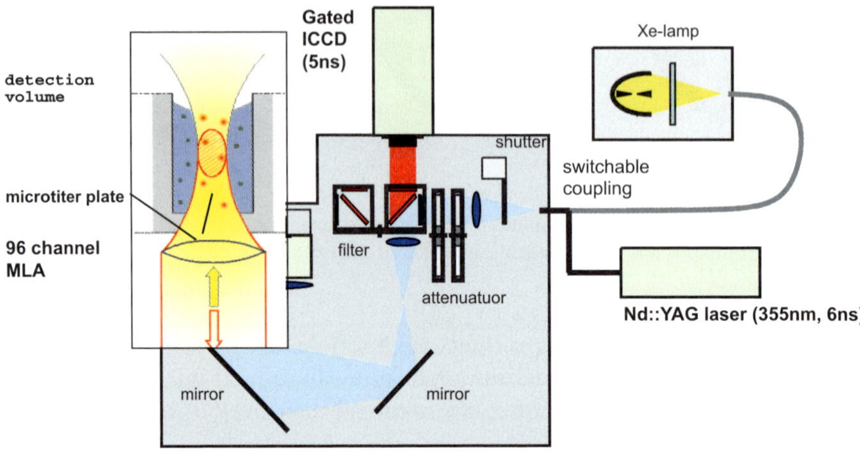

Fig. 28 Schematic view of the Zeiss "Plate::Vision" instrument which is state-of-the art for ultra-high throughput screening (uHTS) for drug discovery. The instrument resembles a 96-well parallel microscope: the light of a excitation source (Xe-lamp or pulsed laser) is expanded to illuminate a microtiter plate. The excitation is structured into 96 channels by a mini-lens array (MLA) and focused into the well with a detection volume of < 100 nL. All 96 channels are read simultaneously by a gated, intensifed CCD. With this fast detector and the pulsed laser excitation, the instrument can be used to carry out miniaturized, 96 parallel lifetime measurements in microtiter plate format with nanosecond time resolution or time-gated detection [190]

europium chelates as donor, and streptavidin conjugates with oxazine or rhodamine dyes as acceptor [190]. Protease assays were also performed with this imaging system. The fluorescence of DBO-labeled polypeptide can be efficiently quenched by a tryptophan residue (DBO = 2,3-diazabycyclo[2.2.2]oct-2-ene) [179]. If the peptide serves as substrate for proteases, the cleavage of the amino acid chain can be monitored by means of a change in the decay time of DBO (Fig. 29) [191]. Competitive immunoassays for cAMP (cyclic adenosine monophosphate) is another challenging application for time-resolved RET measurements.

A further impact on the development of sensitive detection methods for RET arises from single-photon counting modules (SPCM). This technology is based on avalanche photodiodes, which are semiconductor junctions operated under a reversed bias voltage. A charge-free depletion region is thus created, which behaves as insulator, so that almost no current is flowing through the device. Under these conditions, a large electric field can be applied which accelerates electrons generated inside the material (e.g., by single photons) by the so-called process of impact ionization [192]. This method can be combined with confocal microscopes or fluorescence correlation spectroscopy and can measure the fluorescence emitted by analytes in the zeptomole range down to single molecules [193]. Single-photon counting of RET

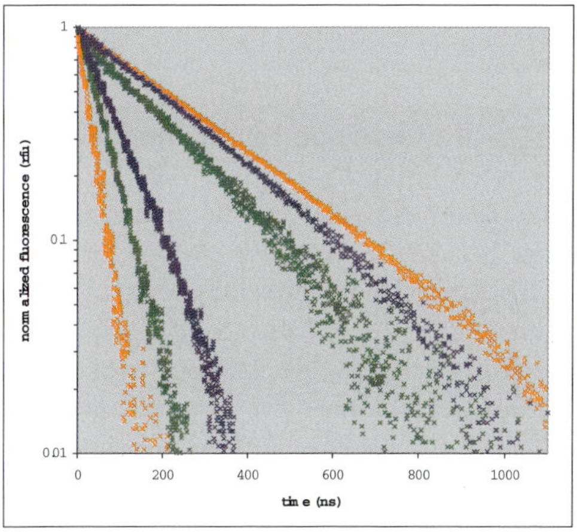

peptide sequence	lifetime	S:B ratio
DBO-ISRRLLGTFTW	59 ns	**43**
DBO-ISRRLL	234 ns	
DBO-RSLKY	95 ns	**11**
DBO-RSLK	287 ns	
WRSLK-**DBO**	35 ns	**1053**
DBO (free label)	315 ns	

Fig. 29 Lifetime measurements with the *Zeiss* "Plate::Vision" of DBO (= 2,3-diazabicyclo [2.2.2]oct-2-ene) for three different oligopeptide protease substrate. For the uncleaved substrates the signal decays fast (τ = 59 ns, 95 ns and 35 ns) due to the collisional, dynamic quenching of the DBO by tryptophan or trypsin. Upon cleavage the quencher and DBO are separated, which results in a lifetime increase (τ = 234 ns, 287 ns and 315 ns). This lifetime change allows a time-gated FTRF detection (delay = 100 ns and gate = 700 ns) of the enzymatic reaction with a high signal-to-background ratio. Data by courtesy of T. Enderle, Hoffmann-La Roche, Pharmaceuticals Division, Assay Development and HTS, Basel/Switzerland [192]

processes was used to analyze single protein–DNA complexes [194], specific DNA sequences in biological samples [195] or homogeneous assays [196], and point mutations [197]. This single-pair FRET (spFRET) technology has some distinct advantages: the processes can be monitored in real time, signals can be detected within a broad dynamic range, and DNA sequence variations can be monitored directly in genomic material without PCR amplification. Furthermore, it is possible to acquire multiple measurements simultaneously with a multi-channel SPCM array. This makes SPCM useful for multi-wavelength systems or the parallel collection of different parameters, such as fluorescence intensity, molecule diffusion times, fluorescence lifetimes, and fluorescence polarization. SPCM could also enable fluorescence detection for HTS instrumentation, using laser scanning optics for consecutive measurements or CCDs for a simultaneous measurement of the sample spots on a microarray. First applications in related systems like capillary electrophoresis [198] or microfluidic devices (e.g., PMMA substrates) [193] could be demonstrated.

4 Conclusion and Outlook

This review illustrates some trends in the development of luminescence lifetime-based or time-resolved ratiometric imaging methods for the quantitative read-out of sensor arrays. Two basic principles can be distinguished. On the one hand, fluorescent dyes can be used as sensitive probes for environmental chemical stimuli like pO_2, pH, (metal) ion concentrations, or H_2O_2 concentrations. On the other hand, luminescent dyes with long decay times can be used as molecular labels for affinity sensors, like DNA or protein arrays and immunoassays. Both applications can be used for a parallel analysis of multiple samples by implementation of microarray and HTS techniques. Among detection methods for HTS, CCD cameras have provoked a great interest because they can be used for fluorescence intensity measurements as well as for time-resolved ratiometric measurements. The latter aspect took the central stage of this review, as time-resolved imaging exhibits particular advantages over steady-state imaging, because background fluorescence and measurement artifacts can be reduced by signal referencing. CCDs provide a high lateral resolution, so miniaturized devices with high spot densities can be screened.

Currently, a large number of molecular sensors are available for different parameters, as described in Sect. 2. They can be incorporated into microwell plate formats and applied either in homogenous solution or immobilized in sensor layers, e.g., polymer hydrogels in the form of ready-for-use sensor arrays. Probes with luminescent lifetimes in the microsecond range, attained by certain metal complexes, are preferred for time-resolved imaging because bottom-of-the-range instrumental set ups are adequate. In this respect, sensor arrays for the determination of one parameter in multiplexed samples have been developed (e.g., pO_2, pH, hydrogen peroxide, or glucose concentrations), or arrays of cross-reactive unspecific sensors for the detection of several parameters (e.g., metal ion concentrations) simultaneously in one sample.

The future perspective within this area will be the preparation of specific multiparameter sensor layers or sensor arrays, which enable the imaging of gradients of several parameters on large area samples or on particular sensor spots in parallel and in real-time. Metabolic imaging is one application in this respect. The indication of regions with increased metabolic rate, as in many tumors, can be achieved by magnetic resonance imaging or by tomographic methods like positron emission tomography (PET). Sensor chips with electrode arrays like ion sensitive field effect transistors (ISFETs), interdigital electrode structures (IDES), or Clark oxygen electrodes can monitor metabolic and respiratory processes in the extracellular environment [31]. In a similar way, fluorescent probes can be arrayed on a sensor chip for a simultaneous online detection of diverse relevant extracellular parameters like changes in pH, oxygen partial pressure, Ca^{2+} concentration, or the consumption of glucose, lactate etc. The combination with advanced time-resolved imaging methods like *t*-DLR could be a successful approach towards a integrated solution. Microarrays based on the immobilization of whole living cells are promising candidates as sensor platforms for the screening of drugs [199, 200], especially for cancer therapy [201], functional genomics [202], expression [203] or transcription studies [204], or for environmental monitoring of pollutants [205] and genotoxins [206]. Expression constructs with GFP or bioluminescent systems can be used as reporter elements for the fluorescent identification and screening. It is assumed that time-resolved imaging methods can also become a useful tool for the screening of cell arrays, due to their intrinsic referencing capabilities. Nevertheless, the large area of whole cell microarrays and their immobilization strategies are out of the scope of this review.

New europium probes have been presented that can indicate changes in hydrogen peroxide concentration and are applicable for time-resolved imaging. As H_2O_2 is a product of the activity of almost all oxidases, quantitative assays for different substrates, e.g glucose, or the determination of the activity of the respective oxidases can be carried out. On the other hand, peroxidases and catalases consume H_2O_2, which can also be monitored via the europium tetracycline probe and can therefore be used for the imaging of ELISAs performed in microwell plates.

There is also a great demand for fluorescent indicators for other reactive oxygen species (ROS) with improved specificity. Oxygen radicals like the superoxide anion radical or the hydroxyl radical play an important role in the peroxidation of unsaturated lipids and thus influence the cell membrane properties and signal transduction pathways [207]. Furthermore, they are of physiological interest, regulating and participating in cellular processes like apoptosis or necrosis. New fluorescent probes, like vanillin-8-aminoquinoline [208] have been introduced for the fluorescent detection of ROS. The dye BODIPY C11 (581/591) contains a long alkyl chain lipohilic moiety and can be distributed especially in cell membranes. It has the advantage that it can be used as a ratiometric dye as it undergoes a red-to-green shift upon oxidation by ROS [209]. The sensitivity of fluorescent europium chelates towards ROS other than H_2O_2 has to be examined and would offer further applications of these probes for time-resolved imaging.

Metal complexes like lanthanide chelates (mainly europium or terbium), ruthenium phenanthrolines or bipyridyls, and platinum porphyrins can be used as fluorescent labels for biomolecules. Their long decay times are perfectly suited for a detection by time-resolved imaging, and the labeled target molecules can be used for the determination of intracellular recognition processes or for the screening of DNA and protein arrays. Ratiometric lifetime-based imaging methods in combination with sophisticated data acquisition and evaluation tools can substantially contribute to the development of straightforward microarray read-out systems, particularly for differential expression analysis or competitive immunoassays. The implementation of RET techniques for the read-out of microarray formats by time-resolved imaging is another intriguing challenge.

As outlined, there has also been a lot of progress in the development of highly sensitive detection systems, culminating in single-photon counting modules, applicable to the analysis of microarrays and related devices. Active-pixel image sensors like CMOS-based digital cameras are one alternative to CCD cameras and can lead to more compact and portable devices [210]. The lifetime-based imaging of sensor arrays as presented here is exclusively performed in the time domain. Nevertheless, phase-sensitive detection systems in the frequency domain exhibit an interesting option by using multichannel photomultplier tubes for the imaging of biosensor arrays [211].

Acknowledgements Special thanks go to Dr. Thilo Enderle, Dr. Gregor Liebsch, Dr. Christian Krause, Dr. Meng Wu, Dr. Zhihong Lin, Dr. Axel Dürkop, Stefan Nagl and Prof. Otto S. Wolfbeis. Their work and support has vitally contributed to the appearance of this article.

References

1. Horobin RW (1980) J Microscopy 119:345–355
2. Sozmen M, Brown PJ, Cripps PJ (1999) Veterin Res 30:99–108
3. Müller M, Tausz M, Wonisch A, Grill D (1999) Free Radical Res 31:S121–127
4. Ruifrok AC, Katz RL, Johnston DA (2002) Appl Imminuhistochem Mol Morph 11:85–91
5. Soini E, Hemmila I (1979) Clinical Chem 25:353–361
6. Wielbo D, Simon A, Phillips MI, Toffolo S (1996) Hypertension 28:147–151
7. Zhang SP, Zhou LW, Morabito M, Lin RC, Weiss B (1996) J Mol Neurosci 7:13–28
8. Guzowski JF (2002) Hippocampus 12:86–104
9. Seisenberger G, Ried MU, Endress T, Büning H, Hallek M, Bräuchle C (2001) Science 294:1929–1932
10. Goulian M, Simon SM (2000) Biophys J 79:2188–2198
11. Kubitscheck U (2002) Single Molecules 5–6:267–274
12. Schäfer B, Gemeinhardt H, Uhl V, Greulich KO (2000) Single Molecules 1:33–40
13. Bastiaens PIH, Verveer PJ, Squire A, Wouters F (2001) Fluorescence lifetime imaging microscopy of signal transduction protein reactions in cells. In: Valeur B, Brochon JC (eds) Springer series on fluorescence 1 (New trends in fluorescence spectroscopy). Springer, Berlin Heidelberg New York p 297–302
14. Szmacinski H, Lakowicz JR (1990) Topics in fluorescence spectroscopy. Probe design and chemical sensing, vol 4. Plenum, New York
15. Wolfbeis OS, Böhmer M, Dürkop A, Enderlein J, Gruber M, Klimant I, Krause C, Kürner J, Liebsch G, Lin Z, Oswald B, Wu M (2002) Advanced luminescent labels, probes and beads, and their application to luminescence bioassay and imaging. In: Kraayenhof R, Visser AJWG, Gerritsen HC (eds) Springer series on fluorescence 2 (Fluorescence spectroscopy, imaging and probes). Springer Berlin Heidelberg New York, p 3–42

16. Valeur B (2002) Molecular fluorescence. Wiley-VCH Weinheim
17. Babilas P, Schacht V, Liebsch G, Wolfbeis OS, Landthaler M, Szeimies RM, Abels C (2003) Br J Cancer 88:1462–1469
18. Liebsch G, Klimant I, Krause C, Wolfbeis OS (2001) Anal Chem 73:4354–4363
19. Mycek MA, Pogue BW (2003) Handbook of biomedical fluorescence. Dekker, New York
20. Gerritsen HC, De Grauw K (2001) Metabol Cell Imaging 309–323
21. Szöllösi J, Damjanovich S, Matyus L (1998) Cytometry 34:159–179
22. Cubeddu R, Comelli D, D'Andrea C, Taroni P, Valentini G (2002) J Phys D: Appl Phys 35:R61–R76
23. Wang XF, Herman B (1996) Fluorescence imaging spectroscopy and microscopy. Chemical analysis series, vol 137. Wiley, New York
24. Gadella TWJ, Jovin TM, Clegg RM (1993) Biophys Chem 48:221–239
25. Wang XF, Periasamy A, Herman B (1992) Crit Rev Anal Chem 23:369–395
26. Hartmann P, Ziegler W (1996) Anal Chem 68:4512–4514
27. Hartmann P, Ziegler W, Holst G, Lübbers DW (1997) Sens Act B 38:110–115
28. Woods RJ, Scypinski S, Cline Love LJ, Ashworth HA (1984) Anal Chem 56:1395–1400
29. Ballew RM, Demas JN (1989) Anal Chem 61:30
30. Ni TQ, Melton LA (1991) Appl Spec 45:983
31. Ehret R, Baumann W, Brischwein M, Lehmann M, Henning T, Freund I, Drechsler S, Friedrich U, Hubert ML, Motrescu E, Kob A, Palzer H, Grothe H, Wolf B (2001) Fresenius J Anal Chem 369:30–35
32. Molls M, Vaupel P (1998) The impact of the tumor environment on experimental and clinical radiation oncology and other therapeutic modalities. In: Molls M, Vaupel P (eds) Blood perfusion and microenvironment of human tumors – implications for clinical radiooncology. Springer, Berlin Heidelberg New York
33. Kellner K, Liebsch G, Klimant I, Wolfbeis OS, Blunk T, Schulz MB, Göpferich A (2002) Biotech Bioeng 80:73–83
34. Shah K, Tung CH, Chang CH, Slootweg E, O'Loughlin T, Breakefield XO, Weissleder R (2004) Cancer Res 64:273–278
35. Weissleder R, Tung CH, Mahmood U, Bogdanov A (1999) Nature Biotech 17:375–378
36. Pottier R, Chow YFA, LaPlante JP, Truscott TG, Kennedy JC, Beiner LA (1986) J Photochem Photobiol 44:679–687
37. Langer S, Abels C, Botzlar A, Pahernik S, Rick K, Szeimies RM, Goetz AE (1999) J Invest Dermatol 112:723–728
38. Rittenhouse-Diakun K, Van Leengoed H, Morgan J, Hryhorenko E, Paskiewicz G, Whitaker JE, Oseroff AR (1995) Photochem Photobiol 61:523–528
39. Kriegmair M, Stepp H, Steinbach P, Lumper W, Ehsan A, Stepp HG, Rick K, Knuchel R, Baumgartner R, Hofstetter A (1995) Urol Int 55:190–196
40. Bedwell J, MacRobert AJ, Phillips D, Bown S (1992) Br J Cancer 65:818–824
41. Schneckenburger H, König K, Kunzi-Rapp K, Westphal-Frösch C, Rück A (1993) J Photochem Photobiol B 21:143–147
42. Vaupel P, Kallinowski F, Okunieff P (1989) Cancer Res 49:6449–6465
43. Orellana G, Garcia-Fresnadillo D (2004) Environmental and industrial optosensing with tailored luminescent Ru(II) polypyridyl complexes. In: Narayanaswamy R, Wolfbeis OS (eds) Optical sensors: industrial, enviromental and diagnostic applications. Springer series on chemical sensors and biosensors vol 1. Springer, Berlin Heidelberg New York, p 309–357
44. Sharma A, Wolfbeis OS (1988) Appl Spec 42:1009

45. Lakowicz JR (1999) Principles of fluorescence spectroscopy, 2nd edn. Kluwer/Plenum, New York
46. Birks JB (1970) Organic molecular photophysics. Wiley, New York, p 545
47. Carraway ER, Demas JN, DeGraff BA, Bacon JR (1991) Anal Chem 63:337–342
48. Mills A, Thomas M (1997) Analyst 122:63–68
49. Mingoarranz FJ, Moreno-Bondi MC, Garcia-Fresnadillo D, de Dios C, Orellana G (1995) Mikrochim Acta 121:107–118
50. Papkovsky DB, Olah J, Troyanowsky IV, Sadowsky NA, Rumyantseva VD, Mironov AF, Yaropolov I, Savitsky AP (1991) Biosens Bioelectron 7:199–206
51. Amao Y (2003) Microchim Acta 143:1–12
52. Khalil G, Gouterman M, Ching S, Costin C, Coyle L, Gouin S, Green E, Sadilek M, Wan R, Yearyean J, Zelelow B (2002) J Porphyrins Phthtalocyanines 6:135–145
53. Spellane PJ, Gouterman M, Antipas A, Kim S, Liu YC (1980) Inorg Chem 19:386–391
54. Zipp AP, Sacksteder LA, Streich J, Cook A, Demas JN, DeGraff BA (1993) Inorg Chem 32:5629–5632
55. Kneas KA, Xu W, Demas JN, DeGraff BA, Zipp AP (1998) J Fluorescence 8:295–300
56. Karpovich DS, Blanchard GJ (1995) J Phys Chem 99:3951–3958
57. Juris A, Balzani V, Barigelleti F, Campagna S, Belser P, Von Zelewsky A (1988) Coord Chem Rev 84:85–277
58. Zhong W, Urayama P, Mycek M-A (2003) J Phys D: Appl Phys 36:1689–1695
59. Eaton K, Douglas B, Douglas P (2004) Sens Act B 97:2–12
60. Liebsch G, Klimant I, Frank B, Holst G, Wolfbeis OS (2000) Appl Spec 54:548–559
61. Kolle C, Gruber W, Trettnak W, Biebernik K, Dolezal C, Reininger F, O'Leary P (1997) Sens Act B 38:141–149
62. Draxler S, Lippitsch ME, Klimant I, Kraus H, Wolfbeis OS (1995) J Phys Chem 99:3162–3167
63. Mills A, Lepre A (1997) Anal Chem 69:4653–4659
64. Gillanders RN, Tedford MC, Crilly PJ, Bailey RT (2004) Anal Chim Acta 502:1–6
65. Mills A, Thomas MD (1998) Analyst 123:1135–1140
66. Holst G, Glud RN, Kühl M, Klimant I (1997) Sens Act B 38–39:122–129
67. Lin J (2000) Trends Anal Chem 19:541–551
68. Schulman SG, Chen S, Bai F, Leiner MJP, Weis L, Wolfbeis OS (1995) Anal Chim Acta 304:165–170
69. Draxler S, Lippitsch ME (1995) Sens Act B 29:199–203
70. Kermis HR, Kostov Y, Rao G (2003) Analyst 128:1181–1186
71. Rao BS, Puschett JB, Mayjaszewski K (1991) J Appl Polym Sci 43:925–928
72. Wolfbeis OS, Offenbacher H (1986) Sens Act 9:85–91
73. Saari LA, Seitz WR (1982) Anal Chem 54:823–824
74. Cajlakovic M, Lobnik A, Werner T (2002) Anal Chim Acta 455:207–213
75. Hanson KM, Behne MJ, Barry NP, Mauro TM, Gratton E, Clegg RM (2002) Biophys J 83:1682–1690
76. Gerritsen HC, De Grauw K (2001) One- and two photon confocal lifetime imaging and its applications. In: Periasamy A (ed) Methods in cellular imaging. Oxford University Press, New York, p 308–323
77. Herman P, Lin HJ, Lakowicz JR (2003) Lifetime-based imaging. In: Vo-Dinh T (ed) Biomedical photonics. CRC, Boca Raton, p 9.1–9.30
78. Vroom JM, De Grauw KJ, Gerritsen HC, Bradshaw DJ, Marsh PD, Watson GK (1999) Appl Environm Microbiol 65:3502–3511
79. Lin HJ, Herman P, Lakowicz JR (2003) Cytometry 52A:77–89
80. Boyer AE, Devanathan S, Hamilton D, Patonay G (1992) Talanta 39:505–510

81. Dybko A, Romaniuk RS, Maciejewski J, Brzozka Z (1992) Int J Optoelectron 7:443
82. Igarashi S, Kuwae K, Yotsuyanagi T (1994) Anal Sci 10:821
83. Kuswandi B, Narayanasway R (1999) Fresenius J Anal Chem 364:605
84. Kostov Y, Tzonkov S, Yotava L, Krysteva M (1993) Anal Chim Acta 280:15–19
85. Wroblewski W, Roznicka E, Dybko A, Brzozka Z (1998) Sens Act B 48:471
86. Wang E, Chow KF, Kwan V, Chin T, Wong C, Bocarsly A (2003) Anal Chim Acta 495:45–50
87. Shahriari MR, Ding JY (1994) In: Klein LC (ed) Sol-gel optics: processing and applications, chap 13. Kluwer, Boston p 279
88. Lobnik A, Oehme I, Murkovic I, Wolfbeis OS (1998) Anal Chim Acta 367:159–165
89. Huiqui H, Gaixia X, Xuesong Y, Ping W (2003) Meas Sci Technol 14:1040–1046
90. Chan CM, Fung CS, Wong KY, Lo W (1998) Analyst 123:1843–1847
91. Hulth S, Aller RC, Engstrom P, Selander E (2002) Limnol Oceanogr 47:212–220
92. Haugland RP (2002) Handbook of fluorescent probes and research products. Molecular Probes, www.probes.com
93. Mortimer RJG, Krom MD, Hall POJ, Hulth S, Ståhl H (1998) Marine Chem 63:119–129
94. Lavigne JJ, Savoy S, Clevenger MB, Ritchie JE, McDoniel B, Yoo SJ, Anslyn EV, McDevitt JT, Shear JB, Neikirk D (1998) J Am Chem Soc 120:6429–6430
95. Mayr T, Igel C, Liebsch G, Klimant I, Wolfbeis OS (2003) Anal Chem 75:4389–4396
96. Mayr T, Liebsch G, Klimant I, Wolfbeis OS (2002) Analyst 127:201–203
97. Leca B, Blum LJ (2000) Analyst 125:789
98. Fähnrich KA, Prawda M, Guilbault GG (2001) Talanta 54:531
99. Saifer A, Gerstenfeld S (1985) J Lab Clin Med 51:448
100. Hirschy LM, van Geel TF, Winefordner JD, Kelly RN, Schulman SG (1984) Anal Chim Acta 166:207–219
101. Wolfbeis OS, Dürkop A, Wu M, Lin Z (2002) Angew Chem Intl Ed 41:4495–4498
102. Wolfbeis OS, Schäferling M, Dürkop A (2003) Microchim Acta 143:221–227
103. Schäferling M, Wu M, Enderlein J, Bauer H, Wolfbeis OS (2003) Appl Spec 57:1386–1392
104. Matsumoto K, Tsukatani T (1996) Anal Chim Acta 321:157–164
105. Metzger A, Anslyn EV (1998) Angew Chem Intl Ed 37:649–652
106. Cabell LA, Best MD, Lavigne JJ, Schneider SE, Perreault DM, Monahan M, Anslyn EV (2001) J Chem Soc Perkin Trans 2:315–323
107. Lin Z, Wu M, Schäferling M, Wolfbeis OS (2004) Angew Chem Int Ed 43:1735–1738
108. Lin Z (2004) Time-resolved fluorescence-based europium-derived probes for peroxidase bioassays, citrate cycle imaging and chirality sensing. PhD thesis, University of Regensburg
109. Moreno-Bondi MC, Wolfbeis OS, Leiner MJP, Schaffar BPH (1990) Anal Chem 62:2377–2380
110. Wolfbeis OS, Oehme I, Papkovskaya N, Klimant I (2000) Biosens Bioelectron 15:69–76
111. Schultz JS, Mansouri S, Goldstein IJ (1982) Diabetes Care 5:245–254
112. Tolosa L, Szmacinski H, Rao G, Lakowicz JR (1997) Anal Biochem 250:120–128
113. DiCesare N, Lakowicz JR (2001) Anal Biochem 294:154–160
114. Schäferling M, Wu M, Wolfbeis OS (2004) J Fluorescence 14:561–568
115. Wu M, Lin Z, Schäferling M, Dürkop A, Wolfbeis OS (2005) Anal Biochem 340(1):66–73
116. Wu M, Lin Z, Wolfbeis OS (2003) Anal Biochem 320:129–135

117. Bucher DJ, Mikhail A, Popple S, Graves P, Meiklejohn G, Hodes DS, Johansson K, Halonen PE (1991) J Clin Microbiol 29:2484–2488
118. Lombardi V, Caniglia M, Scarlatti G, Jansson M, Plebani A, D'Argenio P, Scaccia S, Wigzell H, Rossi P (1993) Eur J Pediatrics 152:484–489
119. Meyer J, Karst U (2001) Analyst 126:175–178
120. Slavik J (1994) Fluorescent probes in cellular and molecular biology. CRC, Boca Raton
121. Wood P, Geoff B (1997) Fluoroimmunoassay. In: Price CP, Newman DJ (eds) Principles and practice of immunoassay, 2nd edn. Macmillan, London, p 391–424
122. Niemeyer CM, Blohm D (1999) Angew Chem Int Ed 38:2865–2869
123. Wang J (2000) Nucl Acids Res 28:3011–3016
124. Jain KK (2000) Pharmacogenomics 1:289–307
125. Pirrung MC (2002) Angew Chem Intl Ed 41:1276–1289
126. Schena M (2002) Microarray Analysis. Wiley, Hoboken NJ
127. Simon RM, Korn EL, McShane LM, Radmacher MD, Wright GW, Zhao Y (2004) Design and analysis of DNA microarray investigations. Springer, Berlin Heidelberg New York
128. MacBeath G (2002) Nature Gen Suppl 32:526–532
129. Angenendt P, Glökler J, Murphy D, Lehrach H, Cahill DJ (2002) Anal Biochem 309:253–260
130. Schäferling M, Schiller S, Paul H, Kruschina M, Pavlickova P, Giammasi C, Kambhampati D (2002) Electrophoresis 23:3097–3105
131. Kambhampati D (2004) Protein microarray technology. Wiley-VCH, Weinheim
132. Preininger C, Sauer U (2004) Design, quality control and normalization of biosensor chips. In: Narayanaswamy R, Wolfbeis OS (eds) Springer series on chemical sensors and biosensors 1 (Optical sensors). Springer, Berlin Heidelberg New York, p 67–92
133. Weller MG (2003) Anal Bioanal Chem 375:15–17
134. www.amershambiosciences.com
135. Oswald B, Patsenker L, Duschl J, Szmacinski H, Wolfbeis OS, Terpetschnig E (1999) Bioconjugate Chem 10:925–931
136. Dürkop A, Lehmann F, Wolfbeis OS (2002) Anal Bioanal Chem 372:688–694
137. Saiki RK, Scharf S, Faloona F, Mullis KB, Horn GT, Erlich HA, Arnheim N (1985) Science 230:1350–1354
138. Rolfs A, Sculler I, Finckh U, Weber-Rolfs I (1992) PCR: Clinical diagnostic and research. Springer, Berlin Heidelberg New York
139. Dirks RW, Van Gijlswijk RPM, Vooijs MA, Smit AB, Bogerd J, Van Minnen J, Raap AK, Van der Ploeg M (1991) Exp Cell Res 194:310–315
140. Cortese JD (2001) The Scientist 15[24]:36
141. Constans A (2003) The Scientist 17[3]:37–38
142. www.api.com
143. Zhu H, Bilgin M, Bangham R, Hall D, Casamayor A, Bertone P, Lan N, Jansen R, Bidlingmaier S, Houfek T, Mitchell T, Miller P, Dean RA, Gerstein M, Snyder M (2001) Science 293:2101–2105
144. Tantra R, Cooper J (2002) Sens Actuat B 82:233–240
145. He H, Nunnally BK, Li LC, McGown LB (1998) Anal Chem 70:3413–3418
146. Lieberwirth U, Arden-Jacob J, Drexhage KH, Herten DP, Müller R, Neumann M, Schulz A, Siebert S, Sagner G, Klingel S, Sauer M, Wolfrum J (1998) Anal Chem 70:4771–4779
147. Valentini G, D'Andrea C, Comelli D, Pifferi A, Taroni P, Torricelli A, Cubeddu R, Battaglia C, Consolandi C, Salani G (2000) Opt Lett 25:1648–1650

148. Waddell E, Wang Y, Stryjewski W, McWorther S, Henry AC, Evans D, McCarley R, Soper S (2000) Anal Chem 72:5907–5917
149. Hemmilä I, Soini E, Lövgren T (1982) Fresenius Z Anal Chem 311:357
150. Degan P, Montagnoli G, Wild CP (1989) Clin Chem 35:2308–2310
151. Markela E, Stahlberg TH, Hemmilä I (1993) J Immun Meth 161:1–6
152. Osman S, Turpeinen U, Itkonen O, Stenman UH (1993) J Immun Meth 161:97–106
153. Härmä H, Pelkkikangas AM, Soukka T, Huhtinen P, Huopalahti S, Lövgren T (2003) Anal Chim Acta 482:157–164
154. Weibel N, Charbonniere LJ, Guardigli M, Roda A, Ziessel R (2004) J Am Chem Soc 126:4888–4896
155. Ge P, Selvin PR (2003) Bioconj Chem 14:870–876
156. Selvin PR (2003) Top Fluor Spec 7:177–212
157. O'Sullivan PJ, Burke M, Soini A, Papkovsky DB (2002) Nucl Acids Res 30:E114
158. O'Riordan TC, Soini AE, Papkovsky DB (2001) Anal Biochem 290:366–375
159. O'Riordan TC, Soini AE, Soini JE, Papkovsky DB (2002) Anal Chem 74:5845–5850
160. Berggren K, Steinberg TH, Lauber WM, Carroll JA, Lopez MF, Chernokalskaya E, Zieske L, Diwu Z, Haugland RP, Patton WF (1999) Anal Biochem 276:129–143
161. www.chromeon.com
162. Kürner JM, Wolfbeis OS, Klimant I (2002) Anal Chem 74:2151–2156
163. Lian W, Litherland SA, Badrane H, Tan W, Wu D, Baker HV, Gulig PA, Lim DV, Jin S (2004) Anal Biochem 334:135–144
164. Tan M, Ye Z, Wang G, Yuan J (2004) Anal Chem 76:513–518
165. Cao Y, Koo Y-EL, Kopelman R (2004) Analyst 129:745–750
166. www.eriesci.com
167. Nagl S (2004) Quantitative determination of proteins on microarrays by time-resolved luminescent imaging. Diploma thesis, University of Regensburg
168. Chan WCW, Gao X, Nie S (2004) Colloids Coll Assem 493–506
169. Heer S, Lehmann O, Haase M, Guedel HU (2003) Angew Chem Int Ed 42:3179–3182
170. Dahan M, Levi S, Luccardini C, Rostaing P, Riveau B, Triller A (2003) Science 302:442–445
171. Gao X, Cui Y, Levenson RM, Chung LWK, Nie S (2004) Nature Biotechnol 22:969–976
172. Ness JM, Akhtar RS, Latham CB, Roth KA (2003) J Histochem Cytochem 51:981–987
173. Chan WCW, Maxwell DJ, Gao X, Bailey RE, Han M, Nie S (2002) Curr Op Biotech 13:40–46
174. www.qdots.com
175. Gerion D, Chen F, Kannan B, Fu A, Parak WJ, Chen DJ, Majumdar A, Alivisatos AP (2003) Anal Chem 75:4766–4772
176. Selvin PR (2000) Nature Struct Biol 7:730–734
177. Jares-Erijman EA, Jovin TM (2003) Nature Biotech 21:1387–1395
178. Suzuki Y, Yasunaga T, Ohkura R, Wakabayashi T, Sutoh K (1998) Nature 396:380–383
179. Hudgins RR, Huang F, Gramlich G, Nau WM (2002) J Am Chem Soc 124:556–564
180. Fang X, Liu X, Schuster S, Tan W (1999) J Am Chem Soc 121:2921–2922
181. Gonzalez-Gomez F, Vergara F, Fernandez A, Pedrosa C, Ramirez JP, Castilla JA, Ruiz R, Galan JJ, Ruiz A, Real LM (2003) Clin Chem Lab Med 41:392–393
182. Summerer D, Marx A (2002) Angew Chem Int Ed 41:3620–3622
183. Shah K, Tung CH, Chang CH, Slootweg E, O'Loughlin T, Breakefield XO, Weissleder R (2004) Cancer Res 64:273–278
184. Kircher MF, Weissleder R, Josephson L (2004) Bioconj Chem 15:242–248
185. Kohl T, Heinze KG, Kuhlemann R, Koltermann A, Schwille P (2002) PNAS 99:12 161–12 166

186. Selvin PR (2002) Ann Rev Biophys Biomol Structure 31:275–302
187. Oswald B, Gruber M, Lehmann F, Probst M, Wolfbeis OS (2001) Photochem Photobiol 74:237–242
188. Augustin CM, Oswald B, Wolfbeis OS (2002) Anal Biochem 305:166–172
189. www.zeiss.de
190. Mueller F, Belik D, Hermann R, König B, Josel H-P (2004) PCT WO0241001 A1
191. Enderle T, Nau W, Roth D, Matile H, Josel HP (2004) MRS Spring Meeting, AA3.7
192. Dautet H, Deschamps P, Dion B, MacGregor AD, MacSween D, McIntyre RS, Trottier C, Webb PP (1993) Appl Opt 32:3894–3900
193. Trottier C, Davies M, Dautet H, Wabuyele M, Soper SA, Kapanidis AN, Lacoste T, Weiss S (2004) PharmaGenomics 4(2):24–34
194. Kapanidis AN, Ebright YW, Ludescher RD, Chan S, Ebright RH (2001) J Mol Biol 312:453–468
195. Castro A, Williams JGK (1997) Anal Chem 69:3915–3920
196. Knemeyer JP, Marmé N, Sauer M (2000) Anal Chem 72:3717–3724
197. Wabuyele MB, Farquar H, Stryjewski W, Hammer RP, Soper S, Cheng YW, Barany F (2003) J Am Chem Soc 125:6937–6945
198. McWorther S, Soper SA (2000) Electrophoresis 21:1267–1280
199. Bailey SN, Wu RZ, Sabatini DM (2002) Drug Discovery Today 7:S113–S118
200. Mousses S, Kallioniemi A, Kauraniemi P, Elkahloun A, Kallioniemi OP (2001) Curr Opin Chem Biol 6:97–101
201. Henning T, Brischwein M, Baumann W, Ehret R, Freund I, Kammerer R, Lehmann M, Schwinde A, Wolf B (2001) Anti-Cancer Drugs 12:21–32
202. Wu RZ, Bailey SN, Sabatini DM (2002) Trends Cell Biol 12:485–488
203. Ziauddin J, Sabatini DM (2001) Nature 411:107–110
204. Van Dyk TK, DeRose EJ, Gonye GE (2001) J Bacteriol 183:5496–5505
205. Belkin S (2003) Curr Opin Microbiol 6:206–212
206. Kuang Y, Biran I, Walt DR (2004) Anal Chem 76:2902–2909
207. Borst JW, Uskova MA, Visser NV, Visser AJWG (2002) Imaging of oxidative stress in plant cells by quantitative fluorescence microscopy and spectroscopy. In: Kraayenhof R, Visser AJWG, Gerritsen HC (eds) Springer series on fluorescence 2 (Fluorescence spectroscopy, imaging and probes). Springer, Berlin Heidelberg New York, p 338–348
208. Tang B, Zhang L, Hu JX, Li P, Zhang H, Zhao YX (2004) Anal Chim Acta 502:125–131
209. Van Borren M, Brady NR, Ravelsloot J, Westerhoff HV (2002) Looking into a living cell. In: Kraayenhof R, Visser AJWG, Gerritsen HC (eds) Springer series on fluorescence 2 (Fluorescence spectroscopy, imaging and probes). Springer, Berlin Heidelberg New York, p 362–372
210. Janesick J, Putnam G (2003) Ann Rev Nucl Part Sci 53:263–300
211. Rabinovich E, O'Brien MJ, Brueck SRJ, Lopez GP (2000) Rev Sci Instrum 71:522–529

Springer Ser Chem Sens Biosens (2005) 3: 93–132
DOI 10.1007/5346_003

Published online: 6 October 2005

Cataluminescence-Based Gas Sensors

Masuo Nakagawa[1] (✉) · Nobuhiko Yamashita[2]

[1]Faculty of Science, Okayama University of Science, 1-1 Ridai-cho, 700-0005 Okayama, Japan
masuo@dap.ous.ac.jp

[2]Faculty of Education, Okayama University, 3-1-1 Tsushima-naka, 700-8530 Okayama, Japan
nobu@cc.okayama-u.ac.jp

Abstract Cataluminescence (CTL) is chemiluminescence emitted in a course of catalytic oxidation. Since 1990, the present authors and coworkers have observed CTL during the catalytic oxidation of various organic vapors in air. This phenomenon has been applied to the CTL-based sensors for detecting combustible vapors. THE CTL response is fast, reproductible and proportional to the concentration of the combustible vapors of ppm orders in air. Based on two types of models of the CTL, the relationship between the CTL intensity and the rate of catalytic oxidation have been investigated analytically. In this article, the effects of catalyst temperature, gas flow-rate and gas concentration on the CTL intensity are demonstrated. Finally, various types of sensing system using the CTL-based sensor are proposed. The results of discrimination and determination of more than ten types of vapors of various concentrations are shown.

Keywords Cataluminescence · Catalytic oxidation · Combustible vapour · Gas sensor · Discrimination

1 Introduction

Generally speaking, there are two gas-detection mechanisms using gas sensors: one utilizing changes in sensor material by interaction with gases, and the other utilizing signals obtained from the gas itself. The latter is favorable in view of the stability of the gas sensor.

The catalytic reaction of gases on a solid catalyst proceeds without changes in the solid. In general, the catalytic oxidation of combustible gases in air causes reaction heat. A very stable gas sensor named "Pellistor" was developed in 1966 utilizing the increase in temperature of the solid catalyst due to the reaction heat [1]. This type of gas sensor has a long history and has been widely used for protection against explosion by leakage of combustible gases in coal mines and chemical plants. This is because of the reliability of this sensor, having no disinformation and a long-life. One disadvantage of this type of sensor is that it is only effective for the detection of combustible gases of higher concentrations, near to the lower explosion limit. This is because the increase in catalyst temperature by reaction heat is less than a few degrees, even in air containing explosive gases of % orders.

The chemiluminescence-based sensing of gases or vapors has been reported by many workers including the present authors and coworkers [2–23] (Table 1). Earlier observations of chemiluminescence during catalysis are reviewed by Claudel et al. [24].

In 1976, Breysse et al. found that chemiluminescence is emitted during catalytic oxidation of carbon monoxide on thoria ThO_2, and named it cataluminescence (CTL) [6]. As the CTL intensity depends linearly on the rate of the catalytic oxidation, this is applicable to the detection of combustible gases.

Since 1990, the present authors and coworkers have observed CTL during the catalytic oxidation of various organic vapors: ethanol, butanol, acetone, *n*-

Table 1 Chemiluminescence from catalytic oxidation of gases or vapors

Catalysts	Gases or vapors	Refs.
Silica-gel (luminol)	Ozone (O_3)	Regener [2]
Silica-gel (rhodamine-B)	Ozone (O_3)	Regener [3]
Silica-gel (rhodamine-B)	Ozone (O_3)	Hodgeson [4]
Gallic acid (rhodamine-B)	Ozone (O_3)	Bersis & Vassiliou [5]
Thoria, ThO_2	CO	Breysse et al. [6]
Thoria, ThO_2:RE	CO	Aras et al. [7]
γ-Al_2O_3	Ethanol and acetone	Nakagawa et al. [8]
$BaSO_4$:Eu	Ethanol and acetone	Nakagawa et al. [9]
γ-Al_2O_3	Ethanol, butanol, acetone, and *n*-butyric acid	Utsunomiya et al. [10–13], Nakagawa [14]
γ-Al_2O_3	Ethanol and acetone	Nakagawa et al. [15, 16]
γ-Al_2O_3	Ethanol, butanol, acetone, MEK(ketone), xylene, and *n*-butyric acid	Nakagawa et al. [17]
γ-Al_2O_3:Dy	Methanol, ethylene, and *iso*-butane	Okabayashi et al. [18]
γ-Al_2O_3:Dy	Linalool, citral, limonene, and α-pinene	Okabayashi et al. [19, 20]
$SrCO_3$, Al_2O_3, TiO_2, MgO, Y_2O_3, and $LaCoO_3$:Sr^{2+}	Ethanol, acetone	Zhu et al. [21]
Nanosized ZrO_2	Ethanol	Zhang Z et al. [22]
Nanosized $SrCO_3$	Ethanol	Shi et al. 2002 [23]

butyric acid, and the fragrance substances of linalool, cital, limonene, and α-pinene [8–20].

The CTL spectrum is different to the incandescent spectrum from solids (black-body radiation), and no luminescence is observed in an atmosphere without combustible gases. This has enabled highly sensitive gas detection together with the recent development of a photodetector with high sensitivity. This very sensitive gas sensor also has a stability and linearity similar to the Pellistor, and we call it the "CTL-based gas sensor".

2
Background of CTL-Based Gas Sensors

Application of chemiluminescence to chemical analysis has been developing since the latter half of the 1950s. This method has many advantages, e.g., high sensitivity, good selectivity, linearity in a wide concentration range, and quick response. It has also been used for the measurement of air pollutants. This method, however, requires a supply of reactants to produce luminescent species through chemical reaction. This is a difficult point for the application of this method to gas sensors.

Regener [2] developed and applied the chemiluminescence-based sensor for determining ozone in 1960 for the first time. The sensor was made of a layer of silica-gel powder containing luminol, which emits luminescence in the course of oxidation by ozone. He improved this sensor to avoid the influence of humidity in ambient atmosphere using rhodamine-B instead of luminol [3]. This sensor had sensitivity of one part per billion (ppb) of ozone, and the luminescence intensity was proportional to the ozone concentration. It also had good selectivity for ozone sensing, and the effects of nitrogen dioxide and sulfur dioxide were about 1/5000 of that of ozone. Although this type of sensor containing a luminescent substance has been improved [4], calibration is necessary because of consumption of the sensor substance (luminol or rhodamine-B) by ozone during the measurement.

On the other hand, Bersis and Vassiliou [5] have attempted to use gallic acid as an ozone acceptor and rhodamine-B as a photon emitter that remains unchanged during the measurement. They used an ethanol solution containing gallic acid and rhodamine-B as reactive reagents, which meant that this method could not be applied to a solid sensor.

In 1976, Breysse et al. [6] observed CTL emission during the catalytic oxidation of carbon monoxide on a thoria surface. This luminescence is ascribed to light emission during catalysis in an atmosphere containing oxygen. This implies that carbon monoxide in the air can be detected continuously without any consumption of sensor substance by the measurement of the CTL emission from a thoria surface.

The present authors and coworkers observed intense CTL emission during the catalytic oxidation of ethanol or acetone vapor on a heated aluminum oxide powder [8]. This phenomenon was applied to the consumption-free CTL-based sensor for detecting combustible vapors. The CTL response was fast and reproducible for a change in concentration of a sample vapor in air. CTL emission has three distinct features:

1. There is an approximately linear relationship between the CTL intensity and the concentration of the combustible vapors, ranging over more than two orders of concentration
2. The profile of the CTL spectrum depends on the kind of combustible vapor
3. The profile of the CTL intensity versus sensor temperature (glow curve) also depends on the kind of combustible vapor

3 Principles of CTL and Working Mechanism of CTL-Based Sensors

There are two types of mechanisms for the CTL emission. One is the recombination radiation (Sect. 3.1) and the other is radiation from the excited species (Sect. 3.2).

3.1 Recombination Radiation

Figure 1 shows a schematic illustration to depict the CTL emission process due to the recombination of carriers in catalyst particles. The figure shows

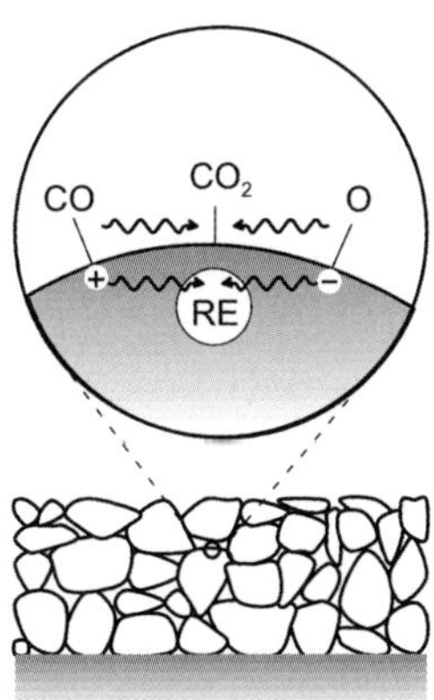

Fig. 1 Schematic illustration of recombination radiation

a layer of the catalyst particles laid on a substrate, and the magnified surface of a catalyst particle in the layer is shown in a circle. Gas molecules in ambient atmosphere of the catalyst diffuse in the gas phase and reach the surface of the particle. For example, in the oxidation of carbon monoxide (CO) on the thoria (ThO_2) surface, CO is chemisorbed to form the CO^+ ion, and O_2 is dissociated and is chemisorbed to form the O^- ion. This means that the chemisorbed CO forms a positively ionized surface state (the surface donor state) accompanied by the localized hole (represented by the symbol + in a circle) and the chemisorbed oxygen forms a negatively ionized surface-state (the surface acceptor state) accompanied by the localized electron (represented by the symbol – in a circle) (Fig. 1). The chemisorbed species CO and O can migrate on the solid surface keeping the chemisorbed state. Surface reaction between these chemisorbed species will occur to form the chemisorbed CO_2. The chemisorbed CO_2 accompanied by the localized electron and hole is a chemisoption surface state bound to an exciton. The desorption of CO_2 is accompanied by annihilation of the exciton, and luminescence is emitted by recombination of electron and hole [6].

This process can explain the detailed events with the help of the electron theory of catalysts advocated by Wolkenstein [25]. Figure 2b shows a conceptual illustration of an ionic crystal consisting of M^{2+} and O^{2-} ions, e.g., a metal oxide

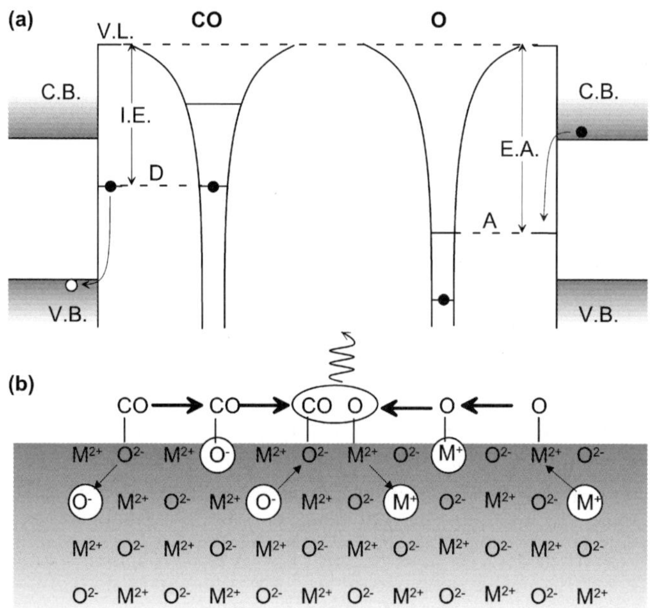

Fig. 2 Conceptual illustration of **a** energy band scheme and **b** an ionic crystal lattice consisting of M^{2+} and O^{2-} ions

crystal. In this figure the shaded region denotes crystal and the upper region denotes gas phase. The outermost shell of the O^{2-} ion is occupied by a valence electron and it forms a valence band in the crystal due to the interaction between atoms in the crystal lattice. When the O^{2-} ion lose one electron, it forms a O^{-} ion. This means that a free hole, which is an unoccupied level in the valence band, is produced. This is expressed by O^{-} in a circle in Fig. 2b. The O^{-} ion can move to the position of the neighboring O^{2-} ion, one after another, by getting an electron from the neighboring O^{2-} ion in the crystal.

On the other hand, when a M^{2+} ion gets an electron, it forms a M^{+} ion. The M^{+} ion can move to the position of the neighboring M^{2+} ion, one after another, by giving an electron to the neighboring M^{2+} ion in the crystal. This means that a free electron is located at the position of a M^{+} ion. This is expressed by M^{+} in a circle in Fig. 2b.

Figure 2a shows the energy band scheme of the ionic crystal; the shaded region V.B. denotes valence band, the shaded region C.B. denotes conduction band, and the energy level V.L. denotes vacuum level. The cone-shaped curves are the potential curves around the chemisorbed particles and the horizontal lines denotes energy levels of the surface state produced by chemisorption of CO and O. The energy difference between the vacuum level and the highest electron-occupied level of chemisorbed CO is ionization energy denoted as I.E. The energy difference between vacuum level and the empty level just above the highest electron-occupied level of chemisorbed O is the electron affinity denoted as E.A. Because of the large value of oxygen E.A., the level A is located at the lower level so that it has an empty level in the forbidden band between C.B. and V.B. Thus, chemisorbed oxygen forms an acceptor level. Because of the small ionization energy of CO, the level D is located at the higher level so that it has an occupied level in the forbidden band between C.B. and V.B. Thus, chemisorbed CO forms a donor level.

The chemisorbed CO is ionized to form CO^{+} by capturing a free hole. This process is denoted by an arrow, which shows that an electron of CO combines with a free hole in the valence band in Fig. 2a. This corresponds to the process that the surface O^{2-} on which CO is chemisorbed gives an electron to the neighboring O^{-} and it becomes a O^{-} ion in Fig. 2b. The CO chemisorbed on O^{-} is shown by CO combined with the mark + in the circle, which represents a localized hole in Fig. 1.

On the other hand, the chemisorbed O is ionized to form O^{-} by capturing a free electron. This process is denoted by an arrow, which shows how a free electron in conduction band moves to a vacant level of O in Fig. 2a. This corresponds to the process that the surface M^{2+} on which O is chemisorbed accepts an electron to the neighboring M^{+} and it becomes an M^{+} ion in Fig. 2b. The O chemisorbed on M^{+} is shown by O combined with the mark – in the circle, which represents a localized electron in Fig. 1.

When these chemisorbed CO and O encounter each other at the crystal surface, they will react to produce chemisorbed CO_2 still having localized electron and hole. When the chemisorbed CO_2 is desorbed from the surface of the crystal, the localized electron should move to the neighboring M^{2+} to form M^+, and a localized hole should move to the neighboring O^{2-} to form O^-. Thus, free electron and free hole are released and they will recombine through a certain energy level in the crystal.

In the case where the catalyst lattice is doped with the rare-earth ion (represented by RE in a circle in Fig. 1), luminescence is ascribed to the electron transition within the RE ion according to the following reaction [7]:

$$\mathrm{RE}^{3+} + p^+ \rightarrow \mathrm{RE}^{4+}$$
$$\mathrm{RE}^{4+} + e^- \rightarrow (\mathrm{RE}^{3+})^* \rightarrow \mathrm{RE}^{3+} + h\nu$$

Figure 3 shows the CTL spectra for the oxidation of CO on undoped thoria (ThO_2) and thoria doped with Pr (ThO_2:Pr).

Figure 4 shows the CTL spectra for catalysts doped with various activators. The peak wavelength of the CTL spectrum for the activated catalyst corresponds to the electronic transition from the ground state to the excited state of the activator.

It should be noted that more complex molecules than CO (e.g., methanol) produce many kinds of intermediates in the course of the catalytic oxidation, and they will chemisorb to form surface states. If the energy of the surface states formed by chemisorption of these intermediates are shallow enough from the delocalized band (conduction band and valence band) edges in the

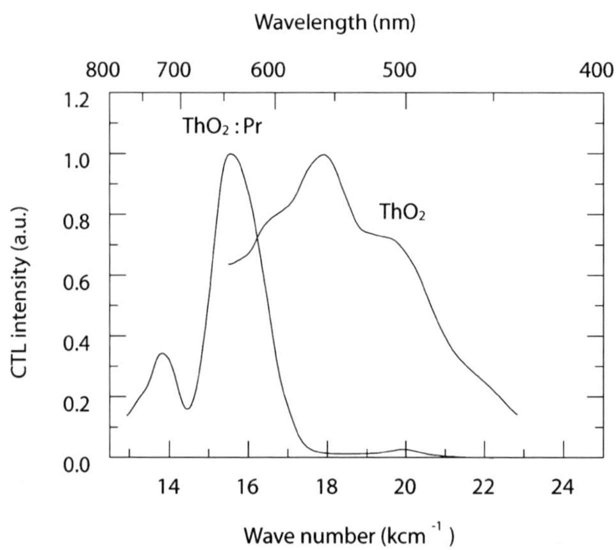

Fig. 3 CTL spectra by oxidation of CO on undoped thoria (ThO_2) and thoria doped with Pr (ThO_2:Pr) [6, 7]

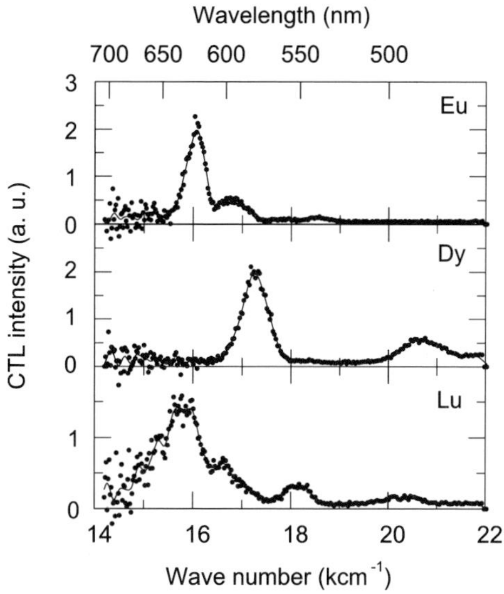

Fig. 4 CTL spectra for catalysts doped with various activators

catalyst, charge carriers will thermally release without surface reaction of the adsorbates, followed by desorption. In either event, the processes mentioned above are characterized as a luminescence by carrier recombination.

3.2 Radiation from the Excited Species

The second CTL mechanism is luminescence from the excited species produced in the course of catalytic oxidation. One of the excited species, formaldehyde (HCHO), is produced during the catalytic oxidation of ethanol, and the reaction process is depicted schematically in Fig. 5. The HCHO is finally oxidized to CO_2 and H_2O in an atmosphere containing oxygen.

Figure 6 shows the CTL spectra observed during the catalytic oxidation of ethanol on γ-alumina, calcium carbonate, and barium sulfate. The profiles of these broad spectrum components are similar to each other, and they peak in the vicinity of 420 nm. The profiles of the CTL spectra from the excited species depend on the kind of catalyst. Fine spectra are observed in the non-porous $BaSO_4$ catalyst. In Fig. 6b, the thin curves denote the fine spectrum components obtained by the peak-fitting technique.

Figure 6d shows the comparison of this spectrum with the theoretical one. The straight lines are the theoretical emission lines from the excited HCHO calculated by Toby [27]. As the peak wavelengths and the intensities for the fine spectra agree with the theoretical lines, we can concluded that the CTL emission results from the excited HCHO produced in the course of

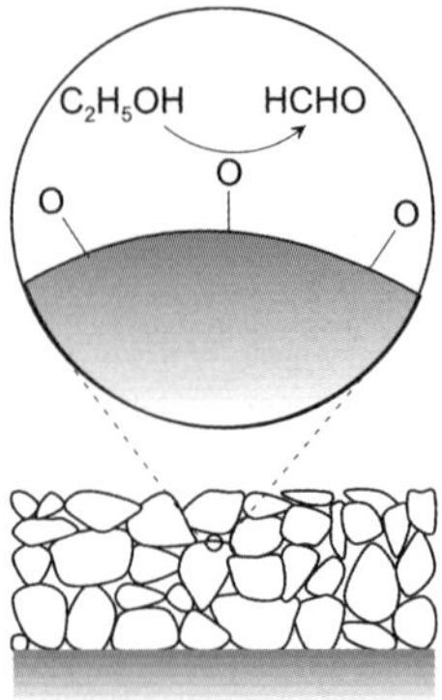

Fig. 5 CTL emission from the excited species

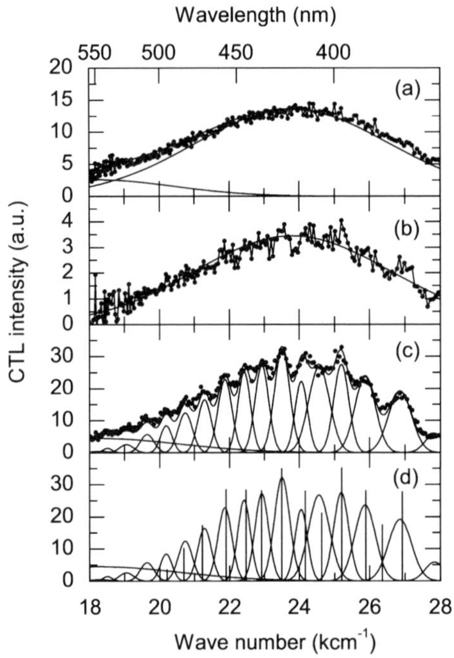

Fig. 6 CTL spectra by the catalytic oxidation of ethanol on **a** γ-Al_2O_3, **b** $CaCO_3$, **c** $BaSO_4$ catalysts at 500 °C [20]. **d** Comparison of the CTL spectrum with the theoretical emission lines from the excited HCHO [27]

the catalytic oxidation. As the fine spectra from the excited HCHO disappear at higher temperatures [28], the fine spectra are not observed for CTL emission at the inner pore of the catalyst particle of higher temperature than the outer surface of the particle, as in the case of CTL emission from porous γ-Al_2O_3 (Fig. 6a). The production process of the excited HCHO is described below.

Ethanol is decomposed to form ether and ethylene on the alumina catalyst. Formaldehyde and formic acid are produced by the catalytic oxidation of ethanol in an atmosphere containing oxygen. In the course of this catalytic oxidation on a certain kind of catalyst, the excited HCHO is produced and the CTL emission is observed in its relaxation process.

Figure 7 shows the CTL response and partial pressure of the desorbed gases during heating (1 °C/s) from the γ-Al_2O_3 catalyst which pre-adsorbed ethanol vapor at room temperature. In Fig. 7a, the mass spectrometer measurements show the desorptions of water (m/z 18), the physisorbed ethanol (m/z 46), diethylether (m/z 59), and ethylene (m/z 25) for the catalyst heated in Ar, but the CTL emission is not observed. In Fig. 7b, the desorptions of ethanol (80 °C), diethylether (230 °C) and ethylene (250 °C) are observed above the same temperature as in Ar, in the course of heating in a mixed gas of 21% O_2 and 79% Ar. The desorption of ethylene in the atmosphere containing oxygen, however, begins to decrease at a lower temperature (320 °C) in Fig. 7b than the peak temperature (340 °C) of ethylene desorption in Fig. 7a. Desorption peaks of CO_2 (m/z 44) and water appear at 340 °C, and the CTL peak is observed at the same temperature in Fig. 7c.

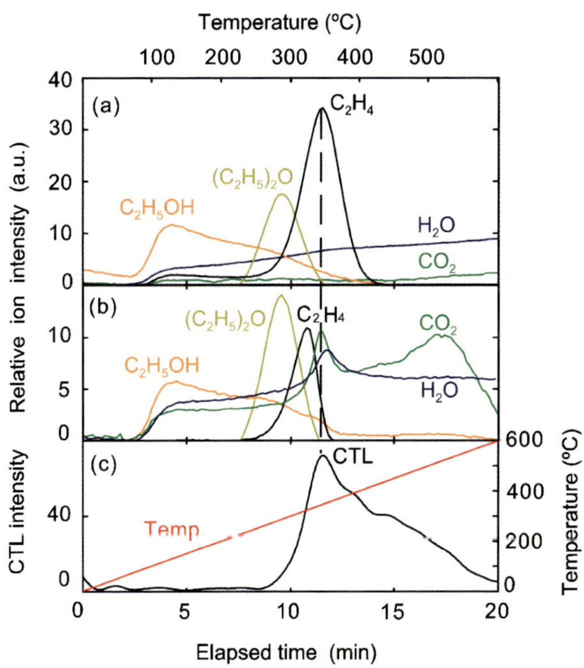

Fig. 7 Desorbed gas and CTL intensity during the heating of γ-Al_2O_3 catalyst on which ethanol is pre-adsorbed **a** in Ar and **b**, **c** in a gas mixture of 21% O_2 and 79% Ar

The results imply that maximum C_2H_4 is produced at 340 °C in the atmosphere containing no oxygen, whereas the excited HCHO is produced instead of C_2H_4 during the course of catalytic oxidation in an atmosphere containing oxygen, and the CTL emission is observed. As HCHO is oxidized to form CO_2 and H_2O, the existence of HCHO cannot be confirmed by mass spectrometry under high oxygen concentrations.

3.3 Working Mechanism of CTL-Based Sensors

As CTL emission is observed in the course of catalytic oxidation, we must consider the overall reaction process in order to describe the working mechanism of the CTL-based sensor. Figure 8 shows a schematic illustration of the catalyst layer to depict the simplified overall reaction processes involving CTL emission on the CTL-based gas sensor.

Oxidation of gas molecules at the surface of the solid catalyst is the heterogeneous catalytic reaction. In this case, the reaction proceeds through the following five stages:

1. Gas molecules R (combustible gas) and O (oxygen) diffuse from the outer gas phase and reach the neighborhood of the catalyst surface.
2. Gas molecules are chemisorbed to form R_{ad} and O_{ad} at the catalyst surface. A part of the adsorbate is desorbed to the gas phase again.
3. Chemisorbed R_{ad} and O_{ad} react to produce chemisorbed RO_{ad} at the surface.
4. The reaction product RO is desorbed from the surface.
5. RO diffuses off to the gas phase.

The total reaction rate r_S depends on the rate-determining step. Now, we assume that the oxygen concentration in the gas phase is large enough to be able

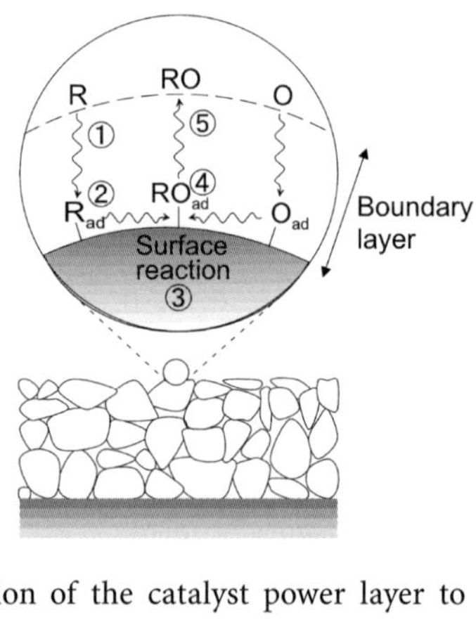

Fig. 8 Schematic illustration of the catalyst power layer to depict the overall reaction processes

to ignore the effect of the oxygen concentration on the rate of total reaction. In the heterogeneous reaction, the reaction proceeds at the interface of the solid catalyst and the gas phase. This reaction is the *surface reaction* because the reaction proceeds on the 2D plane of the catalyst surface. When the surface reaction (3rd step) is the slowest process, i.e., the rate-determining step, the reaction is under the *surface-reaction-controlled condition.* As the both rates of adsorption and desorption of R are faster than the rate of the surface reaction, adsorbates are under adsorption equilibrium and the total reaction rate r_T is equal to the rate of the surface reaction. When the surface reaction of adsorbates R_{ad} and O_{ad} is 1st order, the total rate of reaction is also 1st order. The surface concentration of the adsorbate R_{ad} is given by the Langmuir-type adsorption equilibrium in the gas R of partial pressure p:

$$r_S = k_S\theta \qquad (1)$$

$$\theta = \frac{Kp}{1 + Kp} \qquad (2)$$

where k_S is the rate constant of the surface reaction, and θ is the surface coverage of R_{ad} adsorbates. The equilibrium constant of adsorption K is the ratio k_a/k_d of the rate constant of adsorption k_a to the rate constant of desorption k_d. We obtain the following equation from Eq. 1:

$$r_S = \frac{k_S Kp}{1 + Kp} \qquad (3)$$

When the adsorption rate is slowest in all steps (2nd step), the reaction is under the *adsorption-controlled condition.* That is, the total reaction rate is equal to the adsorption rate as shown below:

$$r_S = k_a p(1 - \theta) \qquad (4)$$

When the desorption rate r'_d of reaction products RO_{ad} at the surface is the slowest rate and the total reaction rate is controlled by the desorption rate r'_d, we obtain:

$$r'_d = k'_d\theta' \qquad (5)$$

where θ' is the surface coverage of reaction products RO_{ad} and k'_d is the rate constant of desorption of RO_{ad}. As the desorption rate is the rate-determining step, θ' is not in equilibrium with the partial pressure of RO in the gas phase. Therefore, the reaction is 0th order and it does not depend on the partial pressure of reactant. This condition is usually observed in a region where p has a large value, and the total reaction rate r_T is inclined to saturate with increase in p.

All of the above mentioned processes are the activation processes and they are characterized by specific activation energies. On the other hand, in some cases, diffusion of gas to the catalyst surface is the slowest rate and it is characterized by a flow-dynamical process.

When the reaction proceeds at the catalyst surface, reactants are transferred from the gas phase to the catalyst surface. This is because the concentration of the reactant in the vicinity of the surface is less than that in the gas phase far from the surface. The diffusion takes place due to the concentration gradient according to Fick's first law. The rate of mass-transfer j through a unit area per unit time to the direction of the x-axis perpendicular to the catalyst surface is given by:

$$j = -D\frac{\partial C}{\partial x} \tag{6}$$

where D is the diffusion coefficient, C is the concentration and $\partial C/\partial x$ is the concentration gradient along the x-axis. Now, we assume that the reactant diffuses from the gas phase to the catalyst surface, and decreases at the surface by the surface reaction. The total reaction rate r_T at steady state is given by:

$$r_T = k_D(C_O - C_S) = k_C C_S \tag{7}$$

where k_D is the mass transfer diffusion coefficient, C_O is the concentration of the reactant in the gas phase far from the catalyst surface, C_S is the concentration in the vicinity of the catalyst surface, and k_C is the rate constant of catalytic reaction process at surface which integrates steps 2–4. k_C is given by:

$$k_C = \alpha \exp\left(\frac{E_C}{k_B T}\right) \tag{8}$$

where α is the frequency factor and E_C is the apparent activation energy of the reaction, k_B is the Boltzmann constant, and T is the absolute temperature.

As the rate of diffusion j equals constant r_T under steady state conditions and D is also constant, the concentration gradient $\partial C/\partial x$ is also constant. From Eq. 7, we obtain:

$$r_T = \frac{C_O - C_S}{1/k_D} = \frac{C_S}{1/k_C} = \frac{C_O}{1/k_D + 1/k_C} \tag{9}$$

$$r_T \cong k_C C_O \quad (k_D \gg k_C)$$

$$r_T \cong k_D C_O \quad (k_C \gg k_D) \tag{10}$$

In this equation, if the rate of diffusion is faster than that of the catalytic reaction at the surface ($k_D \gg k_C$), the Arrhenius plot of r_T gives the apparent activation energy E_C of k_C. This is the *reaction-controlled condition*. On the other hand, if the rate of the catalytic reaction is faster than that of diffusion ($k_C \gg k_D$), the Arrhenius plot of r_T gives the characteristics of temperature dependence of k_D. This is the *diffusion-controlled condition*. Under diffusion-controlled conditions, the transferred reactant decreases at once at the surface ($C_S = 0$) because of the fast catalytic reaction rate. The gas flow along the catalyst surface forms a boundary layer above the surface, and gas molecules diffuse due to the concentration gradient inside the layer in the thickness direction. As the total reaction rate depends only on the diffusion rate of the gas molecule, the reaction rate is kept constant independently of the catalytic activity at the surface.

When the catalyst is a porous solid, most of the surface area of the catalyst is the surface area of the inner surface of the pores. Therefore, most of the reaction proceeds in the pore. Gas molecules are transferred to the outer surface of the catalyst by diffusion. Generally speaking, the diffusion is faster than the diffusion inside the pores. Gas molecules collide with the inner wall of the pore before they collide with another molecule for the porous catalyst having an average pore radius r_{P} of a few nm. Such diffusion is called Knudsen diffusion and its diffusion constant D is given by:

$$D = 9.7 \times 10^3 r_{\mathrm{P}}(T/M)^{1/2} \tag{11}$$

where M is the molecular weight of the gas. If the rate of the surface reaction at the inner surface of the pore is fast enough, the total reaction rate is governed by the rate of Knudsen diffusion. All gas molecules react and are consumed at the portion near the entrance of the pore. In this case, the total reaction rate depends on the particle size even if the surface area of inner pore is larger than the outer surface of the particle.

3.3.1 Equation of CTL Intensity as a Function of Temperature in Laminar Flow

The rate of photon emission in CTL is a function of the rate of formation of chemisorption surface state (Sect. 3.1) or production of excited species (Sect. 3.2) in the catalytic reaction described above, because the CTL process takes place as one step of the catalytic reaction process. Now, we can describe the CTL intensity as a function of temperature according to the mechanism mentioned above. We assume for simplicity that a planar-type CTL sensor is in a laminar flow and the catalyst surface is parallel to the gas flow. A boundary layer, in which the velocity of gas along the surface is distributed, is established above the catalyst surface placed parallel in a laminar flow of a flow velocity u_0. We also assume that the concentration of oxygen O is higher than that of the combustible gas R in the gas phase, and transfer and/or surface reaction of the combustible gas is the rate-determining step. The concentration of the combustible gas R just above the catalyst surface C_{S} is lower than C_{O} in the gas phase because of the consumption of R by the surface reaction. Therefore, the combustible gas R is transferred at an average rate j per unit area by diffusion to the catalyst surface of length l along the gas-flow according to the following equation:

$$j = k_{\mathrm{D}}(C_{\mathrm{O}} - C_{\mathrm{S}}) \tag{12}$$

$$k_{\mathrm{D}} \equiv 0.664 \frac{D}{l} \sqrt{\frac{u_0 l}{\nu}} \left(\frac{\nu}{D}\right)^{\frac{1}{3}} \tag{13}$$

where k_D is the mass transfer coefficient, D is the diffusion constant and ν is the dynamic viscosity. In this equation, k_D scarcely depends on temperature because the value of D and ν have little temperature dependence.

On the other hand, the rate constant of the catalytic reaction at surface k_C increases exponentially with increasing temperature as shown in Eq. 8. Thus in Eq. 9, $k_C \ll k_D$ at low temperatures and $k_D \ll k_C$ at high temperatures. Therefore, r_T increases with an increase in temperature at an activation energy E_C at lower temperatures, and gradually approaches an almost constant value at higher temperatures. The value of r_T is obtained by measuring CO_2 which is one of the final products in the catalytic oxidation of combustible gas containing carbon.

Figure 9 shows the temperature dependence of the CTL intensity and the relative partial pressure of reaction product CO_2 by catalytic oxidation on γ-Al_2O_3 catalyst [16]. The high temperature region where the partial pressure of CO_2 is almost constant is the "diffusion-controlled region" and the low temperature region where it is characterized by an activation energy E_C is the "reaction-controlled region". Although the CTL intensity is governed by the overall rate of the catalytic oxidation, it has a negative temperature dependence in the diffusion-controlled region because of the temperature quenching of the CTL emission. Under constant temperature and constant flow rate in the diffusion-controlled region, CTL intensity as well as r_T are

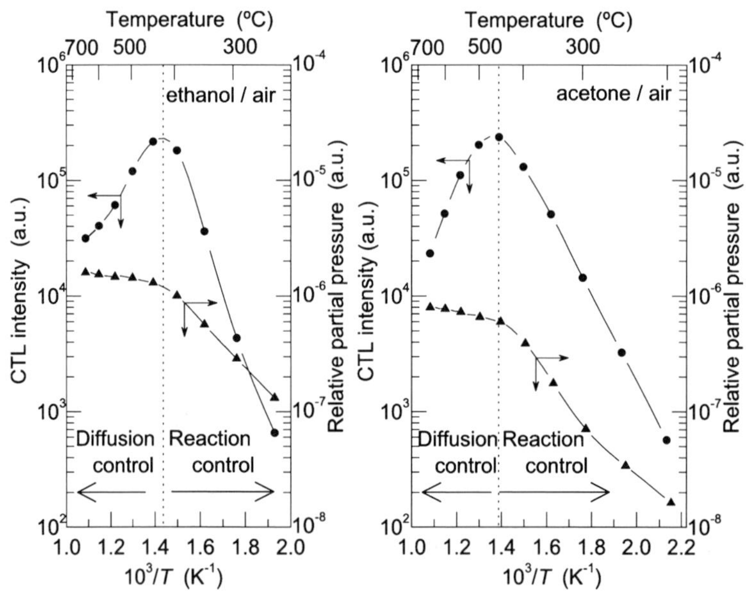

Fig. 9 Temperature dependence of CTL intensity (*circles*) and relative partial pressure of the reaction product CO_2 (*triangles*) by catalytic oxidation on the γ-Al_2O_3 catalyst in **a** ethanol/air **b** acetone/air

proportional to gas concentration C_O and independent of the catalytic activity, as shown in Eq. 9. The difference in activation energy of CTL intensity and CO_2 production may result from a difference in the reaction processes forming HCHO and CO_2 under reaction-controlled conditions. However, the details of the reaction are unknown yet.

If the concentration of reducing gas is rather high and the effect of oxygen concentration cannot be ignored in the catalytic oxidation of reducing gas, we should modify Eq. 1 as follows:

$$r_S = k_S\,\theta_R\theta_O \tag{14}$$

$$\theta_R = \frac{K_R p_R}{1 + K_R p_R + K_O p_O}$$

$$\theta_O = \frac{K_O p_O}{1 + K_R p_R + K_O p_O}$$

where θ_R and θ_O are the surface coverage of adsorbates of reducing gas R and oxygen O, the equilibrium constant of adsorption K_R and K_O is the ratio k_a/k_d for R and O, and p_R and p_O are the partial pressures of R and O in gas phase, respectively. From these equations, we obtain Eq. 15:

$$r_S = \frac{k_S K_R K_O p_R p_O}{(1 + K_R p_R + K_O p_O)^2} \tag{15}$$

Under a constant p_O, this equation has a maximum value of r_S as a function of p_R. The value of r_S increases with increasing in p_R almost linearly in a region where $K_R p_R$ is much smaller than 1. r_S is bound to saturate in a region where p_R has a larger value, and is bound to decreases where p_R is grater than $(1 + K_O p_O)/K_R$, with increase in p_R. Therefore, the concentration dependence of the CTL intensity also deviates from linear characteristics at higher concentration of combustible gas.

4 Fabrication of Sensor Systems

4.1 Sensor Materials

As cataluminescence (CTL) is emitted from the excited species produced in the course of the catalytic oxidation of gas, the CTL intensity is limited by the total reaction rate. As we described in the previous section, the total reaction rate is limited by the rate of the slower process between the surface reaction and the transfer of gases from the gas phase to the catalyst surface. The rate of diffusion is characteristics of the gas itself, so that the maximum gas sensitivity is limited by the flow-dynamical condition and not by the activity of the catalyst. Nevertheless, investigation of the sensor material is very important because the temperature region of the diffusion-controlled condition is lowered when the catalytic activity is high. Therefore, the catalytic activity should be measured at reaction-controlled conditions.

Similar to the catalyst of the catalytic thermometry sensor, the catalytic activity of the CTL-based sensor depends not only on the kind of catalyst material and the surface-to-volume ratio of the powder but also on the preparation procedure of the powder. In considering these conditions, a detailed comparison of the CTL catalytic activity has not been reported so far. The present authors and coworkers observed the CTL emission by ethanol vapor on γ-aluminum oxide, barium sulfate, calcium carbonate, and zirconium oxide at a few hundred degrees. On the other hand, CTL emission is not observed during the catalytic oxidation on metal and semiconductive materials, e.g., tin oxide, zinc oxide, and copper oxide.

Cataluminescence by nano-materials of titanium dioxide and strontium carbonate have also been reported [21–23]. The CTL emission on these nanomaterials is in the diffusion-controlled condition at low temperatures. This means that these materials have high CTL catalytic activity. Systematic research on CTL catalysts should be expected in the future.

In general, CTL intensity depends on the catalytic reaction rate, so that a catalyst with a large surface-to-volume ratio is preferable. In this sense, the catalyst powder or a sintered layer of porous particles is used as the sensor material. As the CTL catalyst should be heated to a working temperature, the catalyst powder is pressed in a ceramic pot with a heating wire, or the sintered catalyst layer is formed on a substrate with an electric heater.

Figure 10 shows the schematic illustration of the sintered layer-type sensor chip. The catalyst powder is ground using an auto-grinder and only fine particles are selected using a mesh filter. When the rare-earth ion is doped into the catalyst, an aqueous solution of rare-earth salt (e.g., ($Dy(NO_3)_3$) is mixed in and then the catalyst is calcined. The rare-earth-activated phosphors used for thermoluminescence (TL) measurements, e.g., $BaSO_4$:Eu, $CaSO_4$:Eu, and $SrSO_4$:Eu, also act as CTL catalysts.

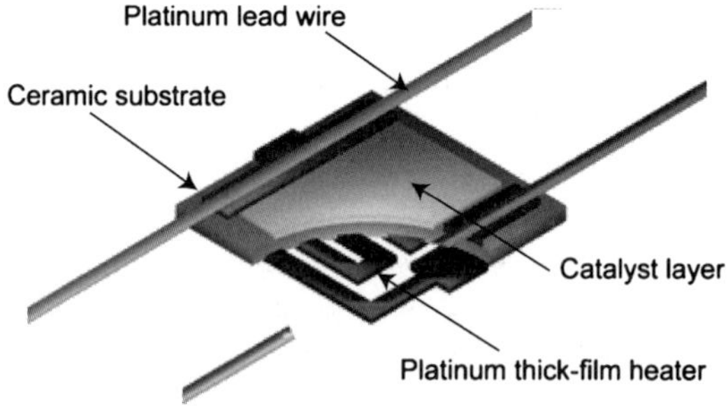

Fig. 10 Schematic illustration of the sintered layer-type sensor chip

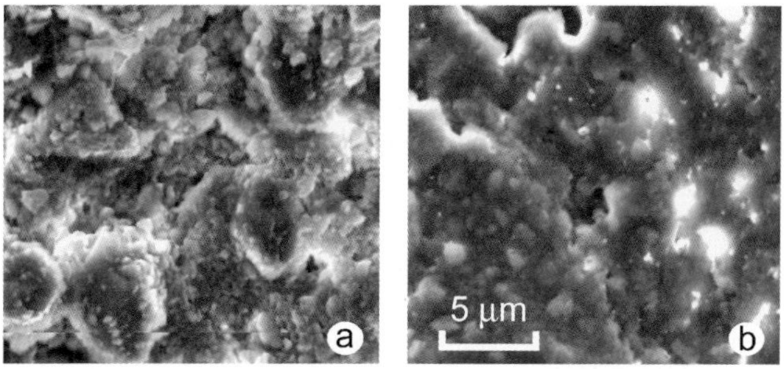

Fig. 11 SEM pictures of **a** γ-Al_2O_3 powder layer and **b** screen-printed layer of the γ-Al_2O_3 powder with colloidal silica binder

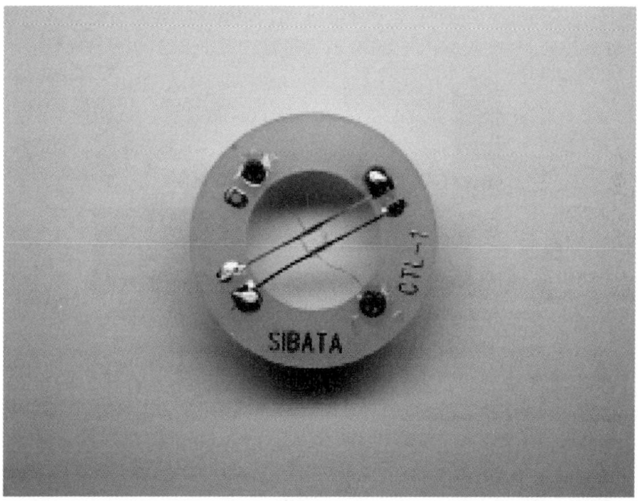

Fig. 12 CTL-based sensor element

The particles are mixed with colloidal silica, and the mixture is screen-printed (0.15 mm in thickness) on a ceramic substrate ($3 \times 3 \times 0.15\,mm^3$). The thick film heater is printed on the other side of the substrate.

Figure 11 shows the SEM pictures of (a) the γ-Al_2O_3 powder layer and (b) the screen-printed layer of the γ-Al_2O_3 powder with the colloidal silica binder. Because the porous structure has space between particles the penetration of gas into the printed layer is allowed.

Figure 12 shows the CTL-based sensor element. The platinum ribbon wire (0.2 mm in width and 0.02 mm in thickness) is spot-welled on the screen-printed substrate as heater lead wires, and the sensor chip is suspended on a plastic frame by the lead wires. In order to measure the catalyst temperature, very thin thermocouple wires are fixed on the substrate using ceramic cement.

4.2 Sensor Cell

The characteristics of the CTL-based sensor depend on the fluid mechanical situation around the sensor because the catalytic oxidation rate depends on the rate of the transfer of gas molecules to the catalyst surface. Therefore, the CTL-based sensor cell must have a structure that keeps a stable flow around the sensing element.

Figure 13 shows schematic illustrations of the two types of sensor cell, the laminar flow type and the impinging jet type.

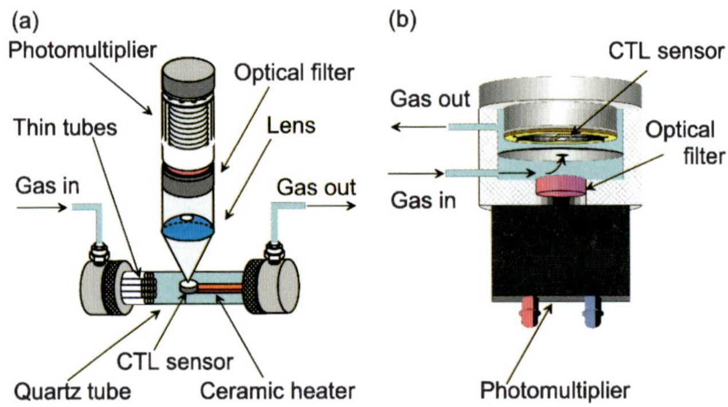

Fig. 13 Schematic illustrations of the two types of sensor cell: **a** laminar flow type and **b** impinging jet type

Fig. 14 Impinging jet type CTL-based sensor cell with a photomultiplier module

In a laminar flow type cell (Fig. 13a), the CTL-based sensor chip is fixed on a planar ceramic heater ($5 \times 1 \times 50\,mm^3$) inside a quartz tube of 10 mm inner diameter. Sample gas flows at a constant rate through a bundle of thin Teflon-tubes to realize a laminar flow around the sensor, and is aspirated by an air pump downstream of the sensor. The CTL intensity is measured by a photomultiplier through a quartz lens and optical filters.

In the impinging jet type cell (Fig. 13b), the sensor cell is separated by a partition with a hole of diameter D. The CTL-based sensor element is placed at H mm just above the hole, and a jet of sample gas is directed to the sensor surface through the hole with a velocity u, and is aspirated by an air pump downstream of the sensor. The CTL intensity through the hole is measured by a photomultiplier through an optical filter. The fast, highly sensitive and stable CTL response is obtained under the condition of $H = 2$ mm, $D = 2$ mm, and $u = 0.5–2$ m/s.

Figure 14 shows the impinging jet type CTL-based sensor cell with a photomultiplier module.

4.3 Sensor System

4.3.1 Fundamental Setup

Figure 15 shows the schematic diagram of a typical sensor system. The CTL-based sensor element is fixed with pins to connectors for heater lead and thermocouple in the sensor cell. The temperature of the sensor is monitored by the thermocouple and is controlled by a digital temperature controller. The sample gas in the sensor cell is aspirated by an air pump downstream of the

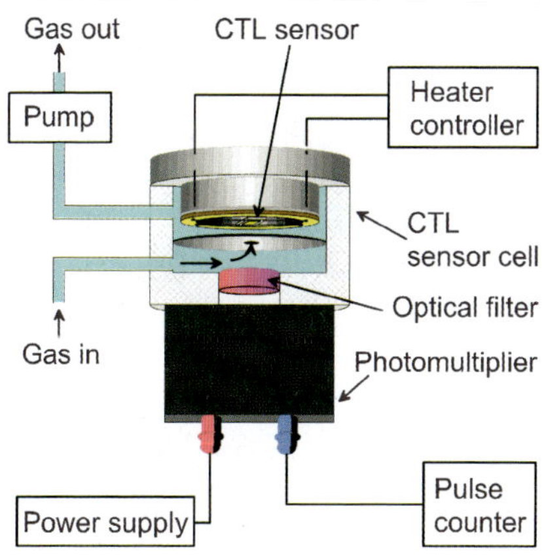

Fig. 15 Schematic diagram of a typical sensor system

sensor, and the jet of sample gas at atmospheric pressure is impinged to the sensor surface. The CTL intensity through the hot-mirror filter and band-bass filter is measured by a photon-counting method with a photomultiplier module. The digitized data are sent to a computer through a RS-232C cable. The lowest detection limit of the photon counting is about 10 photons per second.

Figure 16 depicts the CTL-based sensor system. Figure 16b, shows the CTL-based sensor unit at the center, the flow controller on the left and the air pump on the right. There are electric circuits for the heater controller, the photon counter, and the data communication through the RS-232C cable under

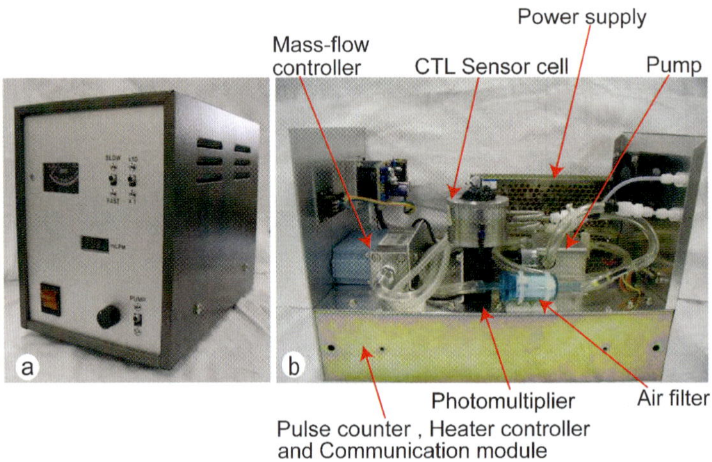

Fig. 16 CTL-based sensor system: **a** outside appearance and **b** inner view

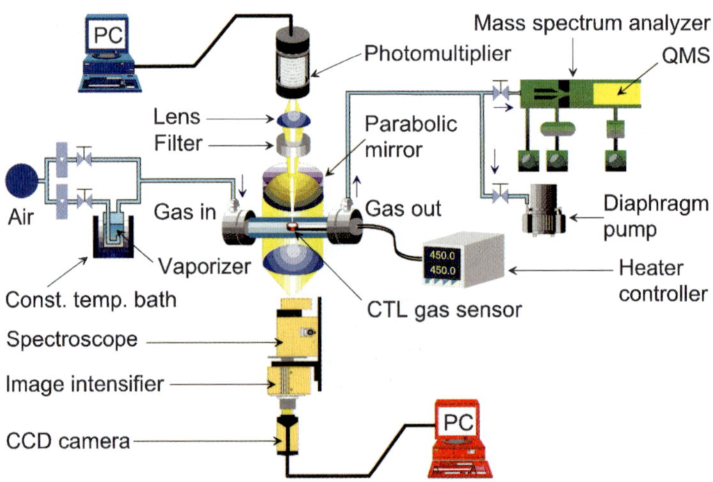

Fig. 17 Experimental setup for measuring the characteristics of the CTL-based sensor

the chassis. The control signal of the system and the measured data are sent from/to a computer through the RS-232C cable.

Figure 17 shows the experimental setup for measuring the characteristics of the CTL-based sensor. The sample gas is prepared using a standard gas generator (Gastec PD-1B). The saturated vapor of organic solvent is mixed with synthetic air (21% oxygen and 79% nitrogen). The sample gas flows through a sensor cell at a constant flow rate. The CTL emission is collected by a paraboloidal mirror and is focused on the entrance slit of a spectroscope. The CTL spectrum is measured using a time-resolved spectrum analyzer consisting of the spectroscope (Ritsu MC-12N), an image intensifier (Hamamatsu C2166), a CCD camera, and an image processor. The CTL emission through the center hole of a paraboloidal mirror is also monitored using a photomultiplier. The sample gas through the sensor cell is introduced into a quadrupole mass spectrometer through a differential evacuation system to measure the changes in partial pressure of reactants and products [18].

5 Characteristics of CTL-Based Gas Sensors

5.1 CTL Spectrum

5.1.1 CTL Spectrum from the Excited Species

This type of the CTL spectrum is independent of the kind of catalyst, but depends on the kind of combustible vapors or gases. Figure 18 shows an example of the CTL spectra during catalytic oxidation of CH_3CHO, C_2H_5OH, and C_2H_4 on nanosized $SrCO_3$ [23]. The CTL spectra consist of three common sub-bands. This means that the same kinds of luminescent species are produced during the catalytic oxidation of these combustible vapors. The peak wavelengths of the CTL spectrum components during the catalytic oxidation of various gases on various catalysts are listed in Table 2.

The luminescent species that shows an emission peak at 420 nm is attributed to HCHO, but the origin of the other sub-bands has not yet been determined. Each amount of the generated luminescent species depends on the kind of combustible gas, so that the intensity of each emission sub-band is different from each other, and the profile of the overall spectrum is also different. This enables us to discriminate the kind of combustible gas from the profile of the CTL spectrum. It should be noted, however, that the intensity ratio of the emission sub-band depends also on the ratio of combustible gas to oxygen and on the catalyst temperature [16, 23].

Table 2 Peak wavelengths of the CTL spectrum components during catalytic oxidation of various gases on various catalysts

Catalyst	Temp. (°C)	Gas	Peak wavelengths (nm)				Refs.
γ-Al_2O_3	450	Ethanol	420		~550 *		[16]
		Acetone	420		~550 *		
$CaCO_3$	500	Ethanol	420				[20]
$BaSO_4$	500	Ethanol	420				
Nanosized TiO_2	410	Acetone	–		~580	~610	[21]
Nanosized ZrO_2	195	Ethanol	425	460	535		[22]
Nanosized $SrCO_3$	400	Acetaldehyde	425	460	530		[23]
	400	Ethanol	425	460	530		
	400	Ethylene	425	460			

* Denotes shoulder

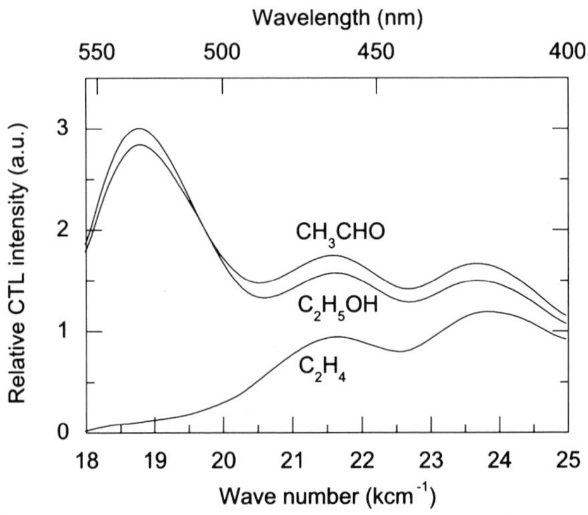

Fig. 18 CTL spectra during catalytic oxidation of CH_3CHO, C_2H_5OH, and C_2H_4 on nanosized $SrCO_3$ [23]

An example is given in Fig. 19, where 3D CTL spectra are shown as a function of temperature for the γ-Al_2O_3 catalyst in air containing ethanol, butanol, acetone, and n-butyric acid [12].

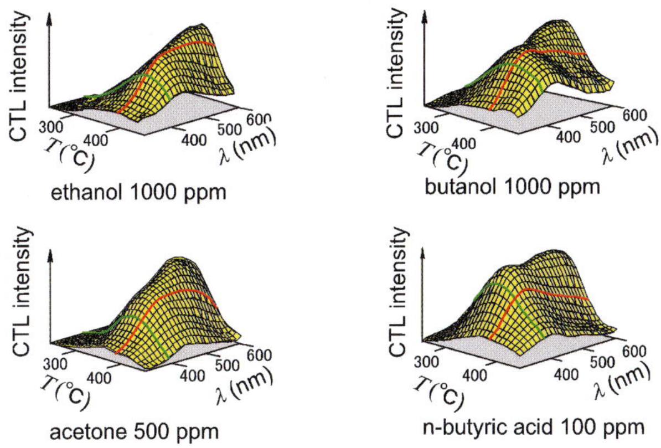

Fig. 19 3D CTL spectra measured in air containing **a** 1000 ppm of ethanol, **b** 1000 ppm of butanol, **c** 500 ppm of acetone, and **d** 100 ppm of *n*-butyric acid at various temperatures

5.1.2 CTL Spectrum Due to Recombination Radiation

CTL emission is also observed during recombination of carriers originating from surface states formed by the chemisorbed adsorbates produced in the course of catalytic oxidation of a gas. As the recombination takes place through the electronic states in solid, the CTL spectrum is independent of the kind of gas, but depends on the kind of catalyst or the doped activator. This type of CTL emission has been reported for ThO_2 [6], ThO_2 activated with Tb or Pr [7], $BaSO_4$ activated with Eu [9], and γ-Al_2O_3 activated with Dy [18].

Figure 20 shows an example of the CTL spectra during catalytic oxidation of methanol, ethylene, and *iso*-butane on γ-Al_2O_3 activated with Dy [18]. Each component of the CTL spectrum from the catalyst activated with the rare-earth ion is attributed to the electronic transition within the activator. The intensity of each component, however, does not agree with that of the corresponding component in the photoluminescence spectrum. Details of the excitation process of the activator by carriers originating from the chemisorbed species have not been clarified yet.

Phosphors doped with rare-earth elements show two types of CTL spectrum, namely emission from the excited species and recombination radiation, simultaneously. Figure 21 shows the CTL spectrum from the TL-phosphor $BaSO_4$:Eu in air containing ethanol vapor. The emission band with fine spectrum components at 420 nm is attributed to the excited HCHO. The line spectrum components peaking at 580 and 615 nm are attributed to the electronic transitions within Eu^{3+} ions.

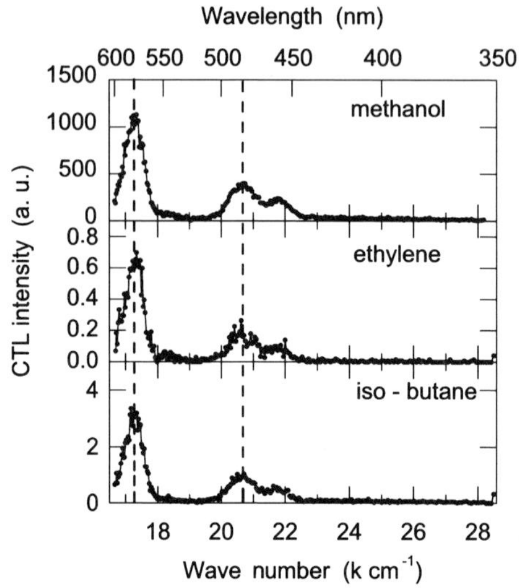

Fig. 20 Examples of CTL spectra during catalytic oxidation of methanol, ethylene, and *iso*-butane on γ-Al_2O_3:Dy

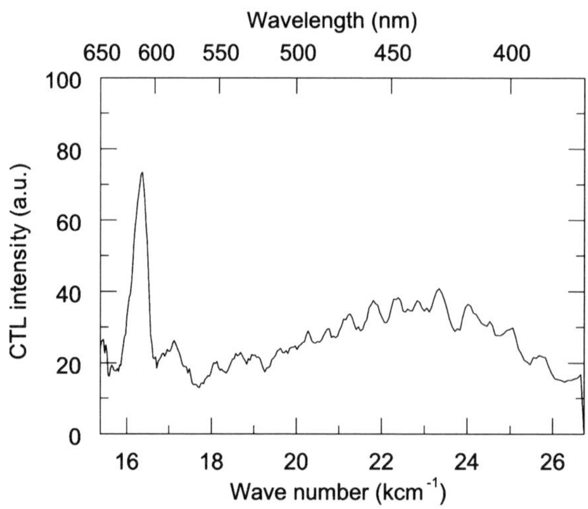

Fig. 21 CTL spectrum from $BaSO_4$:Eu in air containing ethanol vapor

5.2 Temperature Dependence of CTL Intensity

As the gas-phase CTL from the excited species and the solid-phase CTL due to carrier recombination result from different mechanisms, they have different temperature characteristics. Figure 22 shows the temperature dependence of the CTL intensities of the 420-nm band (λ_S) and the 615-nm line spectrum components (λ_L) measured simultaneously in air contain-

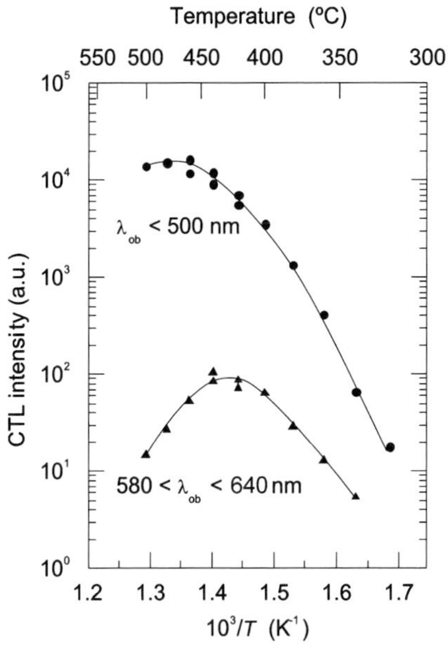

Fig. 22 Temperature dependence of CTL intensity of the 420-nm band (*circles*) and the 615-nm line spectrum components (*triangles*) measured simultaneously in air containing ethanol vapor on the $BaSO_4$:Eu catalyst

ing ethanol vapor on the $BaSO_4$:Eu catalyst. The gas-phase CTL (λ_S) and the solid-phase CTL (λ_L) have different activation energies in the reaction-controlled region below 300 °C. They also have different temperature dependence in the diffusion-controlled region over 400 °C, because the probability of radiative transition of each process is not the same, and they decrease with increasing temperature according to the different temperature dependencies. The characteristics of the temperature dependence of the CTL intensity also depend on the type of combustible gas. This allows discrimination of the different gases by measuring the temperature dependence.

5.3 Gas Flow-Rate Dependence of CTL Intensity

Under reaction-controlled conditions, the total rate of catalytic oxidation is governed by the rate of surface reaction and is independent of the gas transfer rate from the gas phase to the catalyst surface, so that the CTL intensity is also independent of the flow velocity of sample gas around the sensor. Under diffusion-controlled conditions, the rate of catalytic oxidation is independent of the catalytic activity, but depends on the transfer rate of combustible gas in the gas phase, so that the CTL intensity depends on the flow rate of the gas around the sensor. We need a flow dynamics treatment to estimate the transfer rate of combustible gas from the gas phase to the sensor surface through the boundary layer formed by gas flow around the sensor.

In the laminar flow type of sensor cell shown in Fig. 13a, we obtain the reaction rate r_T under diffusion-controlled conditions as follows, applying an approximation of $k_D \ll k_C$ on Eqs. 9 and 12:

$$r_T = k_D C_O$$
$$k_D \equiv 0.664 \frac{D}{l} \sqrt{\frac{u_0 l}{\nu}} \left(\frac{\nu}{D}\right)^{\frac{1}{3}} \quad (16)$$

In the course of catalytic oxidation, the production rate of intermediates that finally generate CO_2 and H_2O is also limited by r_T. Some intermediates are the chemisorbed species that form surface donors and acceptors, and the other intermediates are the excited species. Both the generation rate of carriers originated from the surface states and the production rate of the excited species is governed by r_T. Thus the dependence of CTL intensity on flow velocity should agree with that of r_T in the diffusion-controlled region.

Figure 23 shows the dependence of CTL intensity at 400 and 500 °C on flow velocity. The CTL intensity I at 400 °C is almost independent of the mean flow velocity $\bar{v}$ of the sample gas, i.e., the catalytic oxidation is under reaction-controlled conditions. On the other hand, at 500 °C the curve plotted as $\log I$ vs. $\log v$ has a slope of 1/2. Equation 16 shows that the rate of catalytic oxidation r_T is proportional to the square root of the flow velocity u_0, so that the catalytic oxidation at 500 °C is under diffusion-controlled conditions. As the rate constant of diffusion k_D increases with increasing u_0, the value of r_T

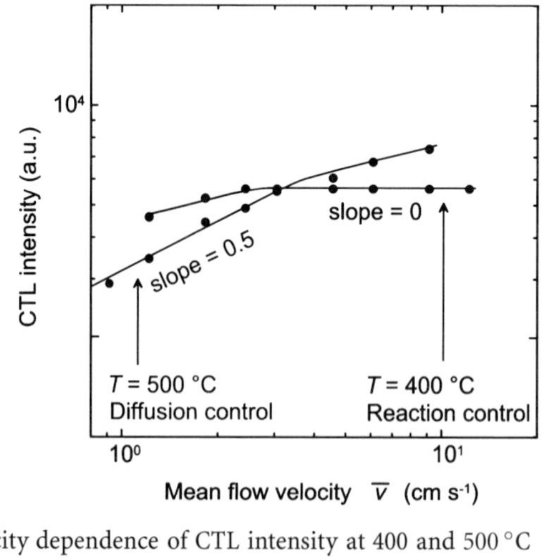

Fig. 23 Flow-velocity dependence of CTL intensity at 400 and 500 °C

approaches the reaction-controlled condition and it saturates in the region of high velocities (Fig. 23).

5.4 Gas Concentration Dependence of CTL Intensity

Equation 16 means that the rate of catalytic oxidation r_T on a catalyst of length l is affected by the diffusion constant D, the dynamic viscosity ν and the flow velocity u_0 except for the gas concentration C_0 under diffusion-controlled conditions. As the parameters D and ν are the physical constants characteristic of a gas, the value of r_T become constant at a constant flow velocity u_0. Thus, under diffusion-controlled conditions, CTL intensity should be proportional to the gas concentration at a temperature at which the CTL efficiency is constant.

Figure 24 shows CTL intensity as a function of (a) the concentration and (b) the flow velocity $\overline{v}$ of ethanol. The CTL intensity I is directly proportional to the concentration of ethanol in the region between 1 and 100 ppm in air (Fig. 24a).

In the $\log I$ vs. $\log \overline{v}$ plots (Fig. 24b), the slope of the flow velocity dependence of CTL intensity is 1/2 in this concentration region. The catalytic oxidation is under diffusion-controlled conditions. As the diffusion rate increases with an increase in the gas concentration, the value of r_T approaches the reaction-controlled condition and it saturates at high gas concentrations.

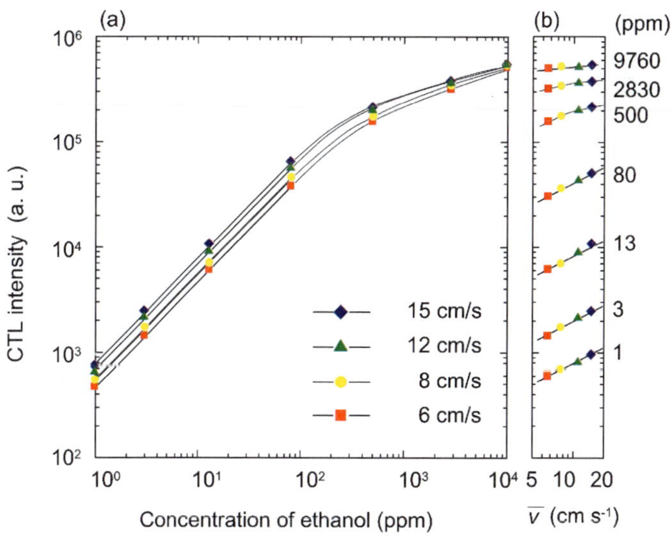

Fig. 24 CTL intensity as a function of **a** concentration and **b** flow velocity of ethanol vapor

In this reaction-controlled region, the CTL intensity is independent of the flow velocity (Fig. 24b).

6
Types of CTL-Based Sensing

6.1
Continuous Measuring System

The CTL emission observed in the course of a catalytic oxidation has a fast response and long-term stability. Thus, it is suitable for continuous monitoring of combustible gases in air. Most of the combustible organic vapors such as alcohols, ketones, aldehydes, and terpenes may emit CTL emission on a catalyst. Figure 25 shows CTL intensity I as a function of the gas concentration C for some gases on γ-Al_2O_3 heated at 450 °C. In Fig. 25, the $\log I$ vs. $\log C$ plots fit into straight lines of unity slope. This is a characteristic of the catalytic reaction under diffusion-controlled conditions.

Under this condition, the response of CTL intensity to an alteration in the ambient atmosphere is fast and reproducible (Fig. 26a). At low temperatures, at which the catalytic reaction is not under diffusion control, the response does not follow the alteration in the gas concentration (Fig. 26b).

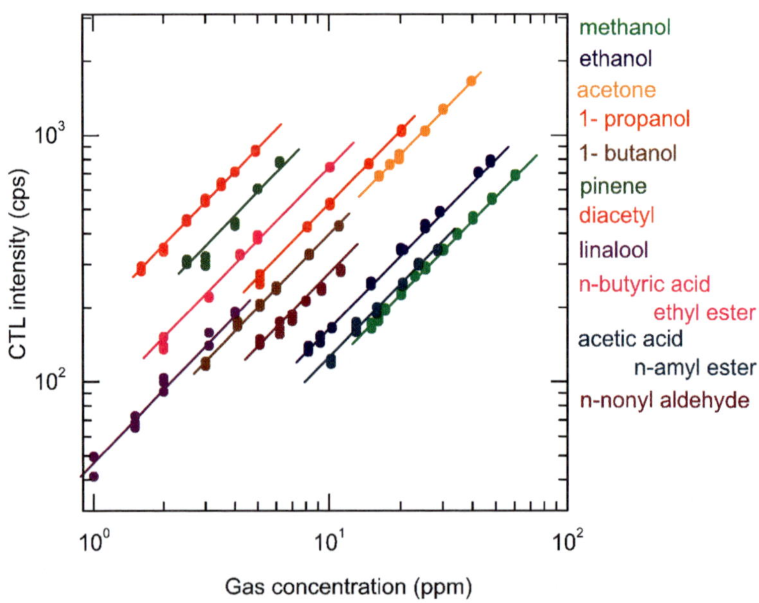

Fig. 25 Concentration dependence of CTL intensity for various gases

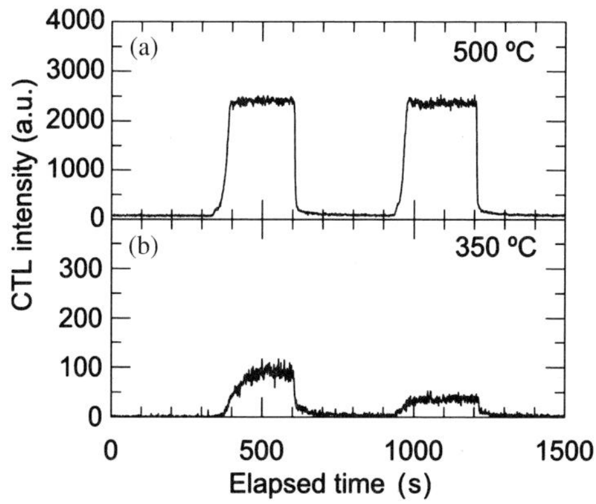

Fig. 26 Response of the CTL intensity on γ-Al_2O_3:Dy to changes in the gas concentration between 0–100 ppm of *iso*-butane at **a** 500 °C and **b** 350 °C [18]

6.1.1
Discrimination of the Type of Vapor Based on the Difference in the CTL Spectrum

As described in Sect. 5.1, the intensity of each CTL spectrum component depends on the vapor type. Because of the broadening of each spectrum component, the profile of the superimposed CTL spectrum depends on the type of vapors, as shown in Fig. 19. This enables discrimination of the vapors. This type of sensor system has been reported in 1995 by Nakagawa [14]. The CTL intensities at four typical wavelengths were measured using a photomultiplier and four band-pass filters mounted on a rotating disk driven by a motor. Ethanol and acetone vapors could be discriminated because the ratio of the CTL intensities at 400 and 500 nm was different for the different vapors, and the CTL intensity at 400 nm has a linear concentration dependence in the 1–200 ppm range for these vapors in air.

6.1.2
Recognition of Organic Vapors by Spectrum-Temperature Imaging

As the temperature dependence of the CTL spectrum has information about the type of vapor, the present authors and coworkers [17] reported a method to recognize organic vapors by means of spectrum temperature imaging. For this purpose, a system to simultaneously measure the CTL spectra at various temperatures was developed (Fig. 27). The sintered layer of the CTL catalyst is laid on a ceramic heater substrate of $5 \times 60\ mm^2$, which has a temperature distribution ranging from 440 to 530 °C along the stream of a sample gas in a quartz tube. A mask with an optical slit of 0.3 mm width is placed on a quartz tube. The CTL emission passing through the slit is focused on

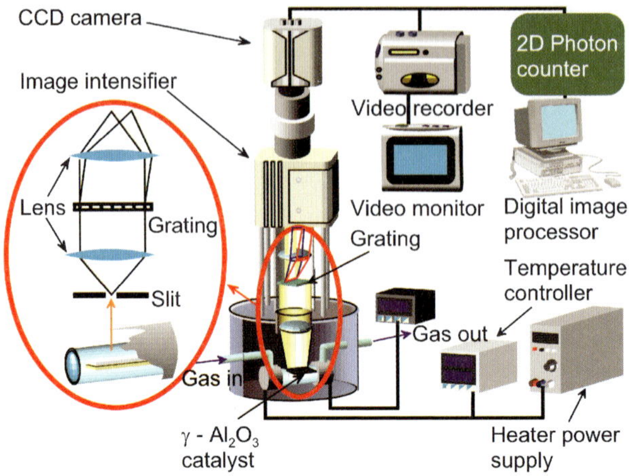

Fig. 27 Schematic illustration of a system to measure the CTL spectrum at various temperatures simultaneously

the incident window of an image intensifier by two lenses, and a grating of 200 lines/mm is placed between these lenses. Both the CTL image and the dispersed CTL image at the slit along the direction of the temperature gradient are intensified by an image intensifier and are monitored by a CCD camera.

Figure 28 shows the spectroscopic images from the CTL-based sensor at 440 to 530 °C for ethanol and acetone in air. These CTL images have different 3D profiles depending on the vapor type. Ethanol, butanol, acetone, MEK, xylene, and butyric acid were discriminated using (i) the ratio of the CTL intensities at 420 and 550 nm estimated from the spectrum dispersion and (ii) the activation energy of the CTL intensity estimated from the temperature profile of these spectral components. Although the CTL image has many

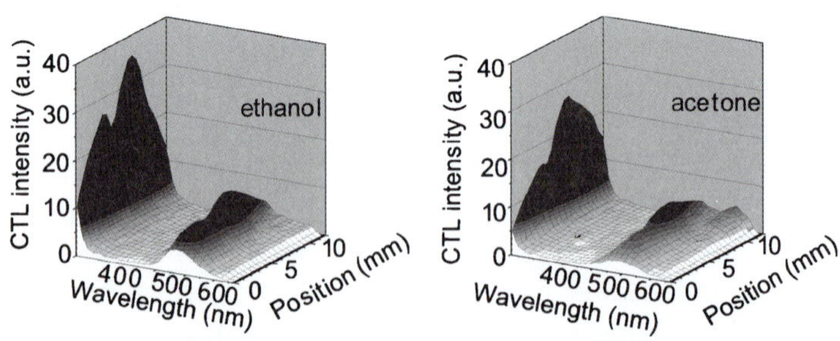

Fig. 28 Spectroscopic images on the CTL-based sensor at temperatures from 440 (0 mm) to 530 °C (12 mm) for **a** ethanol, and **b** acetone in air [17]

advantages for the continuous recognition of vapors, the cost of the sensor system should be lowered for practical use.

6.2 Accumulation-Type Measuring System

When a gas component of very low concentration is measured by means of a gas chromatography system, the gas is condensed in a cold trap. The sample gas is physisorbed and accumulated on the adsorbent surface kept at lower temperatures, and the adsorbent is heated. The desorbed gas is introduced into the chromatography column.

When a porous CTL catalyst is used, we can realize a highly sensitive gas measuring system with a small size because the sensing catalyst itself acts as an accumulator. The sample gas gradually penetrates into the micro-pores of the catalyst powder due to Knudsen diffusion and is physisorbed and accumulated. This process is directly observed as a change in weight of the CTL catalyst when a certain vapor is introduced into the ambient atmosphere after quenching the catalyst at 500 °C in Ar of atmospheric pressure (Fig. 29). In Fig. 29, the ordinate denotes the change in weight M divided by the diffusion constant D_K of Knudsen diffusion. The slow adsorption rate is a characteristic of the adsorption process through Knudsen diffusion in micro-pores. Due to thermal activation during heating of the pre-adsorbed catalyst, the physisorbed molecules are desorbed from the inside of the pore but most of the molecules are adsorbed again to form chemisorbed species, or are changed into the chemisorption state on the catalyst surface. The CTL emission is emitted during the catalytic oxidation of these chemisorbed species. The CTL intensity reflects the amount of adsorbed and accumulated molecules.

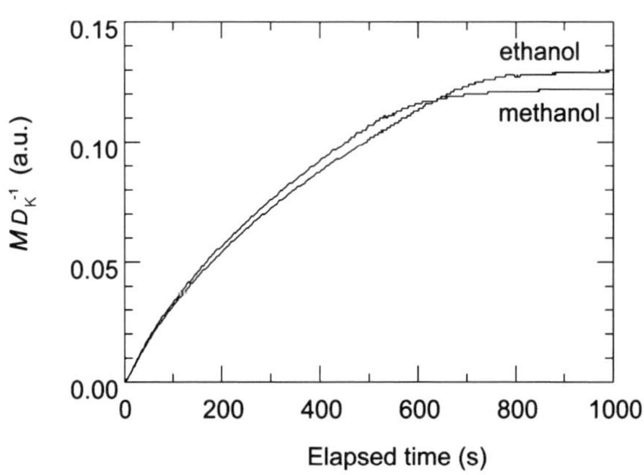

Fig. 29 Change in weight of the CTL catalyst in air containing ethanol and methanol vapor

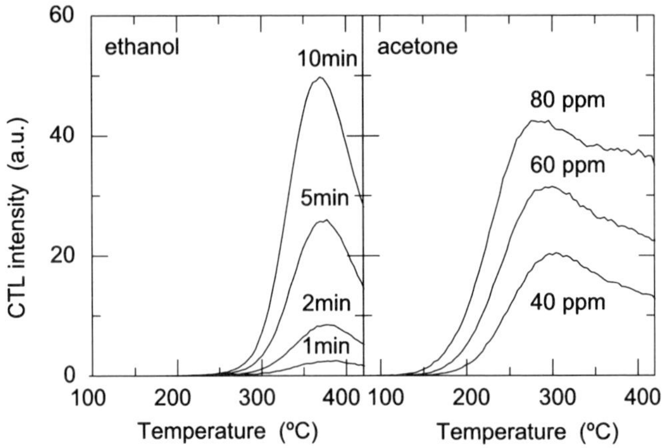

Fig. 30 CTL glow curves from the catalyst heated at a rate of 0.5 °C/s in synthetic air after adsorption of vapor. **a** Adsorption of 100 ppm ethanol in air for 1, 2, 5, and 10 min. **b** Adsorption of 40, 60, and 80 ppm acetone for 2 min

Utsunomiya et al. [11, 12] reported an accumulation-type CTL-based sensor based on this mechanism. The alumina powder (γ-Al_2O_3, 300 mesh in grain size, 30 mg) filled in a glazed ceramic pot (5 mm in diameter) including an electric heater wire was used as a sensor catalyst. First, the catalyst was heated to 500 °C to remove previous adsorbates, then was quenched to room temperature and allowed to adsorb the sample gas for a certain time.

Figure 30 shows the CTL glow curves for the catalyst pre-adsorbed ethanol or acetone vapor. The catalyst is heated at a rate of 0.5 °C/s in synthetic air. The CTL intensity increases at high temperatures, and the total amount of CTL intensity L depends on the adsorption time Δt and the gas concentration C during adsorption. We can measure the gas at a very low concentration by measuring the value of L because L is proportional to the product of Δt and C in a region of low ΔtC where a Henry-type adsorption isotherm holds.

6.3 Temperature-Programming Type System

Temperature-programmed sensing is one of the most convenient methods for realizing a low-cost system for the discrimination and determination of many types of gases because it gives a lot of information for distinguishing gases. This method, however, is effective only for intermittent measurement of the ambient gases whose concentration alternates slowly.

Three kinds of temperature-programmed measurements have been performed. One is CTL measurement during heating after adsorption at room temperature, which was described in the previous section (Fig. 30). The profiles of the CTL glow curves reflect the temperature dependence of the production of luminescent intermediates during the catalytic oxidation, depending on the type of the combustible vapors. Therefore, we can discriminate the type of adsorbed vapor from the profiles of the CTL glow curves. The analytical detection of a vapor mixture by means of this method, however, is difficult because it is hard to avoid the interaction between adsorbed species during the course of heating.

Another temperature-programmed measurement is made during a cycle of heating and cooling in air containing a mixed organic vapor. CTL measurements were made during temperature cycles of ca. 0.5 °C/s between 200 and 740 °C in air containing a mixture of ethanol and acetone vapors [13]. A linear concentration dependence is observed even at concentrations lower than 16 ppm, where interaction between adsorbates can be ignored because most of the reaction products may be desorbed and the catalyst surface may be refreshed at 740 °C. We can measure the concentration of each constituent because the total CTL intensities in the heating and cooling cycles, as a function of the ethanol and acetone concentrations, form grid planes of different gradients. The difference in gradients of the concentration dependence of CTL intensity is attributed to the fact that CTL emission in a heating and cooling cycle has different temperature dependence, as shown by mass spectrum measurements of the desorbed reaction products.

Analytical detection of constituents in a gas mixture is also feasible by the alternating catalyst temperatures in the temperature region of the diffusion-controlled regime. Even under conditions where the total rate of catalytic

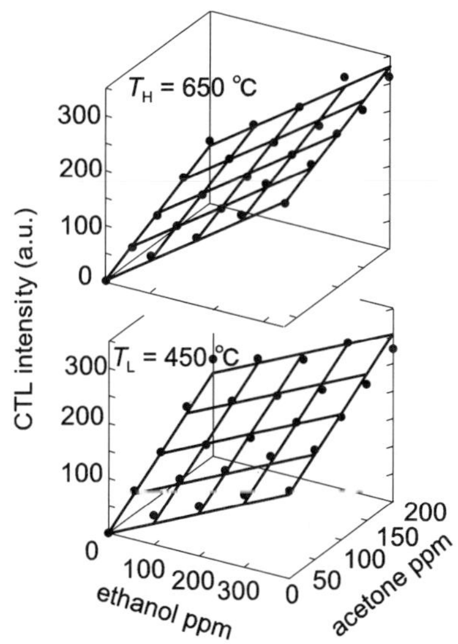

Fig. 31 Total CTL intensities as a function of the concentration of ethanol and acetone in air at **a** 650 °C and **b** 450 °C

oxidation is limited by the rate of gas transfer in the diffusion layer, the formation process of the different luminescent species may depend on the catalyst temperature and the type of the organic vapor. Therefore, the CTL intensities I_H and I_L at temperatures T_H and T_L, respectively, as a function of the constituent concentrations C_A and C_B of a mixed vapor are given by a linear matrix equation as follows:

$$\begin{bmatrix} I_H \\ I_L \end{bmatrix} = \begin{bmatrix} A & B \\ C & D \end{bmatrix} \begin{bmatrix} C_A \\ C_B \end{bmatrix} \tag{17}$$

The sensor system based on this method has been reported by Nakagawa [15]. Concentrations of constituents in a mixed vapor of ethanol and acetone can be measured analytically by alternating the catalyst temperature between 450 and 650 °C (Fig. 31). The concentration dependence of CTL intensity is linear in air containing 1–300 ppm of these vapors.

6.4 Multi-Sensor System

As the CTL-based sensor essentially has a linear concentration dependence, there is an advantage in constructing a multi-sensor system. Figure 32 shows the multi-sensor system using γ-Al_2O_3, γ-Al_2O_3:Dy, $CaCO_3$:Dy, and ZrO_2:Dy for discriminating and determining pinene, limonene, and linalool at the 100 ppb level. These four types of CTL sensors are mounted in a chamber, through which the sample gas flows at a constant flow rate. These sensors are intermittently heated in turn at a certain time interval and the CTL emission from the sensor in a heating cycle is measured using a photomultiplier.

Moreover, the temperature dependence of the CTL intensity of various gases can be modified by addition of an extra catalyst into the matrix cata-

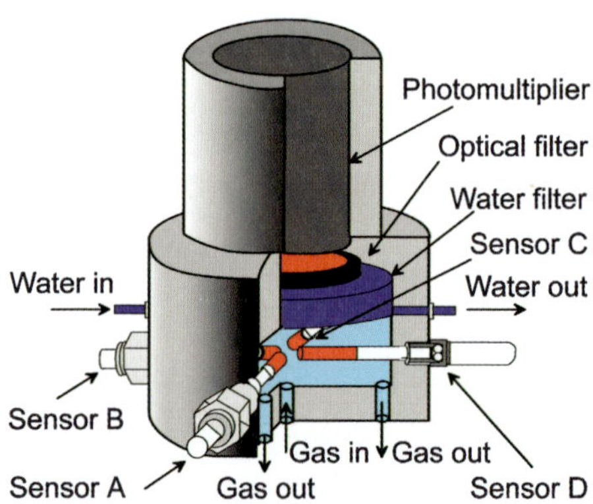

Fig. 32 Multi-sensor system using γ-Al_2O_3, γ-Al_2O_3:Dy, $CaCO_3$:Dy, ZrO_2:Dy for discriminating and determining pinene, limonene, and linalool

lyst. Figure 33 shows changes in the CTL glow curves of 100 ppm benzene on the γ-Al_2O_3:Dy catalyst by adding Sn, Ru, and Pt. These modified catalysts have specific CTL sensitivity on benzene, toluene, and xylene, and they can distinguish these vapors.

Recently, the present authors and coworkers proposed a dual-sensor system for identification and determination of many kinds of gases [28]. Small CTL sensors of $3 \times 3 \times 0.3\ mm^3$ made of the γ-Al_2O_3:Dy (sensor A) and γ-Al_2O_3 catalysts (sensor B) are alternately heated and cooled between 500 and 100 °C in a chamber, in which jets of sample gas are impinged on each sensor through two holes. The CTL intensities from these sensors are measured by a photomultiplier by means of the time-sharing method. During a heating cycle of sensor B, sensor A is kept at 100 °C and the sample gas is physisorbed and accumulated on the surface of sensor A, and the amount of the adsorbed molecules reflects the physisorption property of the gas. During the next cooling cycle of sensor B, sensor A is heated quickly to 500 °C and a certain proportion of the physisorbed gas is chemisorbed by getting thermal activation energy. A sharp CTL peak attributed to carriers from chemisorption surface states and/or reaction between the chemisorbed species is observed. The peak intensity I_p reflects the physisorption characteristics of the sample gas. As the sensor has a fast response, the CTL peak disappears within 30 s according to the oxidation of the pre-physisorbed molecules, and a steady-state CTL intensity (I_{SS}) arises by

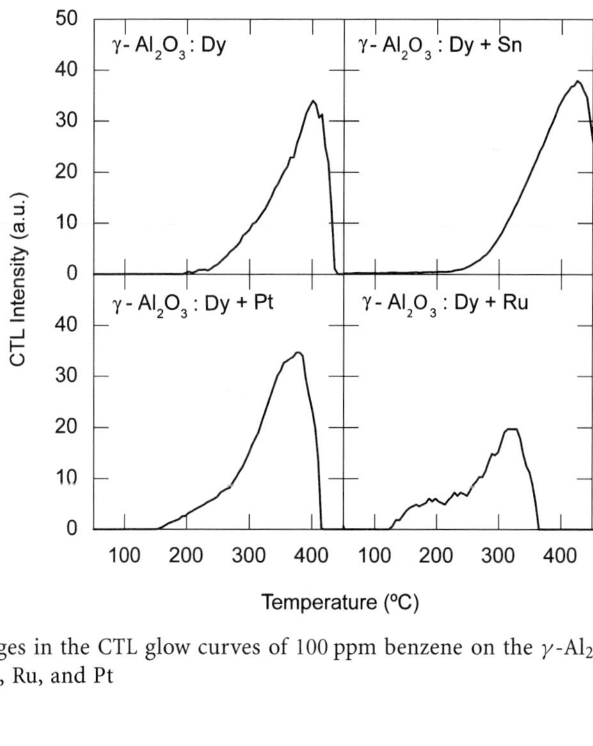

Fig. 33 Changes in the CTL glow curves of 100 ppm benzene on the γ-Al_2O_3:Dy catalyst by adding Sn, Ru, and Pt

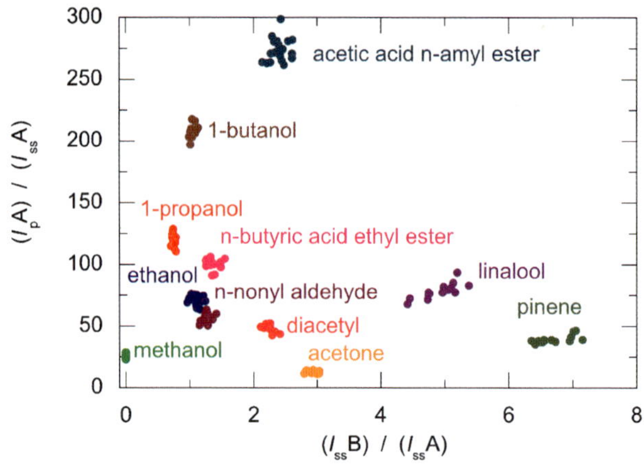

Fig. 34 Identification of the types of flavors of various concentrations using a $(I_{SS}B)/(I_{SS}A)$ vs. $(I_PA)/(I_{SS}A)$ diagram

oxidation of gas transferred from the gas phase to the sensor surface. As the CTL emission from sensor B results from the excited species produced during catalytic oxidation of the gas (as described in Sect. 3.2), the CTL emission mechanism is different from that for sensor A. Thus, the ratio of I_{SS} for sensor A to B ($I_{SS}B/I_{SS}A$) reflects the chemical characteristics of the gas molecule. Therefore, the position of a plot on a graph whose x-axis shows $I_{SS}B/I_{SS}A$ and y-axis shows $I_PA/I_{SS}A$ reflects the physical and chemical characteristics of the gas molecules.

Figure 34 shows the results for alcohol (methanol, ethanol, 1-propanol and 1-butanol), ketone (acetone and diacetyl), terpene (pinene and linalool), aldehyde (*n*-nonyl aldehyde) and ester (acetic acid *n*-amyl ester and *n*-butyric acid ethyl ester) of various concentrations. Because of the linear characteristics of the CTL-based sensor, the plots are located in a similar region for a certain type of gas of various concentrations where the Henry-type adsorption isotherm holds. Thus, we can identify these gases with various concentrations by simple data-processing.

7 Conclusions and Future Development

The elementary working mechanism of the CTL-based gas sensor is summarized as follows: In air containing combustible gases, the catalytic oxidation proceeds on the heated solid catalyst. Two CTL processes proceed during the reaction. One is production of the excited species, e.g. formaldehyde, followed by its relaxation radiation. The other is the recombination radiation of carriers originating from the surface states formed by the chemisorbed species produced in the course of the reaction. (i) When the catalyst is heated to a temperature under the diffusion-controlled condition of sample gases, the rate of catalytic oxidation is limited by the gas transfer rate from the vapor phase to the catalyst surface, and the CTL intensity is essentially proportional to the gas concentration. (ii) When the catalyst is heated after the adsorption of gases at a low temperature, the total CTL intensity is proportional to the amount of the physisorbed gases, which is proportional to the gas concentration and the duration of the adsorption. We can realize the CTL-based sensor because the CTL intensity is proportional to the gas concentration in both cases.

The CTL-based sensor has many advantages:

1. Long-term stability, linear concentration dependence, and fast response due to the catalytic reaction
2. High signal-to-noise ratio and high sensitivity due to the chemiluminescence measurement
3. Memory effect and sensitivity enhancement as a consequence of adsorption into the catalyst pores
4. Discrimination of gases by utilizing the difference in the CTL spectrum and the temperature dependence of CTL intensity

The many advantages of the current CTL-based sensors allow us to develop a gas-monitoring system for product manufacture, e.g., processed foods. The accumulation-type CTL-based sensor may be effective as a film badge to record exposure to harmful gases. It can memorize the physisorption of gases in the past atmosphere, and the total amount of the adsorbed gases can be measured by a read-out system consisting of an optical detector and a heating unit. The multi-sensor system using a CTL-based sensor array may be useful for analyzing various gases because of the linear characteristics of CTL intensity as a function of gas concentration.

The following future developments are expected to improve the sensor characteristics and to expand its use:

1. Catalysts with a better selectivity
2. More active catalysts for lower temperature operation
3. Activators of the catalyst that emit CTL at shorter wavelengths to avoid incandescent radiation
4. Catalysts and/or activators with a higher efficiency of CTL emission to realize a low-cost system using a semiconductor photodetector
5. Many types of catalysts with various sensing characteristics to realize the gas-sensor array for the analytical detection of mixed gases containing many kinds of constituents

The phenomenon of CTL emission has a rather long history since it was first observed in 1976. Unfortunately, however, only a few researchers to date have studied this phenomenon and its application for gas sensors. Although various proposals for measuring techniques using CTL-based sensors have been given in literatures, further research into their practical use is only proceeding slowly. Further investigations and development of CTL-based gas sensors with many potential applications are expected.

References

1. Firth JG (1966) Trans Faraday Soc 62:2566
2. Regener VH (1960) J Geophys Res 65:3975
3. Regener VH (1964) J Geophys Res 69:3795
4. Hodgeson JA, Krost KJ, O'Keeffe AE, Stevens RK (1970) Anal Chem 42:1795
5. Bersis D, Vassiliou E (1966) Analyst 91:499
6. Breysse M, Claudel B, Faure L, Guenin M, Williams RJJ, Wolkenstein T (1976) J Catalysis 45:137
7. Aras VM, Breysse M, Claudel B, Faure L, Guenin M (1977) J Chem Soc Faraday Trans I 73:1039
8. Nakagawa M, Fujiwara N, Matsuura Y, Tomiyama T, Yamamoto I, Utsunomiya K, Wada T, Yamashita N, Yamashita Y (1990) Bunseki Kagaku 39:797
9. Nakagawa M, Matsuura Y, Tomiyama T, Yamamoto I, Utsunomiya K, Wada T, Yamashita N, Yamashita Y (1990) Rep Res Lab Surf Sci, Okayama Univ 6:125
10. Utsunomiya K, Nakagawa M, Chikamori S, Kohata M, Tomiyama T, Yamamoto I, Wada T, Yamashita N, Yamashita Y (1993) Proc Adv Fluor Sens Technol 1885:93
11. Utsunomiya K, Nakagawa M, Tomiyama T, Yamamoto I, Matsuura Y, Chikamori S, Wada T, Yamashita N, Yamashita Y (1993) Sens Actuator B 11:441
12. Utsunomiya K, Nakagawa M, Tomiyama T, Yamamoto I, Matsuura Y, Chikamori S, Wada T, Yamashita N, Yamashita Y (1993) Sens Actuator B13:627
13. Utsunomiya K, Nakagawa M, Sanari N, Kohata M, Tomiyama T, Yamamoto I, Wada T, Yamashita N, Yamashita Y (1995) Sens Actuator B24:790
14. Nakagawa M (1995) Sens Actuator B 29:94
15. Nakagawa M, Kawabata S, Nishiyama K, Utsunomiya K, Yamamoto I, Wada T, Yamashita Y, Yamashita N (1996) Sens Actuator B 34:334
16. Nakagawa M, Yamamoto I, Yamashita N (1998) Anal Sci 14:209
17. Nakagawa M, Okabayashi T, Fujimoto T, Utsunomiya K, Yamamoto I, Wada T, Yamashita Y, Yamashita N (1998) Sens Actuator B 51:159
18. Okabayashi T, Fujimoto T, Yamamoto I, Utsunomiya K, Wada T, Yamashita Y, Yamashita N, Nakagawa M (2000) Sens Actuator B64:54
19. Okabayashi T, Toda T, Yamamoto I, Utsunomiya K, Yamashita N, Nakagawa M (2001) Sens Actuator B 74:152
20. Okabayashi (2002) PhD thesis, Okayama University of Science
21. Zhu Y, Shi J, Zhang Z, Zhang C, Zhang X (2002) Anal Chem 74:120
22. Zhang Z, Zhang C, Zhang X (2002) Analyst 127:792
23. Shi J, Li J, Zhu Y, Wei F, Zhang X (2002) Anal Chim Acta 466:69
24. Claudel B, Breysse M, Faure L, Guenin M (1978) Rev Chem Intermed 2:75
25. Wolkenstein T (1960) The electron theory of catalysis on semiconductors. In: Eley DD (ed) Advances in catalysis. Academic, New York, p 189
26. Toby S (1973) J Lumin 8:94
27. Heusden SV, Hoogeveen LPJ (1976) Z Anal Chem 282:307
28. Okabayashi T, Matsuo N, Yamamoto I, Utsonomiya K, Yamashita N, Nakagawa M (2005) Sens Actuator B 108:515

Springer Ser Chem Sens Biosens (2005) 3: 133–167
DOI 10.1007/5346_004

Published online: 28 September 2005

Hollow Waveguide Infrared Spectroscopy and Sensing

Christy M. Charlton · Bruce T. Thompson · Boris Mizaikoff (✉)

School of Chemistry and Biochemistry, Georgia Institute of Technology,
Atlanta, GA 30332-0400, USA
boris.mizaikoff@chemistry.gatech.edu

Abstract Infrared (IR) hollow waveguides (HWGs) were first utilized in gas sensing applications in the early 1990's and have since been coupled to both FT-IR spectrometers and laser light sources. However, gas sensing with hollow waveguide modules has yet to achieve maturity for widespread use or device commercialization. We review this emerging field with emphasis on technology and application areas where we believe these devices are ideally suited, and discuss the advantages and limitations of using HWGs for chemical sensing, including their optical properties and waveguide losses. As new HWG technologies based on photonic bandgap materials are emerging, relevant applications of

HWG based gas sensing such as in breath diagnostics or process monitoring have the required breakthrough potential to bring the advantages of HWG based sensing concepts to bear.

1 Introduction

1.1 Introduction to Infrared Gas Analysis

IR spectroscopy is routinely used for real-time analysis of gas-phase analytes combining the specificity of fundamental vibrational absorptions with the ease of in situ analysis. Usually, gas-phase IR analysis does not require trapping or extraction of an analyte from its matrix. IR gas cells can be designed to accommodate the sample stream with minimal sample conditioning for either environmental [1, 2] or industrial processes [3, 4]. The specificity of the IR spectral range allows for single-component analysis with minimal interference from other matrix components or multicomponent analysis [1]. Most IR analyzers are used for single-analyte investigations [5]. For example, nondispersive IR (NDIR) analyzers are commonly applied to monitoring carbon monoxide, carbon dioxide, methane, and water in combustion effluent, such as stack gas [6].

IR spectroscopy is widely applicable to many types of environmental analyses. These include the analysis of waste streams, soils, and water. Because of the ease in sample handling, IR analysis of gases, vapors, aerosols, smoke, and exhausts in air is among the commonest applications. The types of compounds under surveillance range from small diatomic molecules, like CO and NO, to complex mixtures of gases and aerosols, like combustion products of organic or petroleum fuels [7, 8]. Likewise, there are a variety of surveillance purposes, such as threshold monitoring to alert to a dangerous condition, and mapping of VOCs (VOCs) emitted from farms and smokestacks by remote sensing [9, 10].

Air quality is important from both a health and a safety perspective. In the USA, the National Institute for Occupational Safety and Health (NIOSH) and the Occupational Safety and Health Administration set limits of exposure to over 2000 different chemicals or classes of respiratory irritants [11, 12]. Many of the compounds that are monitored indoors are of similar interest to outdoor assessment, such as in the atmosphere and stratosphere. VOCs emitted from industrial operations are continuously monitored as required by US and local Environmental Protection Agencies. Power plants and waste incinerators are required to follow emission guidelines for harmful combustion gases, including CO_x and NO_x, as well as other combustion by-products, including polycyclic aromatic hydrocarbons, phenols, and hydrocarbons. Monitoring of smoke stacks at power plants is not only an investigative tool for determining emission compliance, but is also a control diagnostic to optimize the efficiency of the power plant. The major advantage of on-line IR techniques in contrast to other extractive techniques is rapid feedback on the gas composition, usually on the order of seconds. Passive IR sensing can be applied to investigate physical and chemical processes in the atmosphere, to monitor industrial emissions, and to detect climate changes as a result of human activities, such as biomass burning and automobile exhaust analysis. Most notable for this last group of applications are the greenhouse gases, which are contributing to global climate changes. Since many of these compounds have selective IR signatures, they are amenable to qualitative and quantitative IR analysis utilizing the vibrational spectrum.

1.2 Qualitative IR Gas Analysis

The mid-IR (MIR) region of the electromagnetic spectrum is usually considered from approximately 2 to 20 μm. Absorption of IR energy is a result of the vibrational energy levels associated with molecular bonds. Fundamental IR absorptions result from stretching or bending motion of a specific bond or bonds within the molecule under study. Stretching motion results from the movement of the two atomic nuclei towards and away from each other. Diatomic molecules are classic example of such stretching motions. During the motion, the distance between the two atoms decreases and increases from its equilibrium distance. As the internuclear distance decreases, the atoms approach each other, resulting in an increase of the inherent potential energy due to the repulsive nature of their nuclei. As the distance between the two atoms increases, the potential energy approaches zero at equilibrium. Further stretching the bond again increases the inherent potential energy until the dissociation energy is reached and the molecular bond breaks.

The intermolecular motion repeats itself in a periodic fashion modeled after the harmonic oscillator, which is described by

$$\nu_{\mathrm{osc}} = \frac{1}{2}\pi \left(\frac{k}{\mu}\right)^{\frac{1}{2}} , \tag{1}$$

where ν_{osc} is the frequency of vibration, k is the force constant of a hypothetical massless spring connecting the two atoms, and μ is the reduced mass of the two atoms.

In addition to vibrational motions, the molecule can undergo rotational motion perpendicular to the bond axis. For linear molecules, the energy associated with rotational transitions is approximated by the rigid-rotor model.

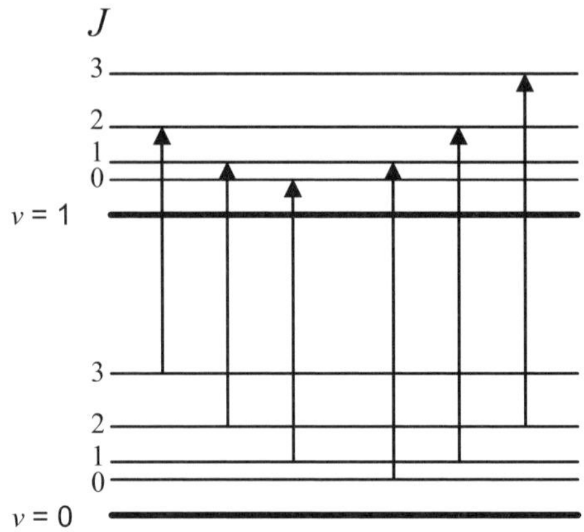

Fig. 1 Schematic of rotational energy transitions at a fundamental IR vibrational absorption

Typically, exciting the rotational energy levels requires much less energy than the energy already available at room temperature. Therefore, the number of rotational levels appreciably populated could be 30 or more.

IR spectra of the fundamental vibrational band of small gaseous diatomic molecules, such as CO and NO, contain a large number of absorption lines that correspond to these vibrational–rotational energy transitions. Since many different rotational levels can be populated at ambient temperature, many different transitions at different energies may occur (Fig. 1). Vibrational–rotational lines are evident only in gas-phase spectra collected at sufficiently high resolution. These lines are not resolved in condensed-phase spectra because of frequent collisions between molecules; hence, condensed-phase spectra are characterized by broad absorption bands occurring at the vibrational transition energies.

Usually, molecules have more than one fundamental IR vibration. These absorptions are divided into two classes, defined by whether they appear at the same frequency in many molecules or not. The absorptions that occur in molecules that have the same functional group are called group frequencies and are evident in the spectral range of approximately 4000–1300 cm^{-1}. The absorptions that arise from combinations of absorptions of the entire molecule appear from approximately 1300 to 400 cm^{-1}. As the absorptions in this region are unique to most organic molecules, this spectral region is called the "fingerprint region." The uniqueness and complexity of the fingerprint region is actually an asset particularly for the IR analysis of organics. In qualitative analysis, the fingerprint region provides specific and unique spectral features for compound identification. For quantitative analysis, these unique absorption bands provide opportunities to measure analytes with minimal interference and to match a molecular IR absorption with a narrow line of an IR laser light source, such as a CO_2 laser or a quantum cascade laser (QCL).

1.3 Quantitative IR Gas Analysis

The Beer–Lambert law expresses the relationship between the sample concentration, c, and the amount of IR radiation absorbed at the measured frequency (in wavenumbers $\overline{\nu}$) $A_{\overline{\nu}}$,

$$A_{\overline{\nu}} = -\log\left(\frac{P}{P_0}\right) = a_{\overline{\nu}}bc\,, \tag{2}$$

where P_0 is the power of the incident radiation, P is the power of the transmitted radiation, $a_{\overline{\nu}}$ is the absorptivity coefficient, which is dependent on the frequency of the IR radiation, and b is the optical pathlength. Although theory indicates a linear relationship, in practice the linearity of this relationship is dependent upon the concentration range and other instrumental factors, such as the instrument bandpass, stray light, and instrument noise. Thus, for a given sample matrix, concentration range, and operating conditions, the absorptivity is usually empirically determined. There are several calibration techniques commonly applied in IR spectroscopy to predict sample concentrations. These include univariate approaches utilizing absorption peak height or integrated absorbance, or approaches that take advantage of more spectral information, such as classical least-squares (CLS) or partial least-squares (PLS) techniques, also referred to as multivariate or chemometric techniques.

Univariate calibrations regress the concentration against an absorbance metric, such as the peak height or integrated absorbance. The peak height is the absorption measurement at a specific frequency, typically the absorption maximum for the fundamental vibrational band. However, depending upon the analyte and the calibration range, shoulders of the absorption band or weaker vibrational–rotational lines may be used for quantitative purposes. Nonlinear relationships between concentration and absorption can be handled with univariate calibrations by performing higher-order regressions, such as quadratic regressions.

Multivariate calibrations have become a commonly applied tool in the field of modern analytical chemistry and, specifically, in quantitative IR analysis [13, 14]. PLS regression is one of several methods that utilize an entire spectral information band present in IR data, often referred to as full-spectrum calibrations. The advantages of full-spectrum calibrations, such as PLS and CLS, are improvements in precision and robustness over univariate calibrations owing to increased signal averaging from including more spectral intensities. The distinction between PLS and CLS manifests in the fact that PLS is a factor-based regression, which means the full spectra for the acquired data are reduced to a smaller number of variables that account for the variability in the original data. Furthermore, PLS calibrations can be developed without knowing the entire composition of the sample, such as for developing a calibration for a single component in a complex matrix.

1.4 Fourier Transform IR Spectroscopy

Fourier transform (FT) IR spectroscopy is one of several nondispersive optical spectroscopies based on interferometry. A two-beam interferometer first proposed by Michelson is the basis of most modern FT-IR spectrometers, as exemplified by the schematic of the Bruker Equinox 55 spectrometer (Bruker Optik, Ettlingen, Germany) in Fig. 2. Simply described, the interferometer comprises a beam splitter and two mirrors. A collimated beam of IR energy is split at the beam splitter into equal halves. Half of the energy travels through the beam splitter to one of the mirrors, which is positioned at a fixed distance away from the beam splitter. The reflected beam travels perpendicular to the incident beam to a moving mirror. IR radiation reflects off the fixed and moving mirrors and recombines at the beam splitter. The recombined IR beam projects from the interferometer towards the detector on an optical path perpendicular to the source beam.

The rate at which data are collected at an interferometer is dependent on the distance the moving mirror travels, which determines the optical resolution, and how fast the mirror travels (i.e., the scanner velocity). Modern

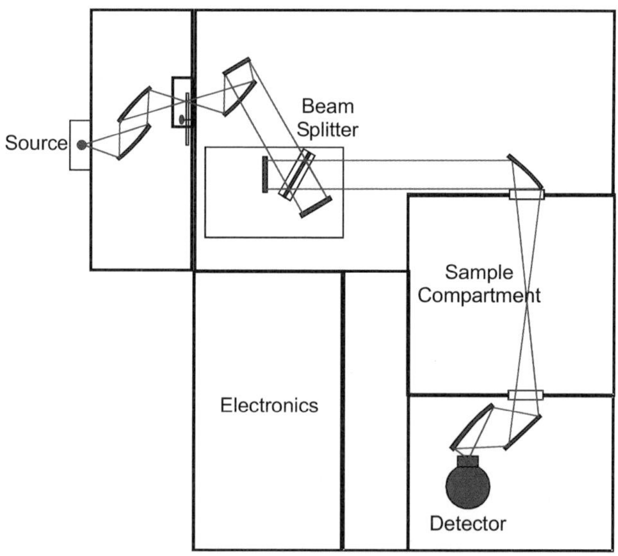

Fig. 2 Block diagram of a Bruker Equinox 55 Fourier transform (*FT*) IR spectrometer

FT-IR spectrometers have interferometers with scanning velocities enabling the collection of tens of spectra per second at a spectral resolution of 8 cm^{-1} or less. With faster data collection capabilities, FT-IR spectroscopy can be used to monitor and observe dynamic gas-phase processes. To observe such a process, interferograms are sequentially collected and stored in the memory of the system. The interferograms are then processed at the end of the data acquisition. The result of this operation is a three-dimensional data cube where each vertical slice of the cube is the spectrum for a time slice in the experiment equal to the interferogram acquisition time.

2 Hollow Waveguide Technology

2.1 IR Waveguides

The first IR transmitting fibers were made of chalcogenide (As – Se – Te) glasses and were produced in the 1960s [15]. Development of other fiber types began in the 1970s pushed by a US Army satellite surveillance program, which needed a fiber that was less brittle and could transmit over the entire MIR band (2–20 μm [15]). At that time, silica fibers were being produced to transmit visible light for telecommunications with intrinsic losses of 0.2 dB km^{-1}, initiating extensive research and development for fibers that could transmit IR light with comparatively low losses. Current solid-core IR fibers can be classified into three main groups: chalcogenide glasses, halide materials (such as silver halides), and crystalline materials (such as sapphire), which all have refractive indices of more than 1 in the MIR, enabling confinement of light by total internal reflection. Several books provide details on IR fiber technol-

Table 1 IR fiber materials. (Adapted from Ref. [18])

Material	Transmission range (μm)	Attenuation (dB m^{-1}) at wavelength (μm)
SiO_2	0.25–2	0.0004 at 1.32
Sapphire	0.4–2.5	5 at 2.9
Fluoride	0.5–3.1	0.015 at 2.5
Chalcogenide	4–11	0.04 at 6
Tellurium halide	3–13	1 at 9
Silver halide	3–18	0.5 at 11
Hollow waveguides	0.9–25	0.5–1.5 at 10.6

ogy [15–17]. Table 1 shows typical transmission ranges and attenuation levels for the commonest types of IR transparent fibers.

As shown in Table 1, hollow waveguides (HWGs) are the only fibers capable of transmitting radiation across the entire MIR band at comparatively low attenuation losses similar to usually more narrow-band solid-core IR fibers.

The main shortcoming of many solid-core fibers was their inability to transmit high-power laser light. This led to the development of air-core fibers, which transmit in the IR regime without concerns of melting the core material at high power levels. HWGs were initially developed for CO_2 laser power delivery at 10.6 μm with main focus on surgical laser applications. Prior to the development of HWGs, CO_2 laser light was predominantly guided by a series of mirrors on a retractable arm which is bulky and difficult to maneuver. HWGs now allow more flexibility to guide light to the point of the surgery.

2.2 History of Hollow Waveguides

Table 2 gives a summary of the key developments in HWG technology. The first HWGs were produced by Garmire [19] in the late 1970s and consisted of two strips of aluminum separated by dielectric spacers. However, not only bending but also twisting of the waveguides caused substantial losses owing to the rectangular cross section. Miyagi [20] then produced the first cylindrical HWGs by sputtering an aluminum pipe with germanium and then plating nickel on top. The aluminum structural tube was then removed, leaving a nickel tube coated with germanium on the inside. Using a dielectric coating on top of the reflective metal was found to increase the reflectivity, and was adopted for all future waveguide structures. The Ni/Ge design was later improved by using a silver tube coated with ZnSe [21], which has much lower losses. Croitoru and his group [22] initiated the use of plastic structural tubes coated with silver and silver iodide on the inside. Using AgI as the dielectric layer provides a distinct processing advantage as it can be formed by passing iodine gas over the silver layer already present on the inside of the waveguide, thereby reducing the number of layers to be deposited.

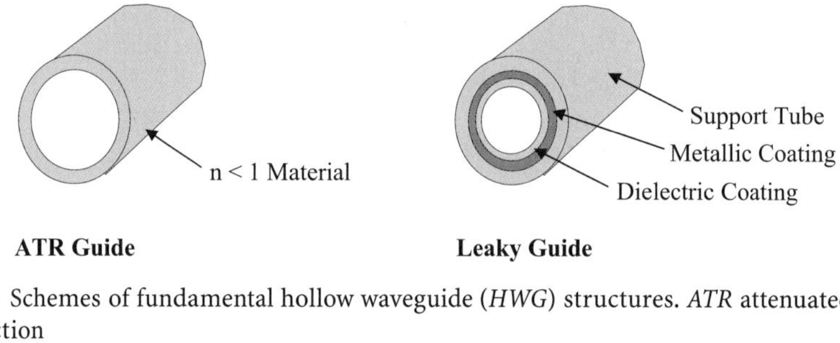

Fig. 3 Schemes of fundamental hollow waveguide (*HWG*) structures. *ATR* attenuated total reflection

Table 2 Milestones in the development of hollow waveguide (*HWG*) technology

Year	Authors	HWG description	HWG schematic	Losses ($dB\,m^{-1}$) (at 10.6 μm)	Improvement
1976	Garmire et al. [19]	Al strip with dielectric spacers; rectangular	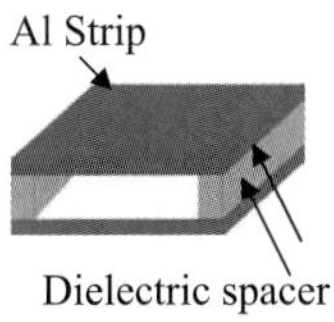	0.22	First HWG; higher powers can be transmitted
1983	Miyagi et al. [20]	Ge-coated Ni Tube cylindrical	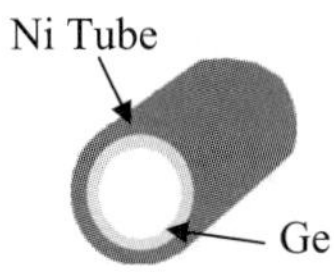	0.5	Losses due to twisting of rectangular guides eliminated
1990	Matsuura et al. [21]	ZnSe-coated Ag tube	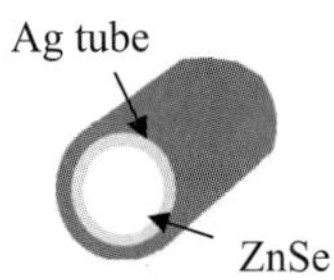	0.021	Much lower losses with silver compared with nickel
1992	Dahan et al. [22]	Ag/AgI-coated plastic tube	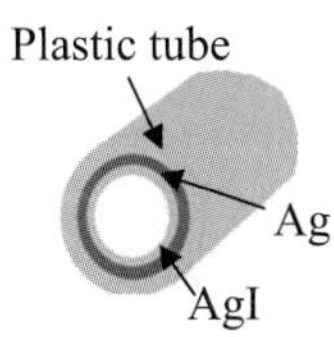	0.5	Very flexible; simple and cheap in fabrication
1994	Abel et al. [23]	Ag/AgI-coated glass tube	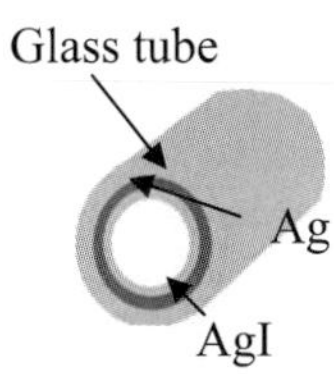	0.15	Smooth inner surface
1999	Fink et al. [24]	OmniGuide (PBG HWG) alternating plastic and glass layers	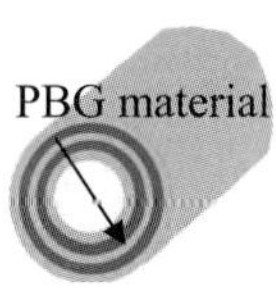		Lower bending losses

PBG photonic bandgap

Harrington and his group [23] introduced glass tubes as the structural support with a Ag/AgI coating. The glass tube provides the advantage of a very smooth inner surface for deposition of the silver layer, giving a lower surface roughness of the final HWG. The latest development in HWG technology is a structure consisting of cylindrical layers of glass and plastic creating a photonic bandgap (PBG) material (OmniGuides [24]; Sect. 2.3.2) promising lower bending losses in contrast to conventional HWGs.

2.3 Hollow Waveguide Technology

The first HWGs confined light to their core either by total internal reflection (attenuated total reflection guides) or by reflection off a metallic surface (leaky guides) as discussed in Sect. 2.3.1. The latest generation of HWGs is based on PBG materials as discussed in Sect. 2.3.2.

2.3.1 Attenuated Total Reflection and Leaky Guides

Attenuated total reflection guides are composed of a tube made from material with a refractive index (n) less than that of the air core ($n = 1$), such as sapphire [25] or special glasses [26]. In contrast, leaky guides consist of a tube coated on the inside with material of $n > 1$. Leaky guides are the more commonly applied HWG type, relying on a reflective coating applied to the inside of a hollow tube to guide IR radiation. Glass, metal, and plastic tubing have been used as a structural support coated on the inside with reflective materials. Among the variety of IR reflective metal layers, silver is the commonest one.

2.3.2 Photonic Bandgap Fibers

PBG materials (also known as photonic crystals) [27] are periodic structures of dielectric materials. The periodicity allows for the control of light inside the crystal in analogy to electrons in a semiconductor crystal. The periodically changing dielectric constant creates diffraction patterns that prohibit photons with a frequency inside the bandgap from propagating through the crystal. This causes a PBG material to act as a perfectly reflecting mirror for a certain band of frequencies, which can be tailored by changing the dimensions of the crystal features. Waveguides made of these materials have the potential to provide much lower losses than HWGs based on light confinement purely by refractive index difference.

PBG fibers are predominantly developed for telecommunications purposes, as they show great potential as a lower-loss alternative to conventional optical fibers. Rather than confining light by total internal reflection as in

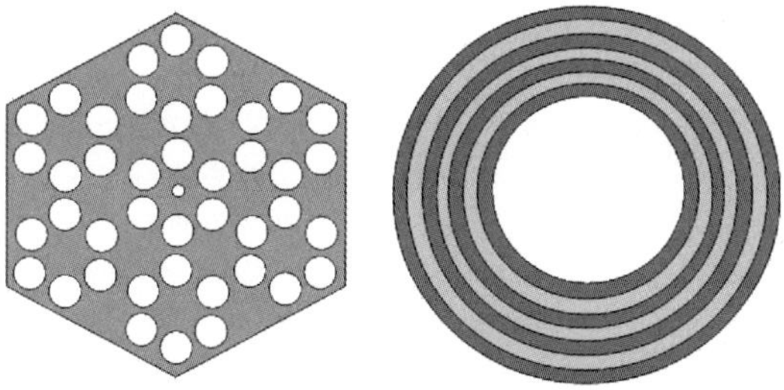

Fig. 4 Schematic photonic bandgap HWG cross section of **a** honeycomb-type structures, where light is confined to the central defect hole, and **b** omnidirectional guide, where light is confined by cylindrical photonic crystal layers

conventional HWGs, these omnidirectional waveguides utilize the bandgap formed by a photonic crystal, confining light in a narrow frequency band regardless of its angle of incidence. Hence, they are in principle able to bend into very sharp turns without losses.

The first PBG fibers with light guided in an air core were developed by Cregan et al. in 1999. These PBG materials consisted of a honeycomb structure made from silica with air gaps similar to the fiber cross section shown in Fig. 4a . Light is confined to a defect air hole in the center of the crystal and guided along the length of the fiber. Omnidirectional guides were developed in 2002 by the Fink group [28, 29] and the Miyagi group [30]. The structure of these materials is shown in Fig. 4b. Omnidirectional waveguides consist of materials with alternating refractive index forming a cylindrical layer stack. Thus, a one-dimensional photonic crystal with a hollow center is created.

The omnidirectional guide is most useful for sensing applications owing to the open hollow core. In honeycomb-type structures [31, 32], light is confined to the small central air hole, which is difficult to fill with analyte gas for sensing owing to considerable back-pressure building up in the hollow structure.

During processing of omnidirectional waveguides, the thickness of the dielectric cylinders can be varied, enabling tuning of the bandgap to confine selected frequencies of light, which renders PBG HWGs ideal for laser-based integrated gas-sensing devices, where the emission frequency of the light source is matched with a selective absorption line of an analyte [33].

2.4
Brief History of Hollow Waveguides as Gas-Transmission Cells

HWG-based light guiding was initiated with the development of the first so-called lightpipes, which is attributed to Azzaraga [34], and to Yang et al. [35]. Designed in the early 1980s in a variety of diameters, lightpipes are gold-coated metal or glass tubes predominantly applied for coupling gas chromatography (CG) with FT-IR spectroscopy. Thermally isolated and surrounded with a heating system, they enable on-line FT-IR spectroscopic detection of gases transferred after separation via heated transfer lines into the lightpipe.

In 1991, Harrington et al. [36] reported the spectrum of 20% CO_2 recorded with a 1.06-mm inner diameter and 150-mm-long hollow sapphire fiber.

Later, Saito et al. [37] detected explosive gases such as CH_4 and n-C_4H_{10}, using ZnS-coated Ag HWGs, and pollution gases such as NO_2 and SO_2 with glass HWGs. The diameter of the hollow core was 1 mm at various lengths ranging from 250 mm to 1 m; a heated SiC rod was used as an incoherent IR light source. The emitted radiation was chopped at 1 kHz, dispersed by a grating monochromator, and detected with a mercury cadmium telluride (MCT) detector element. All gases were detected at percentage concentration levels.

2.5 Optical Considerations for Hollow-Waveguide-Based Gas Sensing

There are various strategies to couple a HWG module to a radiation source and a detector for the purposes of gas sensing. The commonest combination comprises coupling of a FT-IR spectrometer to a HWG-based gas-sensing module. Following the optical requirements for the interferometer, the IR beam exiting the FT-IR spectrometer is nearly collimated and usually up to several centimeters in diameter. Hence, for efficient coupling of this beam to a HWG with an inner diameter usually less than 2 mm, the beam needs to be condensed. Most applications reported in the literature utilize either a single off-axis parabolic mirror or a focusing lens to efficiently couple the source radiation to the HWG sensing module. However, losses due to coupling of up to 80% are reported [38, 39].

At the waveguide entrance aperture, considerable energy is lost owing to the small numerical aperture of the waveguide. As a HWG is considered a nonimaging optical element, radiation emitted at the exit aperture of the HWG is no longer spatially related to the beam geometry at the entrance aperture. Therefore, positioning of the collection optics—usually another off-axis parabolic mirror—is critical and may result in additional losses. A ray-tracing simulation can illustrate these losses. A simple Monte Carlo simulation can be developed by treating all reflective surfaces as ideal reflectors. The geometric surfaces are programmed into the simulation and individual rays are launched from a source and projected to the first surface, usually the off-axis parabolic mirror that focuses the beam into the waveguide. The reflected ray is calculated and propagated to the next surface until it reaches the detector or is lost by, for example, outcoupling. Features of the optical components, such as their geometry and location, can be varied to study their effects. For example, Fig. 5 depicts the simulations of the entrance and exit of a HWG, with and without the approximately 15° acceptance

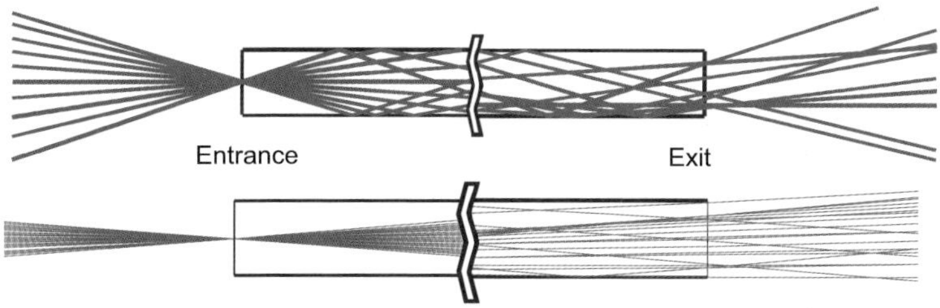

Fig. 5 Schematic of Monte Carlo simulation of IR radiation propagating through a HWG without (*top*) and with (*bottom*) 15° acceptance angle

angle. Radiation is inserted into the HWG via a perfectly focused off-axis parabolic mirror with a focal length of 25.4 mm. Evidently, radiation emitted at the exit aperture does not image the entrance focal point anymore. In addition, a large number of the incident rays would not propagate through the waveguide as their angle exceeds the acceptance angle of the waveguide.

Kozodoy et al. [40] reported results on HWG-based gas sensing for CO_2 by coupling of the HWG to a FT-IR spectrometer. The waveguide was a Ag/AgI-coated silica hollow fiber. The HWG was coupled in a straight (i.e., no bends or loops) and looped configuration to the FT-IR. However, the coiled configuration revealed losses of approximately 50–70% with respect to the straight configuration.

One of the limitations of conventional FT-IR-based gas-measurement systems is the response time. To perform the analyses, the FT-IR spectrometer is typically combined with a long pathlength or multipass gas cell [1, 5, 41]. Conventional gas cells have an internal volume ranging from a few hundred milliliters up to several liters. HWGs usually have an internal volume several orders of magnitude smaller than the multipass gas cell; hence, the response time of the gas-sensing system is dependent upon the optical efficiency of the gas cell defined as the ratio of the optical pathlength to the volume of the gas cell, and the flow response time of the gas cell. For long-pathlength gas cells, the optical efficiency is typically less than 1 [41, 42]. In environmental monitoring, where changes in concentration occur on a comparatively slow time scale, the system response time is usually the most stringent limitation. However, for example in process monitoring and surveillance or for investigating combustion processes, timely feedback is required, rendering gas-sensing systems with long response times less than optimal.

The flow response of a gas cell is an important parameter for sensing applications as it defines the actual response time of the system. The theoretical aspects of determining the time constants from the decay component of the response curve are straightforward. The time constants of the gas cell are calculated from the concentration decay curves (Fig. 6). For example, the time constants for CO are 1.23 and 12.4 s for the HWG and the multipass gas cell,

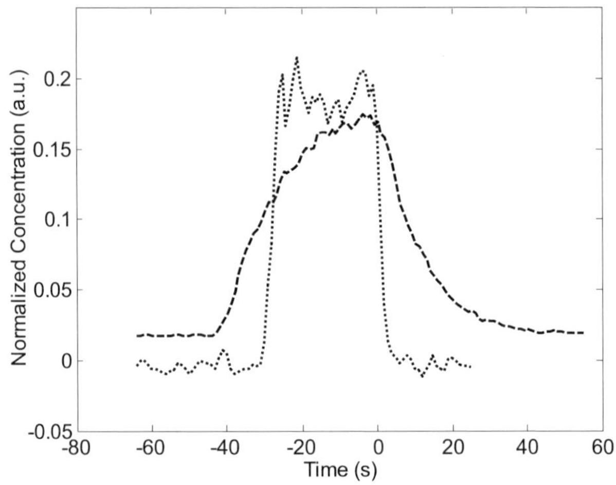

Fig. 6 Flow response decay curves for FT-IR spectroscopy with a HWG gas cell (*dotted line*) and FT-IR spectroscopy with a 3-m multipass gas cell (*dashed line*) [43]

respectively. The response times impressively demonstrate the optical efficiency of the HWG gas cell.

The increase in optical efficiency translates into increased sensitivity for sensing of gas-phase analytes. Plots of the wavenumber ($\overline{\nu}$) versus the sensitivity of the HWG and a 3-m gas cell for CO and NO are shown in Fig. 7. These plots indicate that the HWG reveals increased sensitivity compared with the

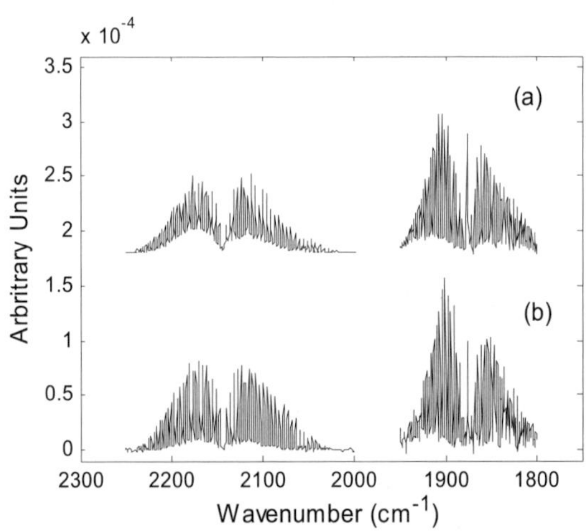

Fig. 7 Wavenumber versus sensitivity per nominal pathlength for **a** 3-m multipass and **b** HWG gas cell

3-m gas cell. Calibrations using certified gas standards show that the increase in sensitivity ranges up to 51% for CO and 59% for NO.

One rationalization for the increased sensitivity is an increase in "photon density" inside the HWG gas cell. In the 3-m gas cell the IR beam occupies approximately 85% of the volume of the cylinder defined by the field mirror and the objective mirrors. Since the HWG is a nonimaging optical device, nearly 100% of its volume is occupied by the IR beam. Assuming all other optical characteristics are equivalent, the HWG should have an increased "photon density" of approximately 125-fold. If the movement of gas molecules and the propagation of photons in the gas are considered random events, the probability of interaction between the two should increase linearly with an increase in either domain. Therefore, a 125-fold increase in sensitivity should be realized by using the HWG; however, the improvement in sensitivity is less than doubled from the experimental findings. Using a three-dimensional Monte Carlo simulation of photons traveling through the HWG with a 15° acceptance angle reveals that only approximately 3% of the photons reach the detector with this optical arrangement. Combined with the estimated attenuation for the HWG of 1 dB m^{-1}, a 1.2-fold increase in sensitivity should still be realized. This estimate, considers only optical efficiencies of the gas cells and simulated throughput of the HWG; however, it is within the same order of magnitude as the measured increase in sensitivity. Simulations also show that by increasing the focal length of the off-axis parabolic mirror focusing the beam into the HWG, fewer photons are rejected because of the HWG's acceptance angle. The percentage of photons reaching the detector increases to 96% if the focal length of the mirror is increase fivefold. Obviously, longer mirror focal lengths diminish the advantages of the compact design of HWG-based sensing modules as the distance from the mirror to the HWG would be nearly half the length of the HWG. Still, since only 3% of the photons reach the detector when using the 25.4-mm focal length mirror, other optical combinations of mirrors or lenses could be applied to achieve more efficient light coupling to the HWG within the 15° acceptance angle, yielding an increased net gain in throughput.

3 Applications of Fourier Transform IR Coupled Hollow Waveguide Gas Sensors

3.1 Analyte Extraction from Intractable Sample Matrices

Frequently, the challenge for the analytical chemist is not only increasing sensitivity, lowering detection limits, or increasing the signal-to-noise of the sensing device, but to reproducibly deliver the analyte to the instrument. This is particularly prevalent in the field of environmental analyses. Two aspects of environmental analytical chemistry that present a challenge are (1) that it is preferable to isolate the analyte from its environment and (2) to preconcentrate the analyte, enabling detection within the analytical range of the instrument. Even in a seemingly trivial case of atmospheric analysis, the challenge is to isolate the analyte from a matrix that is a mixture of potential interferants, including particulates, aerosols, vapors, and gases. Analyses of water or soil samples have obvious challenges to presenting the molecules of interest to an instrument. Neither matrix is suitable for direct introduction into conventional analytical instrumentation used for environmental analysis, such as GC–mass spectrometry, high-performance liquid chromatography, or conventional optical spectroscopic techniques. Thus, extraction and isolation techniques are applied in conjunction with HWG-based sensing techniques to provide fast analysis of intractable samples without sample preparation.

HWGs have successfully been applied to a wide variety of gas-sensing applications [44–52]. Micheels et al. [46] coupled a coiled MIR HWG to a FT-IR spectrometer measuring VOCs in the headspace of water samples. Yang et al. [47, 48] partitioned organics from water or the headspace above a soil sample into the coating of a HWG. The waveguide was then inserted into the sample compartment of the FT-IR.

A membrane-sampling device was coupled to the HWG by de Melas et al. [50] for the analysis of benzene, toluene, and xylenes in water.

Harrington and coworkers [40, 45, 46] have investigated the application of valunit-1m-long coiled HWG to achieve sufficiently long optical pathlengths in a still compact instrumental design. A schematic of the coiled HWG gas-sensing device is shown in Fig. 8. The FT-IR/HWG sensor was applied to the analysis of VOCs in soil. To isolate the VOCs from the soil matrix, the gas-

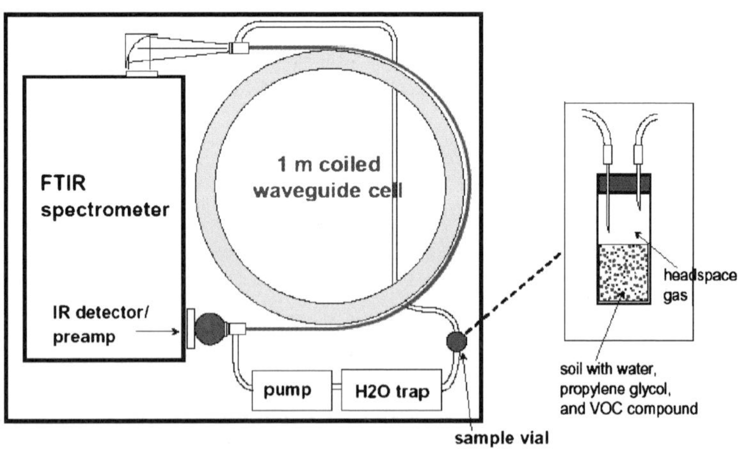

Fig. 8 FT-IR gas analysis system utilizing a 1/2 turn coiled HWG and a headspace sample vial. *VOC* volatile organic compound [46]

sensing system sampled the headspace above the soil sample. The sensing system was calibrated using mixtures of benzene, toluene, *m*-xylene, ethylbenzene, and trichloroethylene from certified gas standards. The concentrations ranged from 20 to 200 ppm (v/v).

Figure 9 shows the recorded spectra of five VOCs studied in this work. Owing to the significant spectral overlap in the group frequencies and fingerprint region, PLS was used to calibrate the instrument. The coefficients of determination ranged from 0.93 to 0.99. This gas-sensing system was used to determine the partitioning ratios between soil and air for the selected VOCs in three types of soil. These ratios range from 19 to 70%, with the majority in the 25–50% range. As expected, trichloroethylene in sandy soil revealed the highest partitioning ratio. In contrast, toluene had the lowest partitioning ratio of 19% in high clay soil. The results indicate that 10 ppm (wt) concentration in soil was equivalent to approximately 100 ppm (v/v) in the gas phase.

Yang et al. [47, 48, 53, 54] developed a HWG sensing system for liquid and soil analyses based on an extractive polymer membrane coated onto the inside of the HWG. The polymer coating performs a solid-phase microextraction of the analyte from the headspace of the sample and preconcentrates the analyte prior to IR analysis.

Two approaches are reported for introducing the sample to the polymer-coated HWG. In the case of soils and liquids, the waveguide was placed above the sample and the gaseous headspace was pulled through the HWG

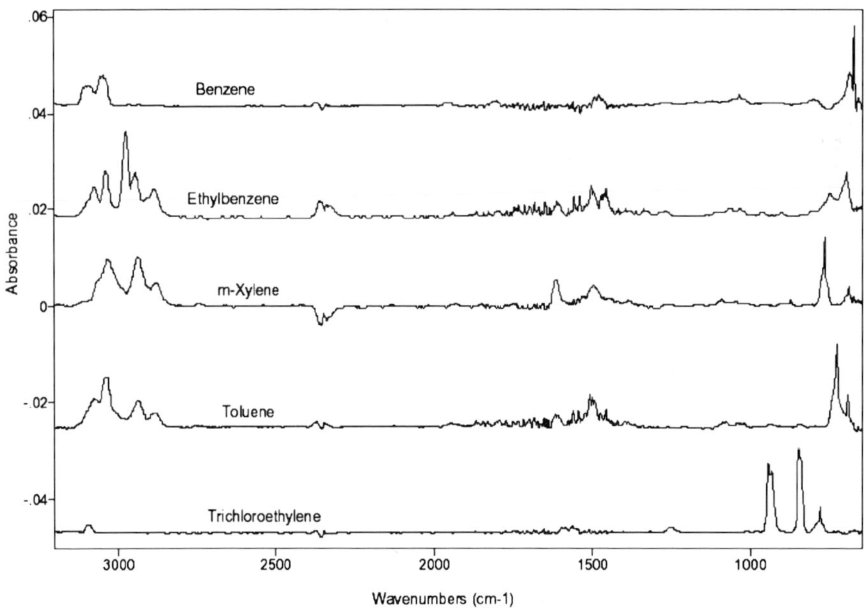

Fig. 9 FT-IR spectra of five VOCs collected with a HWG gas-sensing system [46]

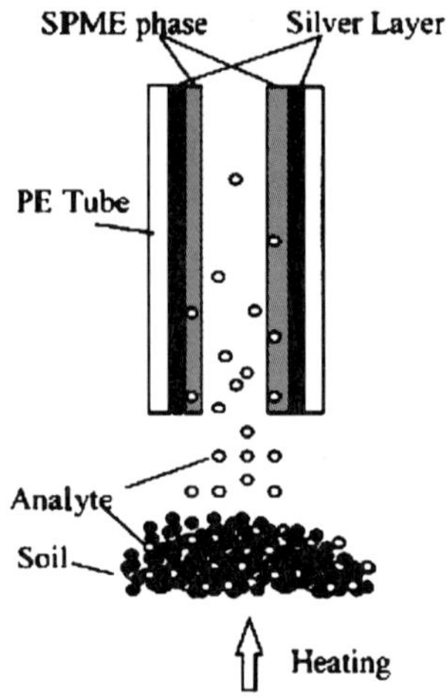

Fig. 10 Schematic of headspace sampling by a HWG internally coated with an extractive polymer membrane adsorbing the analytes investigated [47]

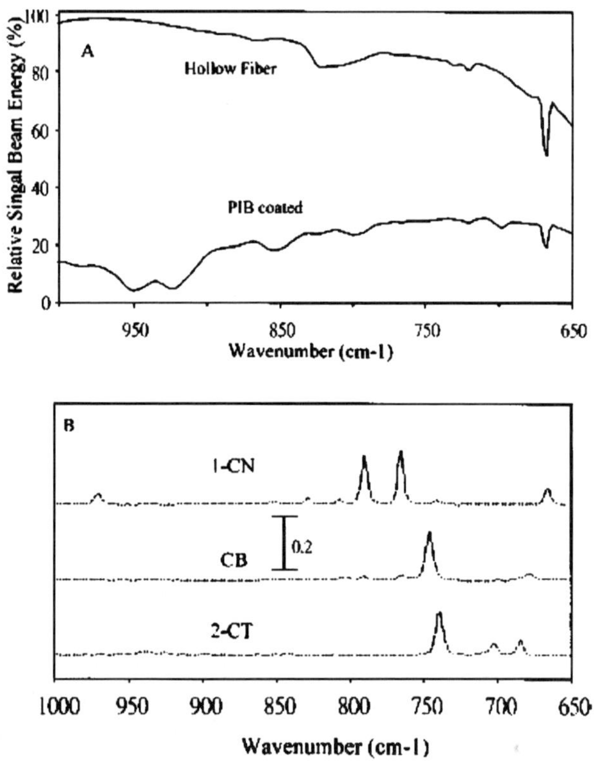

Fig. 11 FT-IR spectra showing **a** the energy throughput of polymer-coated HWGs, and **b** 50 ppm of the chlorinate aromatics 1-chloronaphthalene (*1-CN*), chlorobenzene (*CB*), and 2-chlorotoluene (*2-CT*) [47]

at a low flow rate of approximately 5 mL min^{-1} (Fig. 10). Figure 11 shows spectral data for samples collected under these conditions. The limits of detection for the chlorinated compounds investigated were determined to be less than 1 ppm. Furthermore, chlorinated aromatic amines have successfully been determined in water with the same approach. Limits of detection less than 100 μg mL^{-1} have been obtained for four dichloroaromatic amines investigated.

The second approach was realized by direct injection of microliter quantities of liquid into the internally polymer coated HWG, which is a useful approach if only limited amounts of sample are available. After sufficient exposure time ensuring partitioning of the analyte into the polymer coating, the HWG was purged with air to remove the remaining liquid sample and then inserted into the FT-IR. This approach was used to study several chlorinated hydrocarbons and chlorinated aromatic amines. In this approach, the limits of detection for these compounds were determined to be approximately 1 ppm. However, in both approaches, the analysis is a two-step process requir-

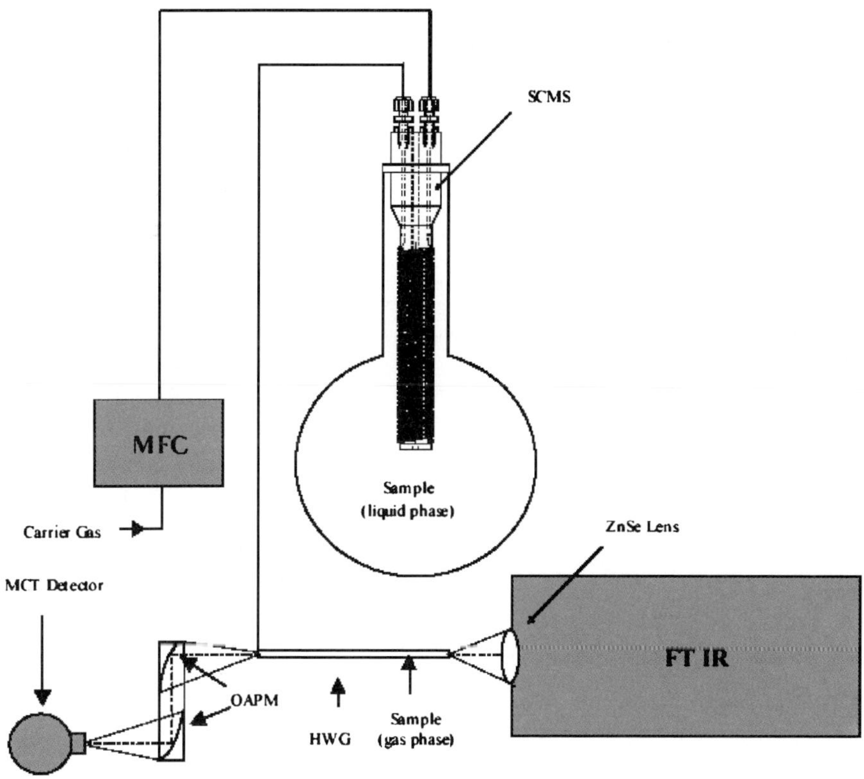

Fig. 12 Experimental setup for a HWG gas-sensing system utilizing a supported capillary membrane sampler, a HWG gas-sensing module, and a compact FT-IR spectrometer [50]

ing a separate sampling step followed by discontinuous analysis. Therefore, this approach is not amenable to continuous real-time analyses.

de Melas et al. [49–52, 55] coupled a supported capillary membrane sampler (SCMS) to a HWG gas sensor to continuously measure parts per billion (v/v) levels of chlorinated hydrocarbons and parts per million (v/v) levels of 1,4-dioxane in real-world samples. The SCMS continuously extracts VOCs from an aqueous matrix into the gas phase [50]. The extraction module consists of a stainless steel probe with a length of 15 cm and a tubular silicone capillary membrane, which is coiled around the steel core and connected to a gas inlet/outlet at the top (Fig. 12). The outer wall of the permeation tubing is in direct contact with the aqueous sample, allowing analyte molecules to permeate from the bulk sample solution through the wall of the polymer

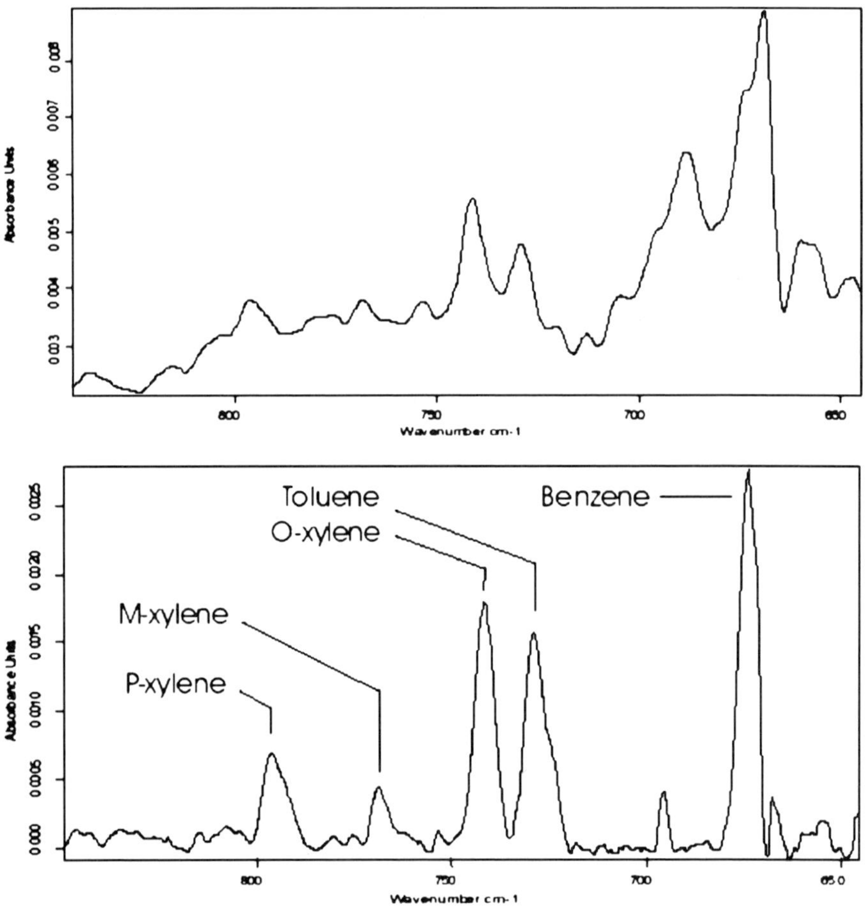

Fig. 13 Raw spectrum (*top*) and CO_2-corrected spectrum (*bottom*) of a sample containing 750 ppm of each component in a mixture of benzene, toluene, and xylenes in water [55]

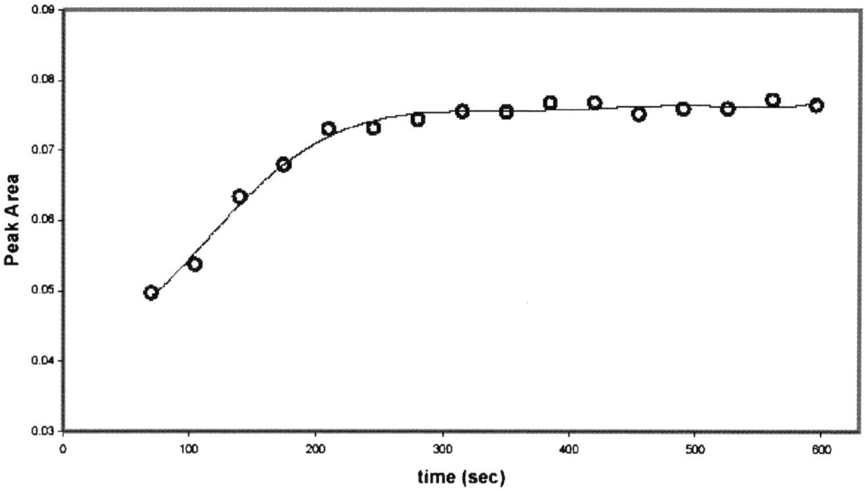

Fig. 14 Equilibrium curve demonstrating the dynamic behavior of the sensor system during a 1-ppm increase in concentration of benzene in aqueous solution, initially present at a concentration of 4 ppm [55]

tubing and to desorb into an inert carrier gas stream. This process is mainly governed by the permeability of the sample molecules through the polymer membrane, the vapor pressure of the analyte molecules, and the temperature of the liquid sample. The length of the permeation tube and the flow rate of the carrier gas are accordingly adjusted, ensuring equilibrium conditions for continuous extraction of analyte molecules from the aqueous phase.

As the analyte absorption features in the fingerprint region are affected by absorptions from CO_2, the CO_2 peak was subtracted and the baseline corrected for quantitative analysis. Figure 13 clearly shows the absorption features assigned to benzene, toluene, and xylenes (BTX) in the mixture. The calibrations yield coefficients of determination of 0.99 or better. With this sampling system, changes in the levels of benzene in water were detected in less than 3.5 min (Fig. 14). Limits of detection ranging from 32 to 116 ppm for individual BTX compounds have been derived using this approach.

Further optimization was performed by adapting the SCMS/HWG sensing system to the determination of several chlorinated hydrocarbons. Calculated detection limits less than 50 and 1 ppm (v/v) were determined for dichloromethane and chloroform, respectively.

3.2 Real-Time Direct Fourier Transform IR Hollow Waveguide Gas Analysis on a Short Time Scale

As discussed in Sect. 2.5, multipass long-pathlength IR gas cells have response times 1 order of magnitude larger than systems utilizing a HWG cell. In order to facilitate analyte determination with a sufficiently short time response, a HWG gas cell offers a suitable substitution for long-pathlength gas cells to achieve the necessary temporal resolution. The application example discussed in the following utilizes a HWG gas-sensing module for time-resolved monitoring of sidestream cigarette smoke. However, this measurement principle is transferable to a wide variety of combustion processes or monitoring situations requiring a time resolution in the second regime.

Various types of IR spectroscopy have been applied to the analysis of cigarette smoke [56–64]. The majority of these studies focus on the analysis of mainstream cigarette smoke, which comprises smoke emitted from the filter end of the cigarette [56–62, 64]. Maddox [58] and Cueto [60, 61] demonstrated analysis of gas-phase components of mainstream smoke by FT-IR spectroscopy. Cole and Martin [63] measured the emission of several gas-phase components of sidestream cigarette smoke, i.e., smoke emitted from the lit end of the cigarette, by FT-IR techniques with univariate calibrations; however, their quantitative analysis relied on unique vibrational–rotational lines and collecting smoke from four cigarettes simultaneously to compute an average amount emitted per cigarette. Consequently, the FT-IR results revealed significant differences for CO and NO in comparison with traditional techniques. Thompson and Mizaikoff [65] demonstrated the application of PLS to the analysis of sidestream cigarette smoke and achieved comparable performance at a resolution of 8 cm^{-1}. However, both studies relied on long-pathlength gas cells with volumes ranging from a few hundred milliliters to 8 L. Consequently, the FT-IR results lack the temporal resolution attainable by the NDIR analyzers for CO. Shi et al. [64] used a QCL and a 0.3-L gas cell under constant vacuum to achieve the desired time resolution for monitoring interpuff concentration profiles for several gas-phase sidestream smoke analytes. The QCL spectrometer was successful in monitoring sidestream smoke, but required independent CO_2 measurements, and emission ratios and emitted amounts for the gas-phase analytes were obtained. However, it is evident from this work that the increased temporal resolution aides in understanding the combustion gas formation during the puff. For sidestream smoke, combustion product emission between products generated from the puff and from smoldering (interpuff period) needs to be temporally resolved on the basis of a single cigarette. These results are of increasing importance as, for example, new types of cigarette papers are used to meet self-extinguishing standards for "fire-safe" cigarettes.

In order to achieve the demanded combination of temporal resolution and accurate yields, FTIR spectroscopy appears a suitable method for collecting data in combination with robust calibrations for quantitative results. Cigarette smoke is known to contain thousands of compounds and the amounts of the vapor phase components in the sidestream smoke are emitted at levels 2–10 times the mainstream amounts [66, 67]. In order to utilize vibrational–rotational lines as reported in previous studies [63, 64], FT-IR spectra would have to be collected at a spectral resolution of 1 cm^{-1} or better. Improved temporal resolution of the sidestream concentration profile could only be attained at an equivalent signal-to-noise level by collecting spectra at lower resolution. However, at lower resolution (below 1 cm^{-1}) the vibrational–rotational lines would no longer be spectroscopically resolved and accessible to quantification.

Multivariate regression techniques, such as PLS, are useful for accurate quantification in the presence of interfering compounds at unknown levels. Multivariate calibrations are more robust to interferences and the demands of dynamically changing systems as they use more spectral information for quantification. FT-IR spectroscopy and PLS have been applied to multicomponent analysis of combustion gases [3, 68–70], process streams [4], and environmental pollutants [2, 71, 72]. Hart et al. [71, 72] studied the effect of spectral resolution and regression techniques on the performance of open-path FT-IR spectroscopy for monitoring environmental pollutants. Qin and Cadet [2] evaluated the combination of low-resolution FT-IR spectra and PLS to characterize effluent from a high-temperature process containing organic compounds. Jaakkola et al. [73] studied the relationship between FT-IR instrument resolution and analytical figures of merit for the analysis of hydrocarbon mixtures. Each of these previous studies achieved successful quantification at lower spectral resolution.

Figure 15 shows a schematic of the sidestream smoke collection apparatus. A fishtail chimney [74] is used to collect the sidestream smoke from the lit end of the cigarette. The chimney design permits cooling of the sidestream smoke and formation of smoke aerosol particles for trapping on the glass fiber filter pad immediately after the chimney. To further reduce the level of volatile and semivolatile compounds in the sidestream smoke, a cold trap immersed in a dry ice/methanol bath (– 78 °C) was place immediately after the filter pad. To prevent ice or other solid material from the trap entering the gas cell, another glass fiber filter pad was placed after the trap.

A comparison of the IR spectra of sidestream cigarette smoke with the 3 m multipass gas cell and the 55-cm HWG is shown in Fig. 16.

Figure 18 shows the real-time traces of the CO concentration for smoke runs of the reference cigarette IM16 using the 3-m multipass and HWG gas cells. Evident in both plots are the breaks between the three cigarettes smoked during the run and the minute but noticeable increases in concentration due to puffing. The improvement of the temporal resolution due to the HWG is

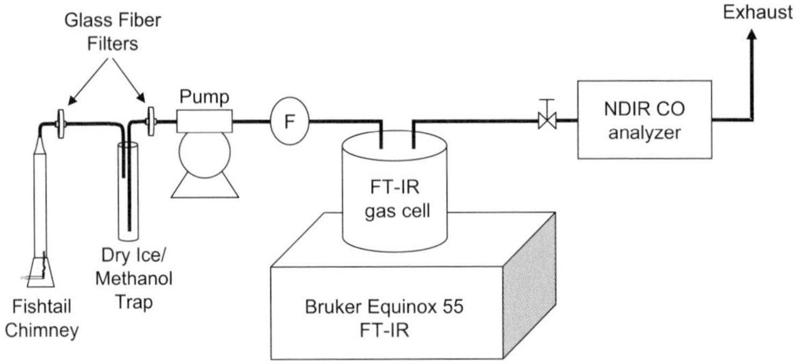

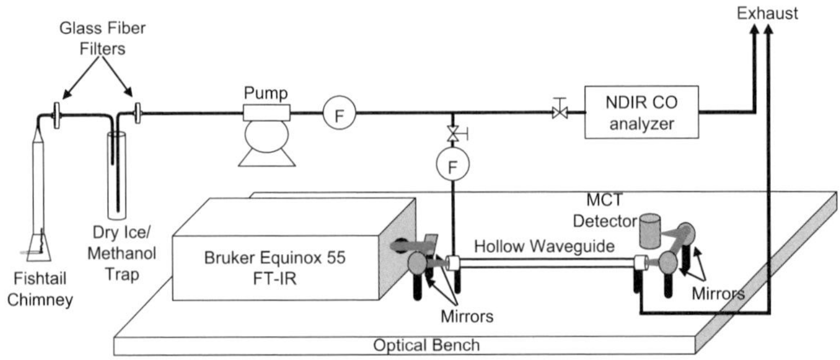

Fig. 15 Schematic layouts of experimental setup for real-time analysis of sidestream cigarette smoke using FT-IR spectroscopy with a 3-m multipass gas cell (*top*) and a HWG gas cell (*bottom*) [43]

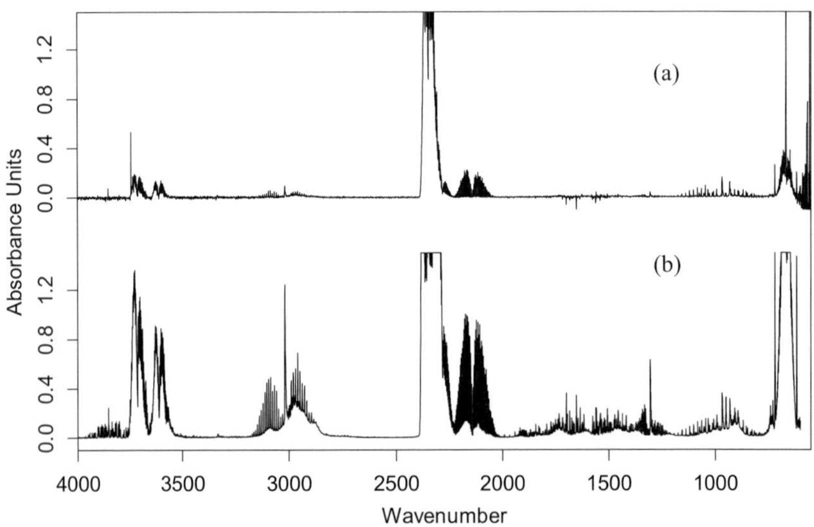

Fig. 16 IR spectra of sidestream cigarette smoke recorded in **a** a HWG cell and **b** a 3-m multipass gas cell [65]

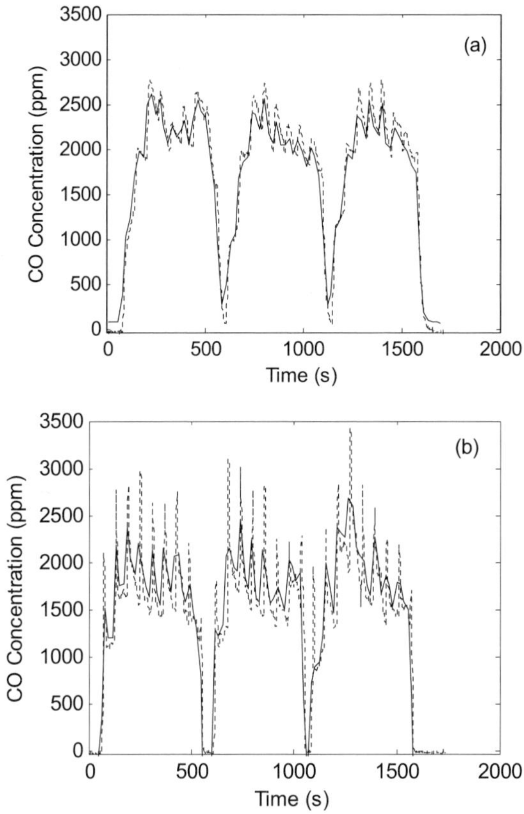

Fig. 17 Carbon monoxide concentration traces for IM16 reference cigarette smoke runs **a** with 3-m multipass and **b** HWG gas cells (*dashed line* nondispersive IR (*NDIR*) analyzer, *solid line* FT-IR spectrometer using partial least-squares calibrations) [43]

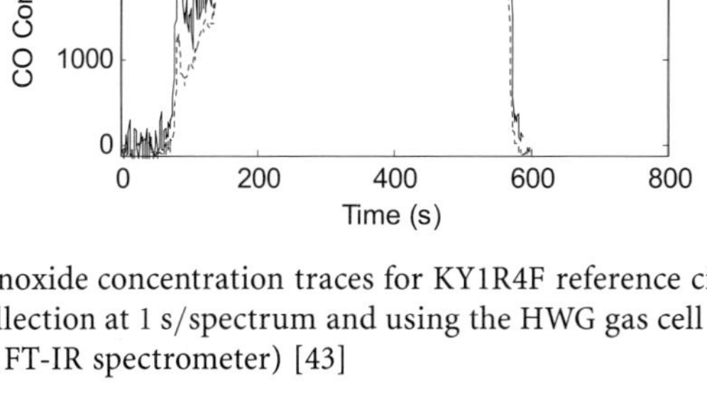

Fig. 18 Carbon monoxide concentration traces for KY1R4F reference cigarette smoke run with FT-IR data collection at 1 s/spectrum and using the HWG gas cell (*dashed line* NDIR analyzer, *solid line* FT-IR spectrometer) [43]

evident by the narrowing of the increases in concentration due to puffing in both the analyzer and the FT-IR traces. The FT-IR trace provides less temporal resolution than the analyzer trace owing to a 21-s sampling interval versus 1 s provided by the analyzer. The scan time of the FT-IR spectrometer at 21 s is a limitation due to the volume and flow response time of the 3-m gas cell and is applied to the HWG data collection to maintain a comparable instrumental signal-to-noise ratio. Even so, the HWG shows better temporal responsiveness owing to its cell-flow response being one tenth that of the 3-m gas cell, as discussed in Sect. 2.5 [43]. Being able to collect analyzer data in parallel to the HWG system shows the true variability in concentration during the smoke run. The variability during the smoke run demonstrates the need to perform real-time analyses for gas-phase smoke constituents with sufficient temporal resolution in order to measure interpuff concentrations. Figure 17 demonstrates the ability of the FT-IR spectrometer based system to perform with the required temporal resolution. The data in this figure shows CO concentration traces for one KY1R4F reference cigarette with both analyzer and FT-IR data collected at intervals of 1 s (FT-IR data collection of 16 scans at a resolution of 8 cm^{-1}). Despite the increased noise due to the reduced scan time, the responsiveness of the HWG system to changes in CO concentration is comparable to that of the NDIR analyzer.

The applicability of HWGs in FT-IR gas-sensing systems is broad. Systems have been developed to handle gas, liquid, or solid samples. The primary utility of the waveguide is its size. Small-bore capillaries used in HWG construction contribute to compact instrument design and greater optical efficiencies. Further applications will materialize as innovative sampling systems are developed. Some, like the capillary membrane, will double as preconcentrators, thus lowering detection limits for some applications.

4 Applications of Laser-Coupled Hollow Waveguide Gas Sensors

Laser spectroscopy is based on coherent radiation with high spectral density and provides a powerful technique for increasing the sensitivity of sensing devices in target analysis. The MIR spectral region was first accessed with CO_2 lasers. Later, tunable semiconductor lead salt lasers (TDLs) operating in the MIR band made from materials such as $Pb_{1-x}Cd_xS$, $Pb_{1-x}S_xSe$, or $Pb_{1-x}Sn_xSe$ were introduced; however, they still required accurate cryogenic cooling [75, 76]. Early works by the research groups of Tacke and Mizaikoff demonstrated improved sensitivity for evanescent field liquid sensing using continuous-wave-operated lead salt lasers in place of broadband FT-IR sources [77]. This section discusses the use of CO_2 lasers and QCLs coupled to HWGs for improved sensitivity in gas analysis.

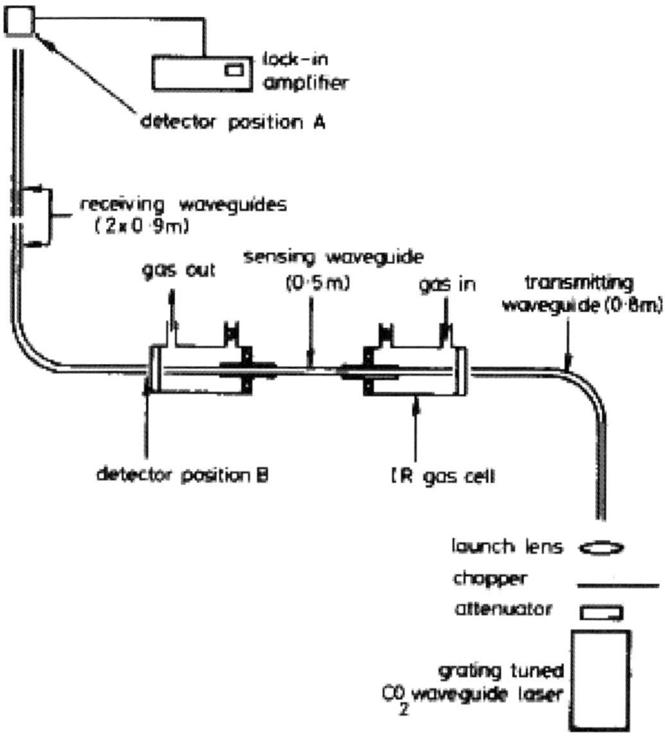

Fig. 19 HWG sensing setup with multiple waveguide sections [78]

4.1
CO_2 Lasers Coupled to Hollow Waveguide Gas-Sensing Modules

Worrell et al. [78] were the first to couple a CO_2 laser to a HWG for sensing purposes in 1992. In this work, the setup was based on four separate sections of 0.7-mm inner diameter HWGs: one transmitting, one sensing, and two receiving waveguides as shown in Fig. 19.

In this approach, ethylene gas was detected using the 944.2 cm^{-1} laser emission frequency down to concentration levels of 100 ppm.

In 1993, Saito et al. [79] introduced the concept of using the HWG as a gas absorption cell. Here, radiation from a CO_2 laser was coupled into a 1.5-mm inner diameter germanium-coated nickel tube waveguide detecting NH_3 gas using the 10.719 μm band of the laser as shown in Fig. 20.

The lowest measured levels of NH_3 were 5 ppm using a 1-m-long HWG; however, the theoretical minimum limit of detection was reported to be at 1 ppb with a HWG providing a length of 5 m.

Later, Worrell and Gallen [44] developed several configurations of HWGs coupled to a CO_2 IR laser for parts per billion level detection of ethylene at

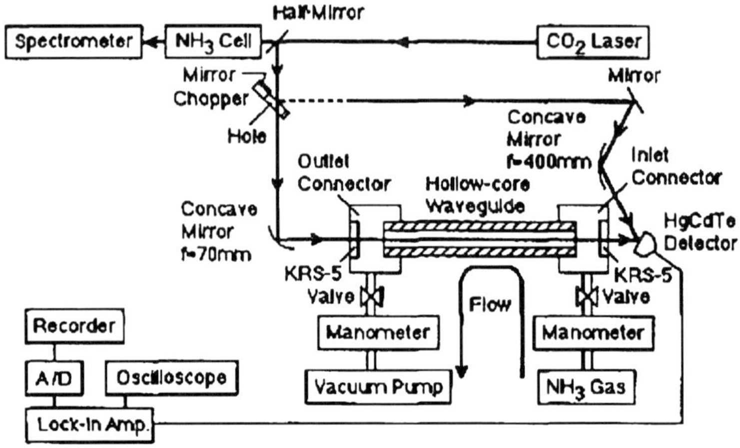

Fig. 20 NH_3 measurements with a HWG-based sensing setup [79]

949.5 cm^{-1}. The applied HWGs were made from both flexible glass with metal and dielectric coatings, and rigid Al_2O_3 guides. The optical configurations included coiled flexible HWGs, as well as several versions utilizing corner mirrors to propagate the radiation through more than one waveguide in the case of rigid waveguide structures. Ethylene was the primary gas investigated in these studies and was detected down to concentration levels of 1.3 ppm at atmospheric pressure and 86 ppb at a pressure of 10.9 atm.

4.2 Quantum Cascade Lasers Coupled to Hollow Waveguide Gas-Sensing Modules

QCLs can be considered the current state-of-the-art light source in the MIR spectral range [80]. QCLs are based on intersubband transitions, which are electronic transitions taking place within the laser heterostructure, rather than electron–hole recombination as in conventional laser diodes. Lasers using heterostructures made from InGaAs/InAlAs/InP have attracted substantial interest as novel light sources for the MIR spectral region [81]. Furthermore, GaAs/AlGaAs QCLs have recently been introduced [82–84]. Since each injected electron generates multiple photons owing to a cascaded active region, output powers up to 300 mW and emission wavelengths extending into the far IR and terahertz regions are achievable [85–88]. Most recently, room temperature continuous-wave operation of InGaAs/InAlAs/InP QCLs has been reported, which can be considered a significant step towards sensor applications of QCLs [89, 90].

A collaboration of the research groups of Strasser and Mizaikoff led to the first combination of QCL technology with HWG gas cells, as reported by

Hvozdara et al. [91–93]. A GaAs/AlGaAs QCL emitting at 10 μm (1000 cm^{-1}) is used to detect the presence of ethylene gas inside a HWG. A Ag/ZnSe-coated silica-based HWG with a length of 434 mm and an inner diameter of 1.13 mm produced by the group of N. Croitoru at Tel Aviv University was applied in this study. Ethylene gas was diluted with helium and the binary mixture was continuously pumped through the HWG gas cell.

Radiation emitted from the QCL was coupled into the HWG and the radiation emitted at the exit aperture was passed through the interferometer of a FT-IR spectrometer and focused onto a MCT detector as shown in Fig. 21. The presence of ethylene gas inside the waveguide gas cell was detected by

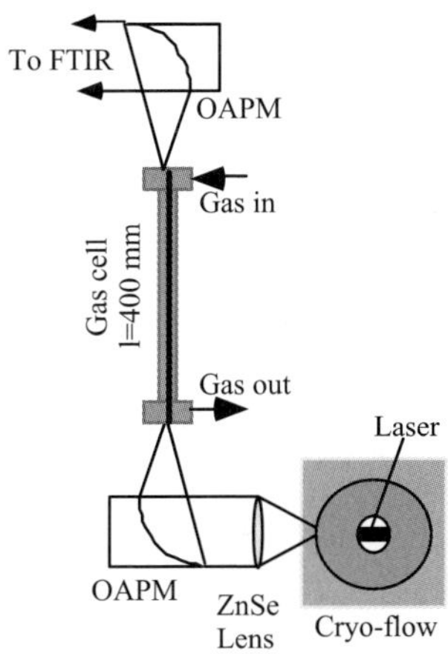

Fig. 21 Experimental setup for quantum cascade laser (*QCL*)–HWG gas sensing [92]

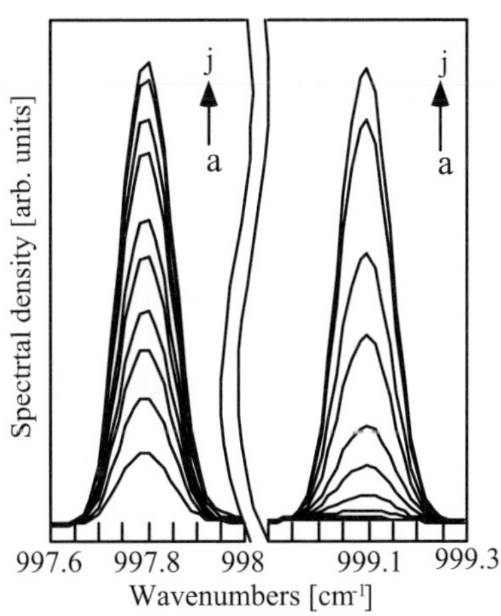

Fig. 22 Damping of two QCL laser modes at 997.8 and 999.09 cm^{-1} with increasing concentration of ethylene; *a* 29 918 ppm ethylene, *j* pure He

damping of the laser line in the IR spectrum as shown in Fig. 22. This sensing system enabled the detection of ethylene concentrations down to levels of approximately 250 ppm.

In 2003, Charlton et al. [51] applied a room-temperature operated InGaAs/AlInAs/InP QCL emitting at 10.3 μm (971 cm^{-1}, Alpes Laser, Neucha-

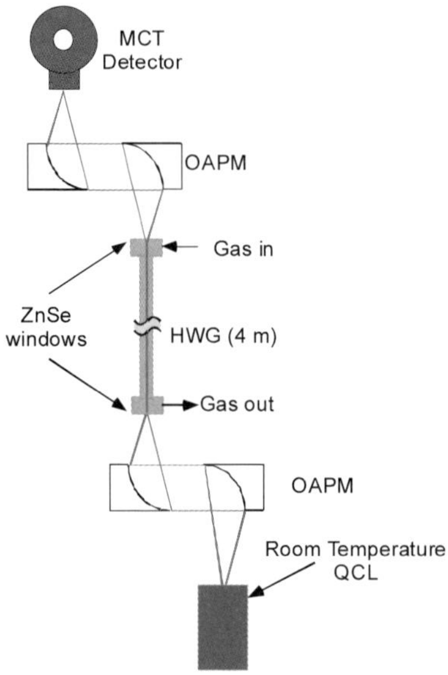

Fig. 23 Experimental setup with a room-temperature operated QCL and a HWG gas cell [94]

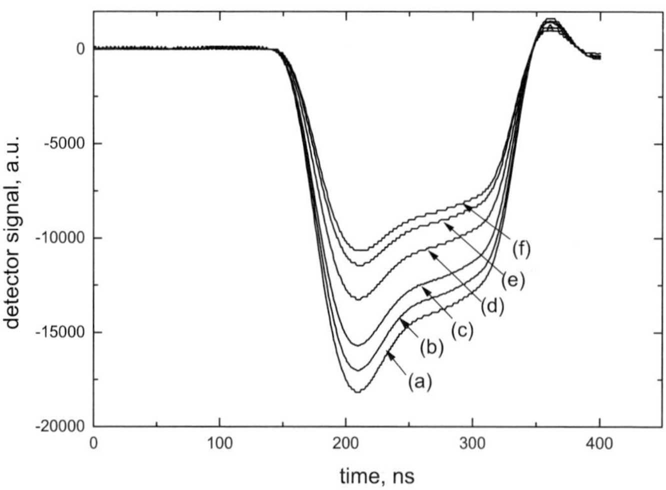

Fig. 24 Damping of the QCL signal due to increasing concentrations (*a–f*) of ethyl chloride gas inside the HWG gas cell

tel, Switzerland) for the determination of ethyl chloride gas in a HWG gas cell. The waveguide used in this study had a length of 4 m and an inner diameter of 0.7 mm. The HWG was based on a silica structural tube coated on the inside with silver and silver iodide (N. Croitoru, Tel Aviv University). Radiation from the QCL was coupled into the waveguide via specifically designed gas cells and radiation at the distal end was focused onto a MCT detector as shown in Fig. 23.

The concentration of ethyl chloride gas present inside the HWG cell was detected by damping of the laser signal due to absorption as shown in Fig. 24. The limit of detection was determined to be 5 ppm.

4.3
Potential for Photonic Bandgap Fibers in Laser-Based Hollow Waveguide Gas Sensing

PBG fibers guiding light in the near IR and MIR are currently being developed predominantly for light guiding of optical telecommunication signals, and for flexible light delivery in medical laser systems at wavelengths of the CO_2 laser emission [28] (10.6 μm) and Nd:YAG laser emission [30] (1.06 μm). However, these fibers also have great potential in sensing applications, which is widely unexplored to date. In particular, omnidirectional waveguides as discussed in Sect. 2.3.2 have the potential to serve as a low-loss HWG gas cell in systems similar to the laser-based sensing concepts tailoring the emission frequency of the laser light source and the PBG of the waveguide to a molecule-specific absorption wavelength of an analyte molecule. Owing to their low attenuation losses in bent or coiled geometries, these waveguides promise long optical pathlengths sufficient for trace gas analysis, while maintaining the demanded compactness of a gas-sensing system. Recently, the research group of Mizaikoff in collaboration with OmniGuide Communications (http://www.omni-guide.com) have demonstrated a limit of detection of 30 ppb (v/v) for ethyl chloride with a response time of 8 s probing a gas sample volume of 1.5 mL in a transmission absorption measurement with an omnidirectional PBG waveguide coupled to a 10.3 μm quantum cascade laser (QCL), which substantiates a new generation of QCL PBG fiber-based trace gas analyzers [33].

5
Conclusion and Outlook

To date, a wide variety of proof-of-concept applications of HWGs in trace gas sensing have been demonstrated—but which areas will be of technological relevance in the future? The main areas of HWG application can be envisaged in process monitoring, environmental analysis, and the biomedical field, which have need for accurate gas analysis and sensing with low limits of detection and fast response times in common. Since HWGs enable combining long optical pathlengths for high sensitivity with small sample volumes facilitating rapid response times, it is anticipated that HWG-based sensors will emerge in the commercial market in the not too distant future. To date, HWG sensing is predominantly related to environmental applications.

Biomedical applications are certainly among the most appealing areas demanding sensitive gas monitoring, such as breath analysis and narcotic gas monitoring [94]. Breath analysis has generated great interest in the gas-sensing community owing to its potential as a noninvasive diagnostic technique in a clinical setting. More than 500 volatile components are present in breath [95]. However, monitoring of these compounds requires sensing devices capable of detecting very low concentration levels (typically below 20 ppb) and concentration changes in small samples, which makes HWGs particularly attractive for this application.

Industrial processes frequently require continuous monitoring of volatile components in process analysis and threshold surveillance. For both gas-phase monitoring, such as the headspace above a tank or the environment around a process line, and for liquid monitoring using continuous sample extraction systems as discussed in Sect. 3.1, HWG gas sensing would be of substantial utility. By providing reliable and rapid sensing methods simultaneously monitoring multiple components, HWGs coupled to FT-IR spectroscopy are among the most promising techniques. Despite numerous potential applications, to date HWG-based sensing has not made it past the proof-of-concept level. The complexity of spectroscopic sensing systems along with the considerably high initial investments still limits the widespread application of this technology. It is expected that component integration into a compact sensing platform utilizing for example, QCL technology in combination with microfabricated waveguides will facilitate overcoming this barriers.

Miniaturization of IR gas-sensing systems incorporating a HWG gas cell presents challenges, in particular related to the required absorption pathlength. To take advantage of a sufficiently long optical path and the resulting increase in sensitivity maintaining a small overall device platform requires a meandered waveguide. However, losses in conventional HWGs dramatically increase with bending of the waveguide, causing the increase in optical pathlength to be negatively compensated by an overall signal decrease. A significant reduction of bending losses is expected by the introduction of omnidirectional mirrors based on PBG materials enabling the demanded optical path length in still compact device format to be achieved.

References

1. Elser MB, Griffith DWT, Wilson SR, Steele LP (2000) Anal Chem 72:206
2. Johansen IR, Lines GT, Honne A, Midtgaard T (1997) Appl Spectrosc 51:1540
3. Martin DM, Medvecz PJ, Nichols KM (1993) Appl Spectrosc 47:1898
4. Qin D, Cadet G (1997) Anal Chem 69:1942
5. Woebkenburg ML (1999) Appl Occup Environ Hyg 14:510
6. Herget WF, Jahnke JA, Burch DE, Gryvnak DA (1976) Appl Opt 15:1222
7. Chaffin CT Jr, Marshall TL, Chaffin NC (1999) Field Anal Chem Technol 3:111
8. Bak J, Clausen S (2002) Meas Sci Technol 13:150
9. Gallagher NB, Wise BM, Sheen DM (2003) Anal Chim Acta 490:139
10. Todd LA, Ramanathan M, Mottus K, Katz R, Dodson A, Mihlan G (2001) Atmos Environ 35:1937
11. Department of Health and Human Services (2003) NIOSH pocket guide to chemical hazards. DHHS (NIOSH) publication no 97–140, DHHS, Washington, DC
12. Occupational Safety & Health Administration (2004) Table Z-1 limits for air contaminants. http://osha.gov/pls/oshaweb/owadisp.show_document?p_table=STANDARDS &p_id=9992, accessed April 27, 2004
13. Lavine BK, Workman J (2002) Anal Chem 74:2763
14. Lavine BK, Workman J (2004) Anal Chem 76:3365
15. Harrington J (2004) Infrared fibers and their applications. SPIE, Bellingham, WA
16. Sanghera J, Aggarwal I (1998) Infrared fiber optics. CRC, Boca Raton, FL
17. Lendl B, Mizaikoff B (2002) In: Chalmers JM, Griffiths PR (eds) Handbook of vibrational spectroscopy. Wiley, New York, p 1541
18. Lopez-Higuera JM (2002) Handbook of optical fibre sensing technology. Wiley, New York
19. Garmire E, McMahon T, Bass M (1976) Appl Opt 15:145
20. Miyagi M, Hongo A, Aizawa Y, Kawakami S (1983) Appl Phys Lett 43:430
21. Matsuura Y, Miyagi M, Hongo A (1990) Opt Laser Tech 22:141
22. Dahan R, Dror J, Croitoru N (1992) Mater Res Bull 27:761
23. Abel T, Hirsch J, Harrington JA (1994) Opt Lett 19:1034
24. Fink Y, Winn JN, Fan S, Chen C, Michel J, Joannopoulos J, Thomas E (1998) Science 282:1679
25. Harrington JA, Gregory CC (1990) Opt Lett 15:541
26. Hidaka T, Kumada J, Shimada J, Morikawa T (1982) J Appl Phys 53:5484
27. Joannoupoulos J, Meade R, Winn J (1995) Photonic crystals. Princeton University Press, Princeton, NJ
28. Hart S, Maskaly GR, Temelkuran B, Prideaux PH, Joannopoulos JD, Fink Y (2002) Science 296:510
29. Temelkuran B, Hart S, Benoit G, Joannopoulos JD, Fink Y (2002) Nature 420:650
30. Katagirl T, Matsuura Y, Miyagi M (2002) Appl Opt 41:7803
31. Barkou S, Broeng J, Bjarklev A (1999) Opt Lett 24:46
32. Laegsgaard J, Mortensen NA, Bjarklev A (2003) J Opt Soc Am B 20:2046
33. Charlton CM, Temelkuran B, Dellemann G, Mizaikoff B (2005) Appl Phys Lett 86:194192
34. Azzaraga LV (1980) Appl Spectrosc 34:224
35. Yang PW, Ethridge EL, Lane JL, Griffiths PR (1984) Appl Spectrosc 38:813
36. Saggese SS, Harrington JA, Siegel GH Jr (1991) Opt Lett 16:27
37. Saito S, Saito M, Miyagi M (1993) Appl Spectrosc 47:1665
38. Matsuura Y, Abel T, Harrington JA (1995) Appl Opt 30:6842

39. Alaluf M, Dror J, Dahan R, Croitoru N (1992) J Appl Phys 72:3878
40. Kozody RL, Micheels RH, Harrington JA (1996) Appl Spectrosc 50:415
41. Hanst PL (1978) Appl Opt 17:1360
42. White JU (1942) J Opt Soc Am 32:285
43. Thompson BT, Inberg A, Croitoru N, Mizaikoff B (2005) accepted Appl Spectrosc
44. Worrell CA, Gallen NA (1997) J Phys D Appl Phys 30:1984
45. Haan DJ, Gibson DJ, Rabii CD, Harrington JA (1998) Proc SPIE Int Soc Opt Eng 3262:125
46. Micheels RH, Richardson K, Haan DJ, Harrington JA (1999) Proc SPIE Int Soc Opt Eng 3540:64
47. Yang J, Her JW (2000) Anal Chem 72:878
48. Yang JS, Tsui CP (2001) Anal Chim Acta 442:267
49. de Melas F, Pustogov VV, Croitoru N, Mizaikoff B (2003) Appl Spectrosc 57:600
50. de Melas F, Pustogov VV, Wolcott DK, Olson DC, Inberg A, Croitoru N, Mizaikoff B (2003) Int J Environ Anal Chem 83:573
51. Charlton C, Mizaikoff B, de Melas F, Inberg A, Croitoru N (2003) IEE Optoelectron 150:306
52. Pogodina OA, Pustogov VV, de Melas F, Haberhauer-Troyer C, Rosenberg E, Puxbaum H, Inberg A, Croitoru N, Mizaikoff B (2004) Anal Chem 76:464
53. Yang J, Her JW, Chen HS (1999) Anal Chem 71:3740
54. Yang J, Chen PC (2002) Anal Sci 18:555
55. de Melas F, Pustogov VV, Wolcott DK, Olson DC, Inberg A, Croitoru N, Mizaikoff B (2003) unpublished data
56. Williams TB, Belk CW (1972) Beitr Tabakforsch 6:210
57. Vilcins G, Lephardt JO (1976) Recent Adv Tob Sci 1:123
58. Maddox WL, Mamantov G (1977) Anal Chem 49:331
59. Williams TB (1980) Beitr zur Tabakforsch Intl 10:91
60. Cueto R, Church DF, Pryor WA (1989) Anal Lett 22:751
61. Cueto R (1990) PhD thesis, Louisiana State University and Agricultural and Mechanical College
62. Parrish ME, Lyons-Hart JL, Shafer KH (2001) Vib Spectrosc 27:29
63. Cole SK, Martin P (1996) Analyst 121:495
64. Shi Q, Nelson DD, McManus JB, Zahniser MS, Parrish ME, Baren RE, Shafer KH, Harward CN (2003) Anal Chem 75:5180
65. Thompson BT, Mizaikoff B (2005) accepted Appl Spectrosc
66. Dube MF, Green CR (1982) Recent Adv Tob Sci 8:42
67. Baker RR, Davis DL, Nielson MT (eds) (1999) Tobacco production, chemistry and technology. Blackwell, London
68. Speitel LC (2002) J Fire Sci 20:349
69. Blackwood TR (2000) J Air Waste Manage Assoc 50:1714
70. Haus R, Schafer K, Bautzer W, Heland J, Mosebach H, Bittner H, Eisenmann T (1994) Appl Opt 33:5682
71. Hart BK, Griffiths PR (2000) Environ Sci Technol 34:1337
72. Hart BK, Berry RJ, Griffiths PR (2000) Environ Sci Technol 34:1346
73. Jaakkola P, Tate JD, Paakkunainen M, Kauppinen J, Saarinen P (1997) Appl Spectrosc 51:1159
74. Cooperation Centre for Scientific Research Relative to Tobacco (2002) CORESTA recommended method no 55
75. Shi Z, Tacke M, Lambrecht A, Böttner H (1995) Appl Phys Lett 66:2537
76. Tacke M (1995) Infrared Phys Technol 36:447

77. Kastner JF, Tacke M, Katzir A, Mizaikoff B, Göbel R, Kellner R (1997) Sens Actuators B 38/39:163
78. Worrell CA, Giles IP, Adatia NA (1992) Electron Lett 28:615
79. Saito Y, Kanaya T, Nomura A, Kano T (1993) Opt Lett 18:2150
80. Faist J, Capasso F, Sirtori C, Sivco DL, Hutchinson AL, Cho AY (1994) Science 264:553
81. Liu F, Zhang Q, Zhang Y, Ding D, Xu B, Wang Z (2001) Solid State Electron 45:1831
82. Gauthier-Lafaye O, Boucaud P, Julien FH, Prazeres R, Glotin F, Ortega JM, Thierry-Mieg V, Planel R, Leburton JP, Berger V (1997) Appl Phys Lett 70:3197
83. Sirtori C, Kruck P, Barbieri S, Collot P, Nagle J, Beck M, Faist J, Oesterle P (1998) Appl Phys Lett 73:3486
84. Strasser G, Gianordoli S, Hvozdara L, Schrenk W, Unterrainer K, Gornik E (1999) Appl Phys Lett 75:1345
85. Capasso F, Gmachl C, Sivco DL, Cho AY (1999) Phys World 12:27
86. Gmachl C, Capasso F, Tredicucci A, Sivco DL, Kohler R, Hutchinson AL, Cho AY (1999) IEEE J Select Top Quantum Electron 5:808
87. Tredicucci A, Gmachl C, Capasso F, Wanke M, Hutchinson A, Sivco D, Chu S, Cho A (2001) Opt Mater 17:211
88. Williams B, Callebaut H, Kumar S, Hu Q, Reno J (2003) Appl Phys Lett 82:1015
89. Beck M, Hofstetter D, Aellen T, Faist J, Oesterle U, Ilegems M, Gini E, Melchior H (2002) Science 295:301
90. Beck M, Hofstetter D, Aellen T, Blaser S, Faist J, Oesterle U, Gini E (2003) J Cryst Growth 25:697
91. Hvozdara L, Gianordoli S, Strasser G, Schrenk W, Unterrainer K, Gornik E, Pustogow V, Murthy C, Kraft M, Mizaikoff B (2000) Physica E 7:37
92. Hvozdara L, Gianordoli S, Strasser G, Schrenk W, Unterrainer K, Gornik E, Murthy C, Kraft M, Pustogow V, Mizaikoff B, Inberg A, Croitoru N (2000) Appl Opt 39:6926
93. Hvozdara L, Mizaikoff B, Gianordoli S, Strasser G, Schrenk W, Unterrainer K, Gornik E, Pustogow V, Murthy C, Kraft M, Inberg A, Croitoru N (2001) Proc SPIE 4204:255
94. Charlton C, Mizaikoff B (2003) Proc SPIE 4957:116
95. Miekisch W, Schubert J, Noeldge-Schomburg G (2004) Clin Chim Acta (in press)

Springer Ser Chem Sens Biosens (2005) 3: 169–188
DOI 10.1007/5346_005

Published online: 28 September 2005

Combinatorial Method for Surface-Confined Sensor Design and Fabrication

Lourdes Basabe-Desmonts · Rebecca S. Zimmerman · David N. Reinhoudt · Mercedes Crego-Calama (✉)

Department of Supramolecular Chemistry and Technology, MESA+ Institute for Nanotechnology and Faculty of Science and Technology, University of Twente, P.O. Box 217, 7500 AE Enschede, The Netherlands
m.cregocalama@utwente.nl

Abstract The procedure for the combinatorial fabrication of new sensing materials for cations and anions based on self-assembled monolayers (SAM) is discussed. A library of different sensitive substrates is generated by sequential deposition of fluorophores and small ligand molecules onto an amino-terminated SAM coated glass. The preorganization provided by the surface avoids the need for complex receptor design, allowing for a combinatorial approach to sensing systems based on individually deposited small molecules. Additionally the sensing system has been miniaturized to the microscale using microcontact printing and integrating the sensory SAMs on the walls of microchannels.

1 Introduction

Combinatorial methods are being widely implemented in the field of optical sensor development. The combinatorial concept is based on the relative ease of production of a large number of potential targets, of which it is hoped that some will exhibit the desired specifications of the researcher. This combinatorial approach is clearly different from the 'classical' rational design and individual creation of specific targets, in that the stress is not on the initial, specific design of the desired system, but on the testing of the large number of resultant targets to determine successful hits. Linked to a proper screening methodology and data processing, it allows for the facile search and optimization of a target lead structure for a certain purpose, for example, drug discovery, catalysis, bimolecular interaction studies or sensitive probe discovery.

Many different types of combinatorial methods have already been employed for obtaining new sensitive probes. Conventional probes that exploit specific interactions between analyte and receptor have been made by combinatorial chemistry. A nice example is the work of Walt et al. in which a combinatorial approach, based on azo coupling of diazonium salts with different phenolic compounds or different aromatic amines, generated a library of azo dyes. Each synthesized compound is a chelate agent for a heavy metal ion and each one showed a different UV–vis absorbance spectrum upon binding to selected metal ions [1]. In the same line multiparallel peptide synthesis was used to make a cyclopeptide library that attached to a glass surface works as an amino acid sensor by reflectometric interference spectroscopy [2]. The split-and-pool method has been used to generate a combinatorial library with more than 4000 different tripetides for the discovery of ATP binding receptors in order to develop an ATP sensor [3]. Other peptide sensors [4–9] have been similarly developed utilizing a wide range of optical detection methods [10]. A different combinatorial approach was used by the group of Brown for the construction of metal binding sites; the method is called *in vitro selection*. With this method they obtained catalytic DNA that can bind a metal ion of choice strongly and specifically. The desired DNA strand was found within a library of 10^{14}–10^{15} random DNA/RNA sequences. By labeling the resulting DNA with a fluorophore–quencher pair they made a new class and metal ion fluorescent (bio)sensor [11]. Additionally, combinatorial methods have successfully been used to generate arrays of nonspecific sensors comprising partially specific molecular receptors [12]. The screening of such libraries yields different patterns of responses for different analytes. This approach, called *differential sensing*, is best illustrated by the work of Dickinson et al. [13] on sensor arrays for odor recognition (electronic noses). Related to differential sensing, Mayr et al. [14] have recently developed cross-reactive sensor arrays in microtiter plate format, in which, owing to the construction of an array of unspecific sensors, determination of mixtures of divalent calcium, copper, nickel, cadmium and zinc ions is performed. Finally, combinatorial chemistry has also been used to produce certain types of molecular imprinted polymers that show specificity to the template molecule used for their formation. Thus, the polymer is made in the presence of a template molecule generating a polymer microenvironment for the imprint molecule during the polymerization process. The resulting polymer network contains then synthetic receptors that are complementary in size, shape and functional group orientation to the template molecule [15]. Libraries of fluorescent polymers have also been generated as has been shown by the work of Dordick, who made a sensor array for Fe^{3+}, Cu^{2+}, Co^{2+} and Ni^{2+} consisting of 15 phenolic homopolymers and copolymers generated from five phenolic monomers.

The sensing process is based on the change of intrinsic polyphenol fluorescence upon addition of a metal ion mixture to an aqueous suspension of a polyphenol [16]. The vast number of targets produced by combinatorial methods begets the need for effective and efficient screening for hit identification. To this end, surface immobilization and individual addressability of chemical sensing systems are advantageous because they allow for facile analyte sensing to be performed in parallel. For that purpose, platforms have been developed that contain or immobilize targets in order to facilitate high throughput screening (HTS) technologies, originally exploited in the field of biosensors [17, 18], such as microtiter plate and microarray technologies, fiber optic tips and solid-phase synthesis [19]. For instance, solid-phase synthesis on the bottom of a microtiter plate has been used by Gauglitz to perform label-free parallel screening of a combinatorial triazine library [20]. To date, surface-immobilized chemical sensing systems have been deposited on such varied platforms as polymer supports [21, 22], silica beads [23] and glass slides [24, 25].

We have been able to synergize the aforementioned concepts to develop a new methodology for sensor fabrication and discovery of self-assembled monolayers (SAMs) on glass [26]. Through the use of a combinatorial approach, we are able to use simple, commercially available fluorophores and small complexing/coordinating molecules directly self-assembled on a glass surface to generate a variety of sensing surfaces able to detect both cations and anions. Additionally, we show that the methodology is fully transferable to the microscale, resulting in individually addressable sensing systems which will allow for the use of HTS techniques to analyze larger libraries in the future.

2 Fluorescent Monolayers on Glass as Probes Suitable for Combinatorial Methods

2.1 General Description

Self-assembly is a valuable alternative to covalent synthesis for the creation of large multifunctional noncovalent assemblies [27–31]. Extensive efforts have been made to develop selective self-assembled receptors [32–36] and ionophores [37–39] in solution. However, further development of these assemblies into sensing systems is complicated since it requires intrinsic readout functions that signal the binding [40, 41]. Therefore, there is a need for self-organization of selective recognition functionalities and transduction systems. Grandini et al. [42] have described a methodology based on comicellar aggregates, where a specific lipophilic ligand, a fluorophore, and a surfactant self-assemble in water to give a fluorescent Cu(II) sensor. However, immobilization of recognition sites on suitable surfaces might prove more practical in terms of actual device implementation [43–47].

SAMs [48–50] provide a convenient way to produce surfaces with specific chemical functionalities that allow the precise tuning of surface properties. Previously, SAMs have successfully been used to demonstrate that the sensing process is feasible at the monolayer–solution interface [24, 26, 44, 51–53]. The advantages of SAMs for surface-confined sensing are ease and reproducibility of synthesis [51], the introduction of additional chelating effects from the preorganization of the surface platform and fast response times [54].

In terms of detection on SAMs, the use of fluorescence is highly desirable. Its high sensitivity and sub-millisecond temporal resolution make fluorescence-based techniques well suited for HTS, and are quickly replacing the more traditional radiometric technology [55]. In spite of these advantages, fluorescence has hardly ever been used to determine host–guest interactions at the monolayer surface, mainly because most of the systems designed so far have been on gold, which causes quenching of the fluorescence [52, 53]. Glass, however, is an appropriate substrate for the purpose of fluorescence detection of chemical sensing [56].

Utilizing these strengths and advantages, we have succeeded in using a combinatorial approach as the basis for the design and fabrication of chemical sensing surfaces of SAMs on glass. After depositing a reactive SAM onto a glass surface [50, 57, 58] to form a preorganized two-dimensional platform, fluorophores and small binding groups are sequentially, covalently attached, resulting in a random distribution of fluorophores and small binding groups across the surface (Fig. 1). This network comprises a dense sensing surface with the fluorophores and binding groups in close proximity to one another. Upon addition of a series of divalent cations or inorganic anions, this network acts synergistically to bind the analyte and translate the binding event into a modulation of the fluorescence intensity. In this way we are able to

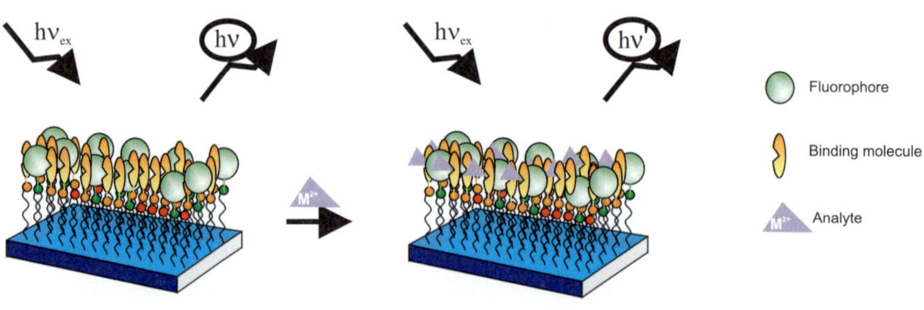

Fig. 1 Schematic representation of a fluorescent self-assembled monolayer (*SAM*) functionalized with binding groups (*orange ovals*) and fluorescent groups (*green spheres*), showing how an analyte (*purple triangles*) can interact with the layer owing to its coordinating properties and be reported by the fluorophore

easily and quickly generate and test a number of different sensing surfaces comprising unique fluorophore-binding group pairs for their binding ability.

Thus, by eliminating the need for the design and synthesis of a specific fluorophore-appended receptor on the basis of geometric and electronic complementarity to match the desired guest, and by employing a combinatorial approach to the deposition of layer components, we are able to quickly generate a number of unique sensing systems easily tested for binding efficacy. We are able to show the viability of a combinatorial approach for the generation of simple sensing surfaces, which could be an important consideration in the still relatively unexplored area of surface chemical sensing.

2.2
Fabrication Methods

Sensitive fluorescent SAMs consist of a SAM made on glass tailored with two building blocks, small molecules that supply different functionalities acting as binding groups (ureas, amides, thioureas, sulfonamides, etc.), and fluorescent probes for reporting the recognition event. Thus, the surface of a SAM on glass is transformed into a fluorescent sensing system for metal ions without any further preorganization of the ligating functionalities other than the directional orientation lent by the surface itself.

The properties of the layer are a result of the combination of various factors, and are directly dependent on the surface functionalization [26]. Neither the nature of the different binding groups nor that of the fluorescent probe is solely responsible for the recognition process. It is the combination of both features plus the intrinsic structure of the monolayer which supplies the final properties to the sensitive monolayer. Such a monolayer can be considered as an enormous macromolecule with an infinite number of sensitive binding pockets and reporters. With this approach, a very large number of different sensing monolayers can be fabricated by making different combinations of the building blocks that comprise them.

This approach has been shown to work with a number of different fluorescent probes such as the short-wavelength fluorophores dansyl sulfonyl chloride and coumarin chloride and the long-wavelength fluorophores tetramethylrhodamine-5-(and-6)-isothiocyanate [5(6)-TRITC], 5-(and-6)-carboxytetramethylrhodamine, succinimidyl ester [5(6)-TAMRA, succinimidyl ester] and lissamine rhodamine B sulfonyl chloride (each in conjunction with different binding functionalities on the SAM surface.

As a general procedure these sensitive fluorescent SAMs are prepared via a three-step procedure. First, a monolayer of *N*-[3-(trimethoxysilyl)propyl]-ethylenediamine is formed on a glass substrate (Fig. 2) [59]. This layer is converted into the fluorescent SAM by reaction with an amino-reactive fluorescent probe (such as lissamine, dansyl or coumarin) [24]. Then the residual amino groups (the steric hindrance renders imposible the reaction of every

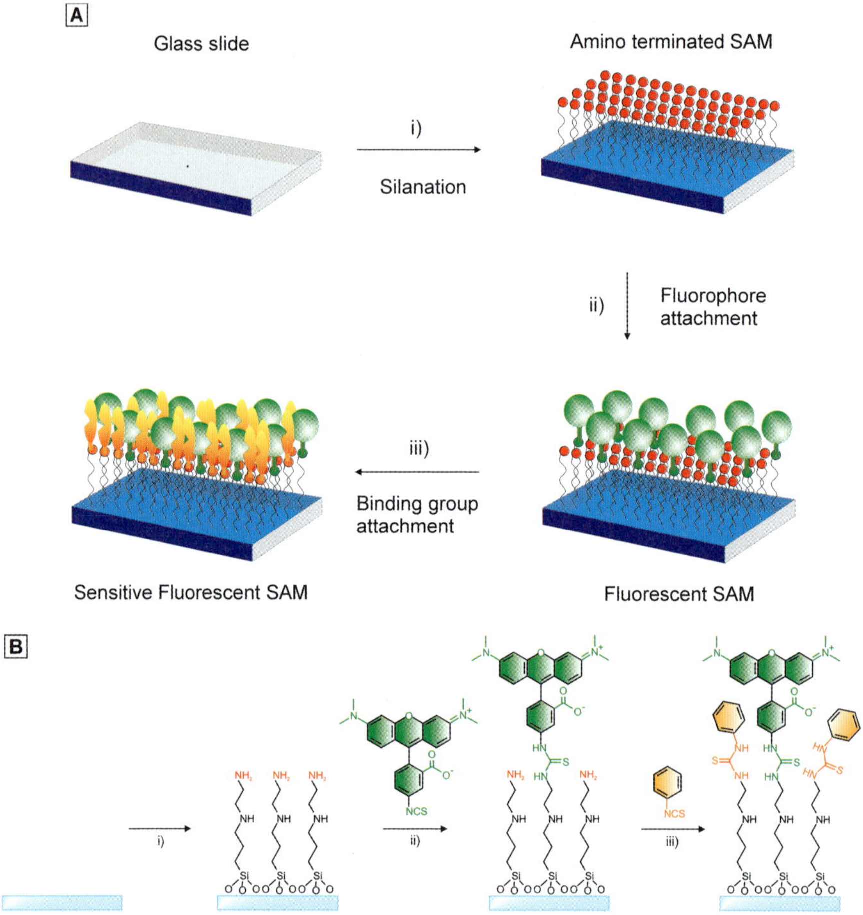

Fig. 2 Schematic cartoon (**a**), and chemical structures (**b**) of the general fabrication procedure of a sensitive fluorescent monolayer on glass: i) silanation of the glass slide with *N*-[3-(trimethoxysilyl)propyl]ethylenediamine to form the amino-terminated monolayer, ii) reaction with an amino-reactive fluorophore, iii) covalent attachment of a binding molecule

surface amino group with a fluorophore molecule) are reacted with a small molecule to form a binding group such as amide, urea, thiourea or sulfonamide, yielding the sensitive fluorescent SAM.

Specifically, two different methods for the fabrication of these sensitive monolayers have been used. These methods differ in the technique used for the covalent attachment of the fluorescent probes (Fig. 2, step ii) while the first and the third fabrication steps (silanation (Fig. 2, step i) and binding molecule attachment (Fig. 2, step iii)) are identical for both methods. The first method is a solution-based procedure and the second method is a microcontact printing (μCP) based procedure. In the solution-based procedure (Sect. 2.2.1), all the steps are carried out by sequential dipping of the glass substrate into the different solutions containing the appropriate reactive compound (silane, fluorophore and binding molecule). In the case of the μCP approach, the covalent attachment of the fluorophore is carried out by soft lithography (Sect. 2.2.2). A poly(dimethylsiloxane) (PDMS) stamp is submersed in the ink solution (fluorophore) and brought into contact with the amino-reactive surface.

Using the solution-based procedure, we obtain fully covered substrates, while by the μCP-based procedure, discrete areas of coverage are patterned on the substrate, which is convenient in terms of microarray fabrication because different fluorophore probes can be placed on specific, discrete regions of the same glass substrate. Such an array could then be exposed to a guest solution and subsequently the surface fluorescence emission scanned by a confocal microscope for the simultaneous acquisition of the optical data from the individually addressable areas [60–63].

2.2.1
Solution-Based Procedure

2.2.1.1
Cation Sensing

Two small libraries of sensitive fluorescent SAMs were fabricated for cation-sensing purposes. The first one (Fig. 3) includes two short-wavelength fluorophores, coumarin and dansyl, and three different complexing functionalities, amino, amido and urea [26]. In this way, each member of the library

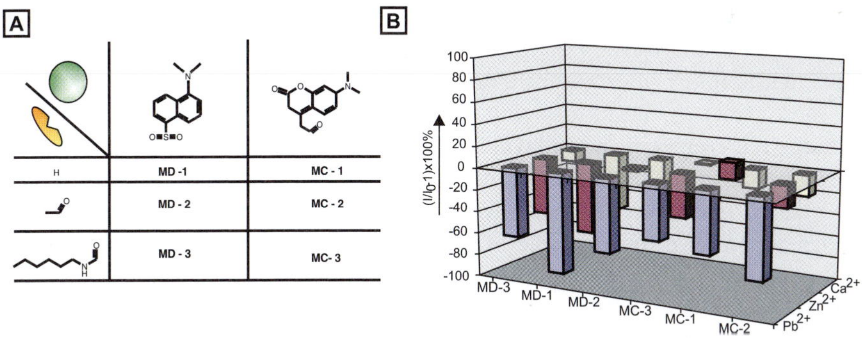

Fig. 3 **a** Schematic representation for the monolayers of the cation-sensing library with short-wavelength fluorophores. **b** Relative fluorescence intensity of surfaces modified with different fluorophores and chemical functionalities in the presence of 10^{-3} M solutions of Pb^{2+}, Zn^{2+} and Ca^{2+} as perchlorate salts. The data have been normalized; in the absence of metal cations the fluorescence emission is set to 0 for the MD-1, MD-2 and MD-3 monolayers and for the MC-1, MC-2 and MC-3 monolayers at 510 and 460 nm, respectively

comprises a different combination of fluorophore and binding molecule randomly distributed on the SAM.

Fluorescence measurements of the differently funtionalized quartz slides in acetonitrile showed the characteristic emission of each fluorophore, confirming the successful attachment of the fluorophore on the monolayer.[1] Each layer was characterized by contact angle, ellipsometry and X-ray photoelectron spectroscopy measurements, confirming the introduction of the different building blocks.

SAMs of dansyl or coumarin adsorbates were prepared via the solution-based procedure by forming a covalently attached monolayer of *N*-[3-(trimethoxysilyl)propyl]ethylenediamine (TPEDA) on the glass substrate [59]. This layer was converted into the dansyl or coumarin SAM by reaction with dansyl chloride or a succinimidyl ester substituted coumarin (Fig. 3) [53]. The residual amino groups were reacted with acetyl chloride or hexyl isocyanate to yield sensing surfaces able to report the binding of various metal cations (Zn^{2+}, Pb^{2+} and Ca^{2+} as perchlorate salts in acetonitrile). This binding was evaluated using the variation in the fluorescence intensity as a function of increasing guest concentration.

For the library functionalized with dansyl, the largest overall response was observed for Pb^{2+}, which almost totally quenches the fluorescence response of MD-1 (Fig. 3) at 9×10^{-3} M. For this surface, with amine as the binding group, 92% quenching of the fluorescence intensity was measured. The other surfaces, MD-2 and MD-3, exhibit 70% quenching of the initial fluorescence. The MD-1 surface is also the best sensing layer for Ca^{2+} and Zn^{2+} with fluorescence quenching of 52 and 65%, respectively. MD-2 would be the material of choice for the optimal discrimination of Pb^{2+} versus Ca^{2+} and Zn^{2+} because it exhibits the largest response differences (70% quenching for Pb^{2+} versus no more than 25% quenching for Ca^{2+} or Zn^{2+}). Nevertheless, the urea-rich monolayer MD-3 could offer the possibility of analyzing Pb^{2+} and Zn^{2+} without any interference from Ca^{2+}.

In contrast to the amino-rich dansyl monolayer MD-1, it is instead the amido-rich coumarin monolayer MC-2 that exhibits the largest fluorescence response toward Pb^{2+}, undergoing 80% quenching. The amino-terminated coumarin monolayer MC-1 shows unique behavior within the library, with a 20% increase in fluorescence in the presence of Zn^{2+}. This enhancement is important for sensor development because it would reduce the likelihood of false-positive signals.

The selectivity of the SAMs was demonstrated by titrating monolayer MC-2 with Pb^{2+} in the presence of Ca^{2+} and Zn^{2+} (1×10^{-3} M) (Fig. 4), which induced a similar response (a large quenching) as in the absence of the competing metal ions (approximately 30 and 40% quenching, respectively, for

[1] When the fluorescence of the layer is measured in air, the emission peak is redshifted, indicating that the solvent induces a polarity effect on the fluorophore.

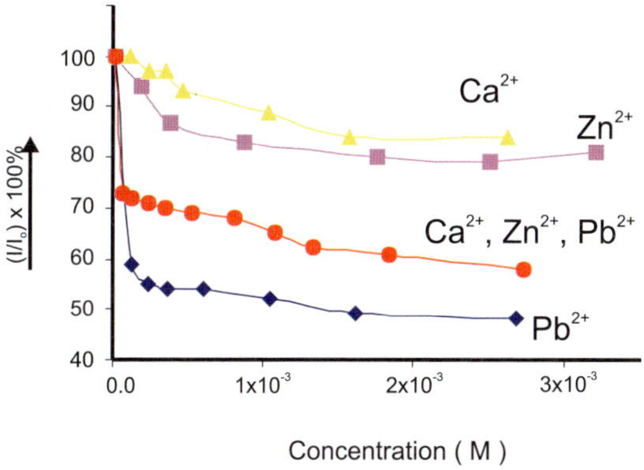

Fig. 4 Dependence of relative fluorescence intensities at $\lambda_{em} = 460$ nm on individual metal ion concentration: Ca^{2+}, Zn^{2+} and Pb^{2+} (1×10^{-3} M) for monolayer MC-2, as well as the response toward Pb^{2+} in the presence of Ca^{2+} and Zn^{2+} (1×10^{-3} M)

$[Pb^{2+}] = 1.2 \times 10^{-4}$ M). This shows that with these very simple systems a substantial selectivity can be obtained.

Basabe-Desmonts et al. [64] have shown the broad scope of this approach in regard to component variation by making a second library with the more useful long-wavelength fluorophores, which are compatible with confocal microscopy measuring techniques. As in the previous library, quartz slides were functionalized with TPEDA. The amino-terminated monolayer was sequentially modified with a long-excitation-wavelength fluorophore (TAMRA or lissamine) and a binding group (i.e., amino, aryl-urea, alkyl-urea, aryl-amide, alkyl-amide) (Fig. 5). Fluorescence measurements were taken in 10^{-4} M acetonitrile solutions of perchlorate salts of Cu^{2+}, Co^{2+}, Pb^{2+} and Ca^{2+}. As observed in the previous library, the different binding group–fluorophore combinations resulted in a range of different responses (Fig. 5).[2]

When the responses within the fluorophore series were compared, changes in the nature of the binding group were seen to significantly affect the binding profiles. For example, both TAMRA layers, the ureido-substituted TM1 and the amido-substituted TM3, showed an increase in fluorescence upon addition of Ca^{2+} and Pb^{2+}, with a greater increase for the TM1 layer than the TM3 layer. More impressive is the comparison of ureido-substituted layers TM1 and TM2. While TM1 shows increased fluorescence for both Pb^{2+} (65%) and Ca^{2+} (90%), TM2, which differs by changing the isopropylphenyl substituent on the binding group for a hexyl group, shows very little change in fluorescence in the presence of these cations. The differences seen between binding groups substituted with an aryl versus an alkyl group are remarkable considering that these substituents, in principle, should not directly coordinate

[2] In each instance the layer response to the analytes occurred within seconds, and furthermore it was regenerated with reproducible results upon washing with 1 N HCl.

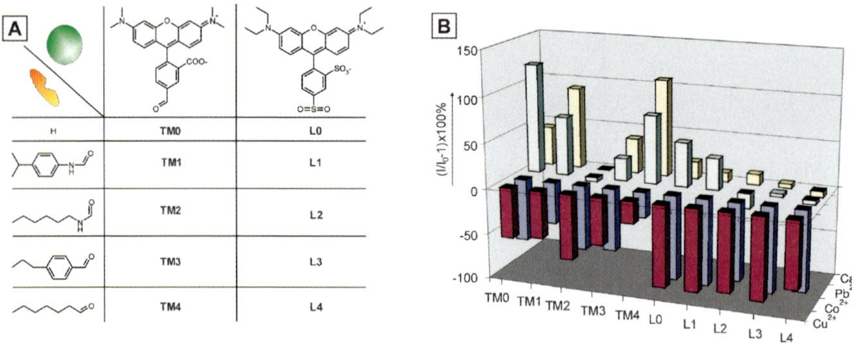

Fig. 5 **a** Schematic representation for the monolayers of the cation-sensing library. **b** Relative fluorescence intensity of surfaces modified with different fluorophores and chemical functionalities in the presence of 10^{-4} M solutions of Pb^{2+}, Ca^{2+}, Co^{2+} and Cu^{2+} as perchlorate salts in acetonitrile. The data have been normalized; in the absence of metal cations the fluorescence emission at 585 and 588 nm for layers TM0–TM4 and for L0–L4, respectively, is set to 0

either Pb^{2+} or Ca^{2+}, and is a case in point of the power of the combinatorial method. Because of the ease of functionalizing different monolayers with different pairs of fluorophore binding molecules, different sensing systems can be easily produced. In this way minimal effort is required to synthesize layers with small variations, which would usually be disregarded in solution-state receptor synthesis.

Within the series of lissamine layers, amino-substituted L0 and ureido-substituted L1 showed substantial increases in response to Pb^{2+}, while virtually no response was seen from other amido and ureido layers, L2–L4. On the other hand, systems L2–L4 showed very large fluorescence quenching upon addition of Co^{2+} and Cu^{2+} (approximately 80%), while no response was seen for Ca^{2+} and Pb^{2+}.

As also observed for short-wavelength functionalized libraries, significant differences can be seen when comparing responses between the two fluorophores. For example, hexylamido TAMRA system TM4 shows fluorescence intensity increases for both Ca^{2+} (107%) and Pb^{2+} (75%), while the corresponding hexylamido lissamine system L4 exhibits virtually no response for these cations. Overall, all systems exhibited a marked fluorescence intensity decrease for Co^{2+} and Cu^{2+}, although more quenching was seen in the lissamine layers than in the corresponding TAMRA layers. Additionally, TM0 is the best for the sensing of Pb^{2+} (121%) versus the other ions owing to the large response differences (Ca^{2+} 43%, Co^{2+} 68% and Cu^{2+} 7%).

Several successful competition studies were performed with these layers. For example, addition of 10^{-4} M Ca^{2+} to layer L2 induced a 20% fluorescence intensity increase (Fig. 6). A further addition of 10^{-4} M Cu^{2+} induced 80% quenching. When reversing the addition, 10^{-4} M of Cu^{2+} caused a 90% de-

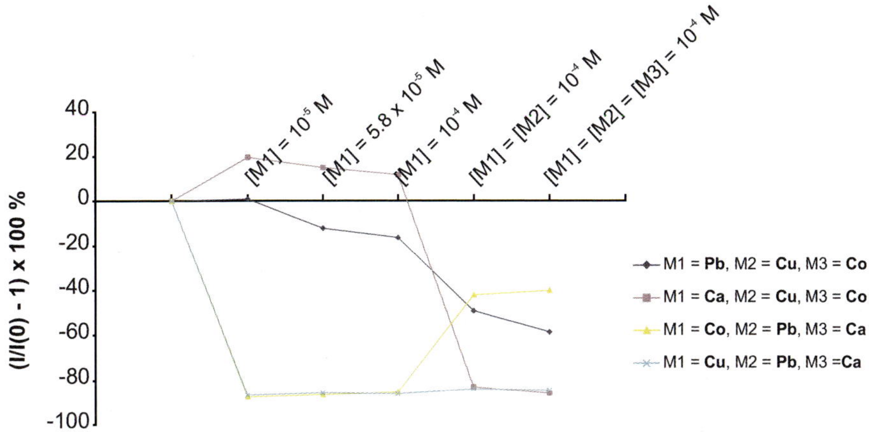

Fig. 6 Competition studies with the alkyl ureido layer L2 upon addition of perchlorate salts of Pb^{2+}, Cu^{2+}, Co^{2+} and Ca^{2+} in acetonitrile were performed. The first metal (*M1*) was added in concentrations ranging from 10^{-5} to 10^{-4} M, followed by subsequent additions of the second and third metals (*M2* and *M3*) at 10^{-4} M

crease in signal, and further addition of 10^{-4} M Ca^{2+} had no effect, indicating that the layer is selective for Cu^{2+} in the presence of Ca^{2+}. Additional studies showed that the detection limit of these layers for the different cations is $10^{-7}-10^{-6}$ M.

Thus, the fluorophores, the binding groups and their substituents each have a significant impact on the sensitivity and selectivity of the sensing system toward a series of cations. The binding abilities of individual fluorophore–binding group pairs is likely due to a combination of factors such as the cation binding site, whether the binding is shared between various surface functionalities and whether there are steric constraints or additional intermolecular surface interactions induced by non-coordinating substituents, such as changes in monolayer packing, van der Waals forces, cation–π interactions or π–π interactions between the monolayer substituents.[3]

2.2.1.2
Anion Sensing

Anion recognition in biological systems is achieved via hydrogen bonds by highly preorganized proteins. The complexation properties of these proteins

[3] When quenching by the analyte dominates, reduction of the fluorescence is observed. Increased fluorescence, however, is a delicate interplay between quenching by both the analyte and the binding groups. If the binding groups are already quenching the fluorescence, resulting in a lowered initial fluorescence, it is possible that unquenching can be induced upon addition of the analyte, which could lead to increased fluorescence.

can be mimicked by chemically sophisticated macromolecules with highly preorganized binding sites. Good selectivity in anion binding has also been achieved with more structurally simple cyclic receptors. However, achieving high binding strengths and specificities for anionic guests remains challenging owing to complications such as high heats of hydration, geometric complementarity factors and larger size-to-charge ratios [65].

The ease and success of the method developed by Crego-Calama and coworkers for cation sensing thus inspired an attempt to apply it to the more difficult field of anion sensing. In this way, a small library of sensitive fluorescent SAMs was fabricated for anion-sensing proposes [64]. An amino-terminated SAM on a quartz slide was functionalized sequentially, in the manner previously described for the cation sensing, with one fluorophore (TRITC or lissamine) and one known anion binding functional group (i.e., amino, amide, sulfonamide, urea or thiourea) (Fig. 7a). The fluorescence response of the layers to 10^{-4} M acetonitrile solutions of tetrabutylammonium salts of HSO_4^-, NO_3^-, $H_2PO_4^-$ and AcO^- anions was recorded (Fig. 7b). As in the case of sensing the cations, each layer revealed a different response pattern to the anions.

As a general trend, the TRITC systems T0–T4 exhibited a larger magnitude of response to all of the anions than did the corresponding lissamine systems L0 and L5–L8, which illustrates the differences between the two fluorophores. This difference is likely due to the attachment point functionality. TRITC forms a thiourea bond upon reaction with the layer, while lissamine forms a sulfonamide bond. An especially significant response is the large sensitivity of layers T1, T2 and T3 toward $H_2PO_4^-$, showing fluorescence quenching from 35 to 50%.

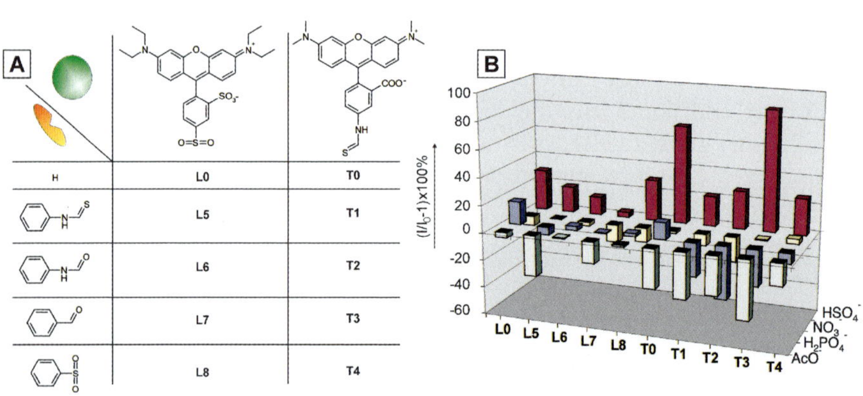

Fig. 7 **a** Schematic representation for the monolayers of the anion-sensing library. **b** Relative fluorescence intensity of surfaces modified with different fluorophores and chemical functionalities in the presence of 10^{-4} M solutions of HSO_4^-, NO_3^-, $H_2PO_4^-$ and AcO^- as tetrabutylammonium salts in acetonitrile. The data have been normalized; in the absence of anions the fluorescence emission at 585 and 575 nm for layers L0, L5–L8 and T0–T4, respectively, is set to 0

Within the lissamine series, L5 functionalized with a thioureido group undergoes 35% quenching in the presence of AcO^-, while the same layer has a moderate increase of fluorescence in the presence of HSO_4^- (20%).

Noteworthy and important for sensing purposes is the response of the TRITC-functionalized layers T1–T3. The fluorescence response of these layers increased between 24 and 87% in the presence of HSO_4^-, while the same layers showed a fluorescence intensity decrease in the presence of $H_2PO_4^-$ between 35 and 56%.

As mentioned for the previous libraries, these varied responses (especially the increases in fluorescence) across the library help decrease the chances of false-positives for the individual analytes. Additionally, amino-functionalized T0 is an excellent sensor for HSO_4^-; not only is the magnitude of the fluorescence increase quite large (72%), but it is the only anion that induces such an increase. The addition of AcO^- induces a quenching of 34%, while NO_3^- and $H_2PO_4^-$ result in virtually no response. These results support those found for the previous cation systems, wherein making a "library" of fluorophores and binding groups results in a unique array of responses to the anions.

Also for the anion library two selectivity experiments for the binding of HSO_4^-, NO_3^- and AcO^- were conducted with T3. In both experiments, first 10^{-4} M NO_3^- was added to a T3 layer, showing no response as expected. Then, two separate experiments were performed reversing subsequent additions of solutions of HSO_4^- and AcO^-, ranging from 10^{-6} M to 10^{-3} M. In each case, the addition of AcO^- caused 70% fluorescence quenching, but *only* when the HSO_4^- was added *first* did the HSO_4^- result in a fluorescence intensity in-

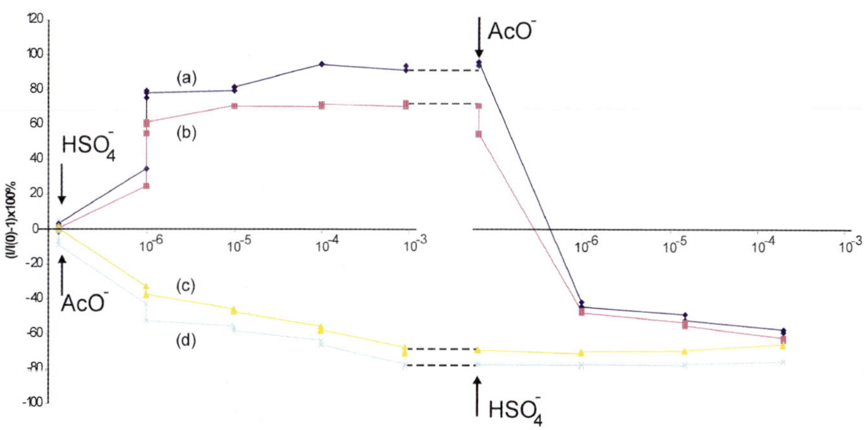

Fig. 8 Competition studies with the amido layer T3 in acetonitrile. Two experiments are shown in duplicate (*a*, *b* and *c*, *d*). In experiments *a* and *b*, three additions of HSO_4^- were made ranging from 10^{-6} to 10^{-3} M. Three additions of AcO^- were then made ranging from 10^{-6} to 10^{-3} M. In experiments *c* and *d*, the order was reversed. First, AcO^- was added from 10^{-6} to 10^{-3} M, followed by HSO_4^- from 10^{-6} to 10^{-3} M

crease. Thus, it appears that the system T3 is able to selectively sense 10^{-5} M AcO^- in the presence of 10^{-4} M NO_3^- and 10^{-3} M HSO_4^- (Fig. 8). Additionally, the detection limit of these surfaces for anion sensing has been found to be lower than 10^{-6} M.

These findings show that with this methodology it is possible to obtain a degree of sensitivity and selectivity in the detection of a series of inorganic anions, and it could represent an important step forward in the field of anion binding, which often requires significant synthetic effort in order to achieve binding affinity and selectivity.

2.2.1.3
Miniaturization: Sensing Self-Assembled Monolayer Integrated in a Microchannel

The new approach described for the fabrication of cation- and anion-sensitive fluorescent SAMs is particularly amenable to the transition toward miniaturization and subsequent array format. Analytically, the system allows both for a fingerprint type of response for detection of analyte mixtures, as well as for hand-picking of particularly selective fluorophore–binding group pairs for a specific analyte [66].

One way to achieve miniaturization using this solution-based procedure is through incorporation of the system into a microchannel. Microchannels are particularly attractive for sensing purposes because they provide a convenient platform for rapid analysis and detection, as has been shown both for biological samples [67] and, more recently, for chemical analytes [68–70]. Furthermore, it eliminates problems such as solvent evaporation in the preparation of the monolayer and in the analysis, as well as pollution or contamination.

A series of solutions were flowed through a glass microchannel to generate a sensing surface comprising a fluorescent probe and a binding molecule. The synthetic protocol of flowing the monolayer components through the channel intrinsically results in the coating of the entire microchannel. Additionally, having three inlets available allowed us to fill them all with the same solution simultaneously, thus preventing evaporation of the small solvent volumes used. First, the channel was hydroxylated with boiling piraña (solution of 1 : 4 30% H_2O_2 and concentrated H_2SO_4), followed by sequential deposition of a toluene solution of TPEDA, an acetonitrile solution of TAMRA, and a chloroform solution of *p*-isopropylphenyl isocyanate to form the TM1 layer inside of the microchannel (Fig. 9a). Addition of a 10^{-4} M acetonitrile solution of the perchlorate salt of Pb^{2+} resulted in a 50% increase in fluorescence intensity when compared with the acetonitrile-filled channel in the absence of the analyte, as imaged using confocal microscopy (Fig. 9b, c). This showed not only that the sensitive system was successfully incorporated on the walls of the microchannel, but that it displayed a response analogous to that of the macroscale TM1 layer.

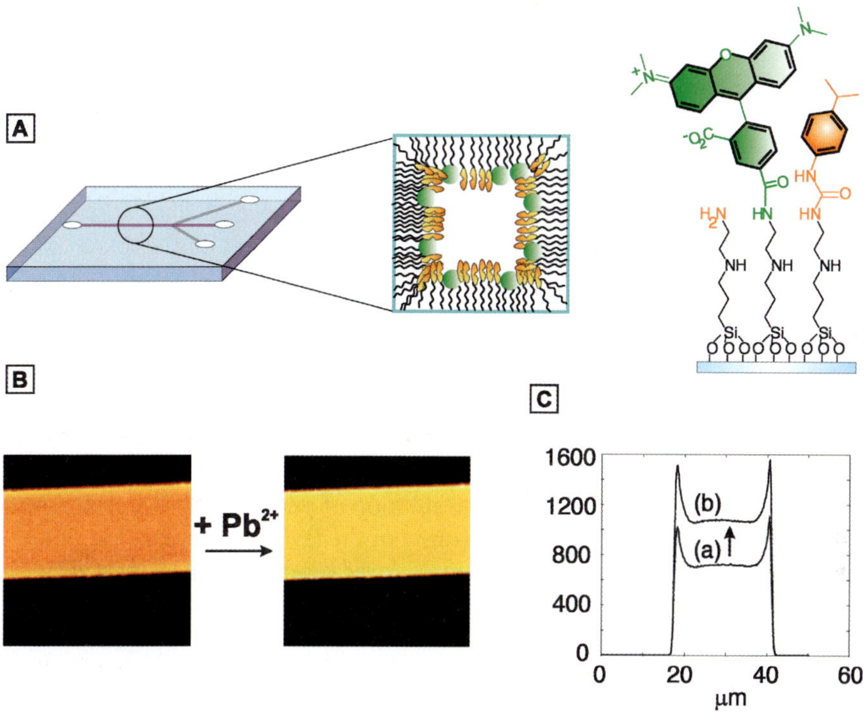

Fig. 9 **a** Schematic representation and enlarged cross section of the microchannel funtionalized with the system TM1. **b** Confocal microscopy images (70 µm × 70 µm) of the channel: filled with acetonitrile (*left*) and filled with a 10^{-4} M solution of Pb^{2+} in acetonitrile (*right*). **c** Fluorescence intensity profiles of the images: *a* in acetonitrile; *b* in 10^{-4} M solution of Pb^{2+} in acetonitrile. The *y – axis* is the fluorescence intensity (counts per pixel). The confocal microscopy measurements were imaged in the *xy* plane. The intensity of the laser was focused on only one wall of the microchannel, resulting in higher intensity on the edges

These results showed the initial proof-of-principle for the miniaturized format. In terms of future applications with HTS analysis, it is possible to envision a network with different fluorophore–binding group mixtures in each channel, allowing for the generation of a fingerprint of the entire network with a single fluorescence "snapshot" for a large number of analytes, and additionally, the possibility of regenerating the channel activity for the sequential testing of multiple analytes.

2.2.2
Microcontact Printing Based Procedure

μCP has been used in the past decade to print proteins [71, 72], peptides [73, 74] and enzymes [75] but, surprisingly, the area of small-molecule printing remains virtually unexplored [76–78]. In an earlier report we demonstrated the use of μCP as a delivery tool and not simply as a patterning technique, in order to achieve the first microcontact-printed surface on SAMs for the purpose of chemical ion sensing [64]. In this context, μCP is an amenable technique for the creation of discrete functionalized areas on the glass substrate for the generation of a sensitive surface array.

An amino-functionalized TPEDA SAM was synthesized onto a glass surface as described previously. Then a PDMS stamp previously inked in an acetonitrile solution of the fluorophore (TAMRA or lissamine) was brought into contact with the SAM for a few seconds, resulting in the covalent attachment of the fluorophore to the layer. The slide was subsequently immersed in an acetonitrile solution of a reactive molecule for the attachment of the binding groups onto the surface (i.e., urea or amide) at the sites of the unreacted surface amino groups (Fig. 10).

The slides were first imaged in acetonitrile by laser confocal microscopy, then exposed to a 10^{-4} M acetonitrile solution of perchlorate salts of different cations. Finally, the fluorescence changes of the layers upon addition of analyte were recorded (Fig. 11). For example, layer L3, microcontact-printed with the lissamine fluorophore and functionalized with a *p*-propylbenzoyl amide

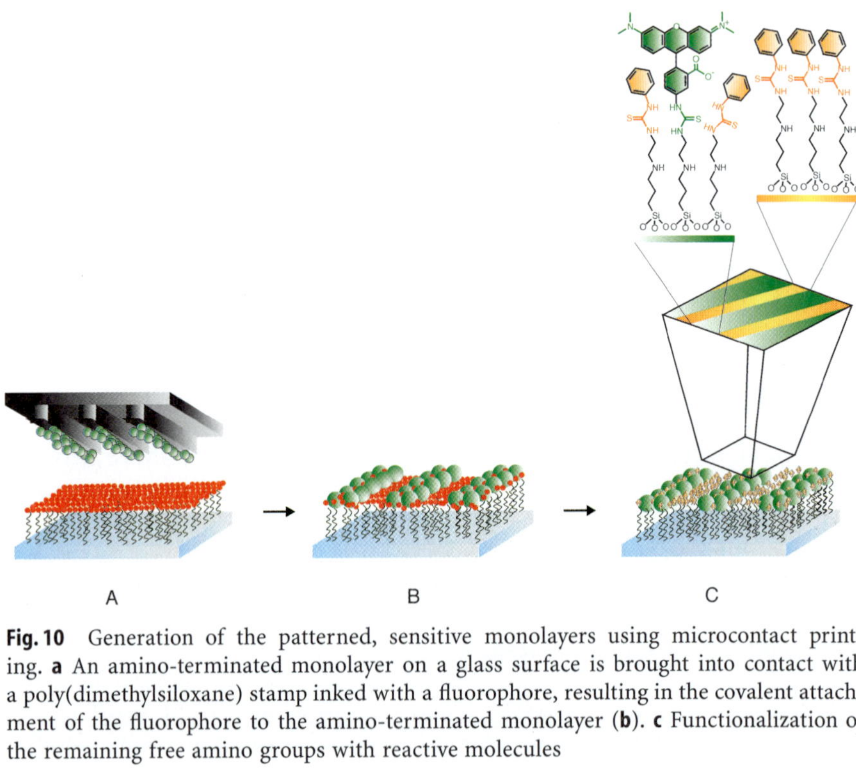

Fig. 10 Generation of the patterned, sensitive monolayers using microcontact printing. **a** An amino-terminated monolayer on a glass surface is brought into contact with a poly(dimethylsiloxane) stamp inked with a fluorophore, resulting in the covalent attachment of the fluorophore to the amino-terminated monolayer (**b**). **c** Functionalization of the remaining free amino groups with reactive molecules

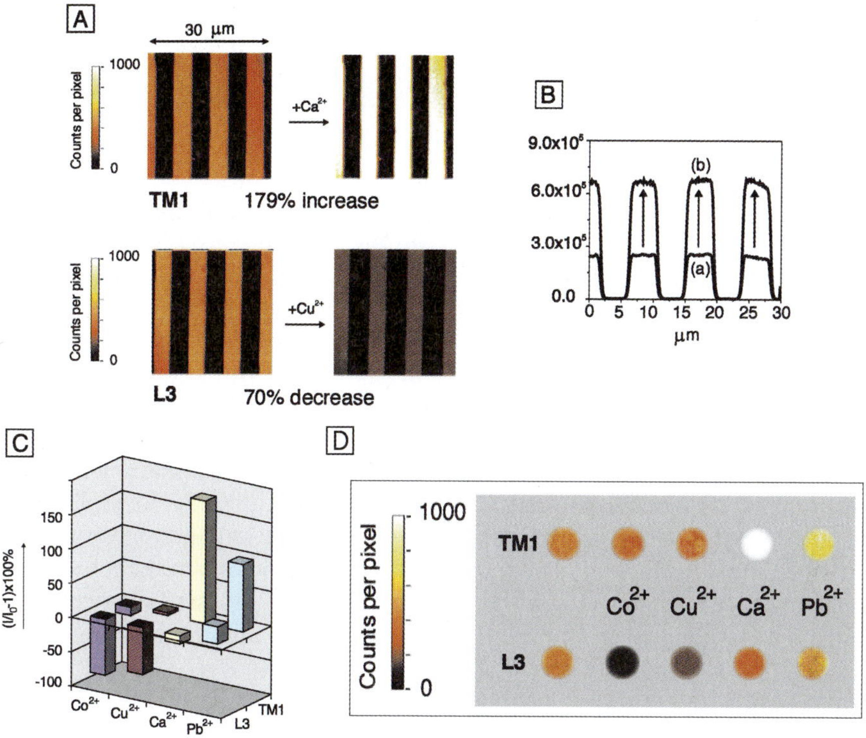

Fig. 11 **a** Fluorescence confocal microscope images of the systems TM1 and L3 in contact with acetonitrile (*left*) and in contact with a 10^{-4} M Ca^{2+} and Cu^{2+} solution in acetonitrile, respectively (*right*). **b** Profile of cross sections of the image of TM1 before (*a*) and after (*b*) addition of Ca^{2+}; *y – axis* is the fluorescence intensity (counts per pixel) and the *x – axis* is in microns. **c** Relative fluorescence intensity of systems TM1 and L3 in the presence of 10^{-4} M acetonitrile solutions of Pb^{2+}, Cu^{2+}, Co^{2+} and Ca^{2+} as perchlorate salts. **d** Array of fluorescence confocal microscope images of the systems L3 and TM1 in contact with acetonitrile (*first spot*) and the metal salt solutions. The printed surface is not perfectly homogeneous, and the *array spots* depict only a very small area and are meant to be a pictorial representation only, whereas the quantitative values for the fluorescence intensity changes are taken as an average over the whole measuring surface [79]

binding group, resulted in quenching of 70% for Cu^{2+} and 82% for Co^{2+}, and little response toward Pb^{2+} and Ca^{2+}. On the other hand, for the TAMRA layer TM1 with a ureido binding group, responses were completely different. A dramatic 179% increase was seen in the fluorescence for Ca^{2+} (Fig. 11a, b) and a 98% increase for Pb^{2+}, while there was very little response for both Co^{2+} and Cu^{2+}.

In the future, it would be possible to generate a microsensing array comprising different printed fluorophores and either cross-printed or reacted in solution with reactive molecules to obtain certain surface binding groups. The advantage of this approach, as illustrated in Fig. 11d, is the direct qualitative visualization of the sensing process, which is amenable to and compatible with HTS.

2.3 Conclusions

A new combinatorial approach where the sequential deposition of a fluorophore and a binding group randomly distributed onto a SAM on glass results in a sensing system for cations and anions has been developed. Through a combinatorial approach to the generation of sensitive surfaces from commercially available, simple small molecules, and by means of a simple analytical protocol, the possibility to generate in situ a huge number of sensing probes exists without the necessity of independent design and synthesis of receptor molecules. Furthermore, labeling or library deconvolution to determine successful sensing systems is rendered unnecessary. The overall simplicity of the system is its strength, in that it imparts generality to the methodology in terms of both sensing system generation and exploration of microanalysis technology. The transfer of the methodology to the microscale was evidenced by producing the first covalently bound sensing system in a glass microchannel and through a new use of μCP. Both of these miniaturization protocols open the possibility for integration with HTS techniques for analysis of larger libraries. Furthermore, if insufficient selectivity is obtained and, given a data set containing a number of general trends, the overall responses can be taken as a whole fingerprint representing a unique response for each analyte.

Acknowledgements The research of M.C.-C. has been made possible by a fellowship of the Royal Netherlands Academy of Arts and Sciences.

References

1. Wu X, Kim J, Dordick JS (2000) Biotechnol Prog 16:513
2. Leipert D, Nopper D, Bauser M, Gauglitz G, Jung G (1998) Angew Chem Int Ed Engl 37:3308
3. Schneider SE, O'Neil SN, Anslyn EV (2000) J Am Chem Soc 122:542
4. Iorio EJ, Shao YF, Chen CT, Wagner H, Still WC (2001) Bioorg Med Chem Lett 11:1635
5. Chen CT, Wagner H, Still WC (1998) Science 279:851
6. Hioki H, Kubo M, Yoshida H, Bando M, Ohnishi Y, Kodama M (2002) Tetrahedron Lett 43:7949
7. Wang W, Gao SH, Wang BH (1999) Org Lett 1:1209
8. Davis CJ, Lewis PT, Mccarroll ME, Read MW, Cueto R, Strongin RM (1999) Org Lett 1:331
9. James TD, Sandanayake Kras, Shinkai S (1996) Angew Chem Int Ed Engl 35:1911
10. Gauglitz G (2000) Curr Opin Chem Biol 4:351

11. Lu Y, Liu JW, Li J, Bruesehoff PJ, Pavot CMB, Brown AK (2003) Biosens Bioelectron 18:529
12. Lavigne JJ, Anslyn EV (2001) Angew Chem Int Ed Engl 40:3119
13. Dickinson TA, White J, Kauer JS, Walt DR (1996) Nature 382:697
14. Mayr T, Igel C, Liebsch G, Klimant I, Wolfbeis OS (2003) Anal Chem 75:4389
15. Batra D, Shea KJ (2003) Curr Opin Chem Biol 7:434
16. Szurdoki F, Ren D, Walt DR (2000) Anal Chem 72:5250
17. Entzeroth M (2003) Curr Opin Pharmacol 3:522
18. Khandurina J, Guttman A (2002) Curr Opin Chem Biol 6:359
19. Lam KS, Renil M (2002) Curr Opin Chem Biol 6:353
20. Birkert O, Tunnernann R, Jung G, Gauglitz G (2002) Anal Chem 74:834
21. Bearman KR, Blackmore DC, Carter TJN, Colin F, Wright JD, Ross SA (2002) Chem Commun 980
22. Rowe HM, Xu WY, Demas JN, Degraff BA (2002) Appl Spectrosc 56:167
23. Brasola E, Mancin F, Rampazzo E, Tecilla P, Tonellato U (2003) Chem Commun 3026
24. Flink S, Van Veggel FCJM, Reinhoudt DN (1999) Chem Commun 2229
25. Kanoh N, Kumashiro S, Simizu S, Kondoh Y, Hatakeyama S, Tashiro H, Osada H (2003) Angew Chem Int Ed Engl 42:5584
26. Crego-Calama M, Reinhoudt DN (2001) Adv Mater 13:1171
27. Conn MM, Rebek J (1997) Chem Rev 97:1647
28. Linton B, Hamilton AD (1997) Chem Rev 97:1669
29. Philp D, Stoddart JF (1996) Angew Chem Int Ed Engl 35:1155
30. Lawrence DS, Jiang T, Levett M (1995) Chem Rev 95:2229
31. Whitesides GM, Simanek EE, Mathias JP, Seto CT, Chin DN, Mammen M, Gordon DM (1995) Acc Chem Res 28:37
32. Crego Calama M, Timmerman P, Reinhoudt DN (2000) Angew Chem Int Ed Engl 39:755
33. Lehn JM (1999) Chem Eur J 5:2455
34. Timmerman P, Reinhoudt DN (1999) Adv Mater 11:71
35. Ganesan A (1998) Angew Chem Int Ed Engl 37:2828
36. Gennari C, Nestler HP, Piarulli U, Salom B (1997) Liebigs Ann Rec 637
37. Sasaki DY, Shnek DR, Pack DW, Arnold FH (1995) Angew Chem Int Ed Engl 34:905
38. Kolusheva S, Shahal T, Jelinek R (2000) J Am Chem Soc 122:776
39. Cai MM, Sidorov V, Lam YF, Flowers RA, Davis JT (2000) Org Lett 2:1665
40. Czarnik AW (1992) In: Fluorescent chemosensors for ion and molecular recognition, chap 1. American Chemical Society, Washington, DC
41. de Silva AP, Gunaratne HQN, Gunnlaugsson T, Huxley AJM, Mccoy CP, Rademacher JT, Rice TE (1997) Chem Rev 97:1515
42. Grandini P, Mancin F, Tecilla P, Scrimin P, Tonellato U (1999) Angew Chem Int Ed Engl 38:3061
43. Steemers FJ, Walt DR (1999) Mikrochim Acta 131:99
44. Kaifer AE (1996) Isr J Chem 36:389
45. Kepley LJ, Crooks RM, Ricco AJ (1992) Anal Chem 64:3191
46. Sugawara M, Kojima H, Sazaya Y, Umezawa Y (1987) Anal Chem 59:2842
47. Wolfbeis OS, Schaffar BPH (1987) Anal Chim Acta 1:198
48. Nuzzo RG, Allara DL (1983) J Am Chem Soc 105:4481
49. Ulman A (1991) (ed) Ultrathin organic films: from Langmuir–Blodgett to self-assembly. Academic, Boston
50. Chechik V, Crooks RM, Stirling CJM (2000) Adv Mater 12:1161
51. Crooks RM, Ricco AJ (1998) Acc Chem Res 31:219

52. Motesharei K, Myles DC (1998) J Am Chem Soc 120:7328
53. Flink S, Van Veggel FCJM, Reinhoudt DN (2000) Adv Mater 12:1315
54. Major RC, Zhu XY (2003) J Am Chem Soc 125:8454
55. Hertzberg RP, Pope AJ (2000) Curr Opin Chem Biol 4:445
56. Van Der Veen NJ, Flink S, Deij MA, Egberink RJM, Van Veggel FCJM, Reinhoudt DN (2000) J Am Chem Soc 122:6112
57. Sagiv J (1980) J Am Chem Soc 102:92
58. Ulman A (1996) Chem Rev 96:1533
59. Kurth DG, Bein T (1993) Langmuir 9:2965
60. Albert KJ, Lewis NS, Schauer CL, Sotzing GA, Stitzel SE, Vaid TP, Walt DR (2000) Chem Rev 100:2595
61. Lavigne JJ, Savoy S, Clevenger MB, Ritchie JE, Mcdoniel B, Yoo SJ, Anslyn EV, Mcdevitt JT, Shear JB, Neikirk D (1998) J Am Chem Soc 120:6429
62. Pirrung MC (1997) Chem Rev 97:473
63. Cremer PS, Yang TL (1999) J Am Chem Soc 121:8130
64. Basabe-Desmonts L, Beld J, Zimmerman RS, Hernando J, Mela P, García Parajó, MF, van Hulst NF, van den Berg A, Reinhoudt DN, Crego-Calama M (2004) J Am Chem Soc 126:7293
65. Beer PD, Gale PA (2001) Angew Chem Int Ed Engl 40:487
66. Phizicky E, Bastiaens PIH, Zhu H, Snyder M, Fields S (2003) Nature 422:208
67. Andersson H, Van Den Berg A (2003) Sens Actuators B 92:315
68. Sato K, Hibara A, Tokeshi M, Hisamoto H, Kitamori T (2003) Adv Drug Delivery Rev 55:379
69. Guijt RM, Baltussen E, Van Der Steen G, Schasfoort RBM, Schlautmann S, Billiet HAH, Frank J, Van Dedem GWK, Van Den Berg A (2001) Electrophores 22:235
70. Munro NJ, Huhmer AFR, Landers JP (2001) Anal Chem 73:1784
71. Bernard A, Renault JP, Michel B, Bosshard HR, Delamarche E (2000) Adv Mater 12:1067
72. Kane RS, Takayama S, Ostuni E, Ingber DE, Whitesides GM (1999) Biomaterials 20:2363
73. Csucs G, Michel R, Lussi JW, Textor M, Danuser G (2003) Biomaterials 24:1713
74. Sprossler C, Scholl M, Denyer MC, Krause M, Nakajima K, Maelicke A, Knoll W, Offenhausser A (2001) Synth Met 117:281
75. Wilhelm T, Wittstock G (2002) Langmuir 18:9485
76. Andersson H, Jonsson C, Moberg C, Stemme G (2001) Sens Actuators B 79:78
77. Lahiri J, Ostuni E, Whitesides GM (1999) Langmuir 15:2055
78. Yan L, Zhao XM, Whitesides GM (1998) J Am Chem Soc 120:6179
79. Basabe-Desmonts L, Beld J, Zimmerman RS, Hernando J, Mela P, Parajo MFG, van Hulst NF, van den Berg A, Reinhoudt DN, Crego-Calama M (2000) J Am Chem Soc 126:7293. DOI 10.1021/ja049901o

Springer Ser Chem Sens Biosens (2005) 3: 189–225
DOI 10.1007/5346_006

Published online: 27 September 2005

The Interplay of Indicator, Support and Analyte in Optical Sensor Layers

Guillermo Orellana[1] (✉) · Maria C. Moreno-Bondi[2] · David Garcia-Fresnadillo[1] · Maria D. Marazuela[2]

[1]Department of Organic Chemistry, Faculty of Chemistry, Universidad Complutense de Madrid, 28040 Madrid, Spain
orellana@quim.ucm.es

[2]Department of Analytical Chemistry, Faculty of Chemistry, Universidad Complutense de Madrid, 28040 Madrid, Spain
mcmbondi@quim.ucm.es

Abstract It has been recognized since the pioneering times of fiber-optic sensing development that the best indicator dye is worth nothing without a (polymer) support fitted to both the determinand species and the indicator itself. However, the task of selecting an organic or inorganic polymer for manufacturing a sensitive head among the myriad of materials available nowadays may seem daunting to the researcher or technologist. Moreover, if we also incorporate a biological recognition element to develop an ultrasensitive or specific biosensor, the multifaceted problem appears even more puzzling. This chapters aims to guide the reader through the current world of both organic and inorganic materials and their effect on (bio)chemical sensing. Selected examples illustrate the diversity of solid supports and composites and their effect on the indicator response, photostability, interaction with the analyte, stability of the different biological elements, and ease of preparation, among other factors, shedding some light on the complex interaction between the key components of chemical sensors and biosensors.

Keywords Biosensors · Dyes · Fluorescence · Glass · Immunosensors · Luminescence · Nanocomposites · Optodes · Optosensors · Ormosils · PVC · Silica · Siloxanes · Sol-gel

Abbreviations

BNHS	biotinyl *N*-Hydroxysuccinimide ester
CCD	Charge-coupled device
CPG	Controlled-pore glass
D	Diffusion coefficient
DMMP	Dimethyl methylphosphonate
DPO	4-(*p*-*N*,*N*-Dimethylaminophenylmethylene)-2-phenyl-5-oxazolone
dppz	Dipyrido[3,2-a:2′,3′-c]phenazine
ELBL	Electrostatic layer-by-layer assembly
FRET	Fluorescence resonance energy transfer
GMBS	*N*-Succinimidyl 4-maleimidobutyrate
I	Luminescence intensity
ISO	Ion-selective optode
K_d	Dissociation constant
k_q	Quenching rate constant
K_{SV}	Stern–Volmer constant
MB	Methylene blue
MBe	Molecular beacon
NR	Nile Red
ormocers	Organically-modified ceramics
ormosils	Organically-modified silicates
PAH	Polycyclic aromatic hydrocarbons
PAMAM	Polyamidoamine
PEBBLE	Probes encapsulated by biologically localized embedding
phen	1,10-Phenanthroline
PrA	Protein A
PrG	Protein G
PV	Pyrocatechol Violet
PVC	Poly(vinyl chloride)
RB_3^{2+}	Tris(2,2′-bipyridine)ruthenium(II) cation
RDP_3^{2+}	Tris(4,7-diphenyl-1,10-phenanthroline)ruthenium(II) cation
RTP	Room temperature phosphorescence
S	Solubility
SPR	Surface plasmon resonance

τ	Luminescence lifetime
TBP	Tributyl phosphate
TOA	Tetraoctylammonium
TPP	Tripropyl phosphate

1 Introduction

Both organic and inorganic polymer materials have been used as solid supports of indicator dyes in the development of optical sensors for (bio)chemical species. It is known that the choice of solid support and immobilization procedure have significant effects on the performance of the optical sensors (optodes) in terms of selectivity, sensitivity, dynamic range, calibration, response time and (photo)stability. Immobilization of dyes is, therefore, an essential step in the fabrication of many optical chemical sensors and biosensors. Typically, the indicator molecules have been immobilized in polymer matrices (films or beads) via adsorption, entrapment, ion exchange or covalent binding procedures.

The polymer materials not only act as supports for the dye and other necessary additives in the sensing phase, providing protective covering for the transduction element; polymers also play various roles in chemical sensors. They provide a compatible environment for the indicator molecules, maintaining or improving the appropriate photophysical features (compared to those observed in homogeneous solution) on which the sensing principle is based. In many cases they collect and concentrate the analyte molecules on sensor surfaces. In addition, the polymer can play an important role in the sensitivity and selectivity of an optical sensor, and its interactions with indicator and analyte molecules influence the analytical performance of the device.

So far, the selection of polymers for chemical sensors based on fluorescent dyes in polymer matrices has been largely empirical, based on the accomplishment by the polymer of a number of requisites that are desirable for any support aimed to be used in optical sensing. These features are the following:

- Chemical inertness
- Affordability
- High thermal stability
- Optical transparency
- Low intrinsic fluorescence or absorbance
- Compatibility with the indicator dye
- Homogeneity
- Good mechanical properties (strength)
- Flexibility or porous structure to allow fast analyte diffusion

- Permeability to analyte
- Reproducibility of film/bead production
- Controllable film thickness (compromise between sensitivity and response time of the sensor system)

It is not easy to predict the performance of a sensor membrane containing an immobilized indicator because the indicator dye can change a number of its characteristics after immobilization. Many papers in the optosensor literature discuss key features of the polymer support and its interaction with the indicator dye and/or the analyte that have to be considered in order to optimize the response. In this chapter we review a number of papers where considerations concerning the role played by the polymer matrix and its relation to the indicator/analyte couple are addressed in terms of the optical sensing layer performance. Examples covering widely studied indicators such as organic fluorophores and phosphors, and luminescent coordination compounds applied to optosensors for oxygen, pH, ions, humidity and organic analytes have been selected.

2 Organic Polymer Supports

Many of the optical sensors developed during the last few decades utilize organic polymers as the supporting material of the indicator. Among the most widely used organic supports we can find poly(styrene), poly(vinyl chloride), poly(methyl methacrylate), poly(dimethyl siloxanes), poly(tetrafluoroethylene); cellulose derivatives; different copolymers used to impart hydrophilicity or hydrophobicity to the polymer matrix; ion exchange resins, etc. This popularity is due to their availability for manufacturing sensing layers since they fulfil most of the desirable requisites mentioned above for any optosensing polymer matrix: they are commercialized in various molecular sizes at low prices since they are used for many different ordinary applications, they possess homogeneous structures, they display sufficient chemical, thermal and photochemical stability, they can be easily manufactured as thin transparent films or (micro)beads with high reproducibility, their (micro)porous structure enables high indicator loading and analyte diffusion, and they show compatibility with many different organic/inorganic molecules that may be used as necessary additives in the sensing phase. The plasticity and antibiofouling features of many organic polymers (silicones, poly(tetrafluoroethylene), polyolefins, etc.) are a bonus in the preparation of sensing layers and keep microorganisms at bay.

2.1
Effect of the Support on the Indicator Response

In general, indicator dyes are chosen on the basis of their high absorption coefficients and/or emission quantum yields in solution, long emission lifetimes, their solvatochromism, large Stokes' shift and, very importantly, because a measurement principle can be established. The interaction of indicator and analyte molecules in the sensing phase must lead to modulation of the optical properties of the former (modification of absorption/emission band maximums, changes in emission intensities or lifetimes, etc.) which can be correlated to the analyte concentration. The proposed sensing principle is firstly tested in homogeneous solution and then, following immobilization of the dye, tested in a (micro)heterogeneous polymer matrix. If the sensing principle still works after indicator immobilization, optosensing analytical applications can be developed.

2.1.1
Modulation of the Light Absorption/Emission Features of the Indicator

Our first example is an optical sensor membrane based on the metal ion indicator Pyrocatechol Violet (PV), a non-specific indicator and water-soluble dye [1]. The authors prepared a plasticized PVC membrane based on a lipophilic form of the PV (an ion pair with tetraoctylammonium, TOA). A 1 : 2 PV/TOA ratio in the membrane was optimal to prevent leaching of the dye. Compared to what is normally observed in solution, a drastic improvement in the indicator selectivity and sensitivity was achieved after its immobilization, which together with a significant shift of the working wavelength in the near infrared region, made the system very promising for Cu(II) sensing. Improved selectivity after immobilization is possibly due to a combination of effects: a reduced complexing ability of the immobilized PV and the different permeation selectivity provided by plasticized PVC. The extraction of Cu(II) ions into the membrane is coupled with release of protons from the membrane into the sample solution. Therefore, on exposure to Cu(II) ions at pH 6, the indicator membrane changes its color from yellow to green (with considerable increase in the absorption coefficient), while the effects of other interfering metal ions such as Sn(II) and Al(III) are less pronounced. The presence of a cationic surfactant such as TOA causes intensification of the chelate absorption band as well as a significant bathochromic shift of the chelate absorption maximum, providing a response to Cu(II) in the 1–100 μM range. The immobilization process restricts formation of complexes with a high indicator : analyte ratio (e.g., 2 : 1 with Sn(II) and Al(III)), thereby making the indicator specific to complexes with low stoichiometric ratios (1 : 1 with Cu(II)).

Azlactone molecules can be protonated non-reversibly in acidic solution due to ring opening. A fully reversible pH sensor based on an azlactone derivative, 4-(*p*-*N*,*N*-dimethylaminophenylmethylene)-2-phenyl-5-oxazolone, DPO, embedded in a plasticized PVC polymer film has been described [2]. The sensor membrane exhibits a pH-induced color change which is a function of pH in the 1–7 range. The azlactone derivative displays enhanced fluorescence quantum yield (two orders of magnitude) and emission lifetime (one order of magnitude) when immobilized in the PVC film compared to solution. The Stokes' shift decreases by 50% in the plasticized PVC matrix compared to acetonitrile medium. The singlet state energy DPO in PVC film, which is $1.1-1.4\,\mathrm{kcal\,mol^{-1}}$ lower than that in solutions, is taken as evidence of less singlet–triplet crossing in the excited state of polymer-supported DPO. Therefore, fluorescence is enhanced. The smaller Stokes' shift is suggested to arise from rigidochromism in the immobilized species with respect to solutions.

The solvatochromic dye Nile Red (NR), with basic functional groups that may participate in hydrogen bonds when dispersed in selected acidic polymer matrices, demonstrates strong *fluorescence enhancement* in the presence of dimethyl methylphosphonate (DMMP) vapors [3]. Sorption of basic vapors may compete with the dye and displace it from hydrogen-bonding sites within the polymer, altering the local environment and fluorescence properties of the former. Two acidic polymers were examined as dye matrices, one with fluorinated alcohol groups on a poly(styrene) backbone and the other with fluorinated bisphenol moieties alternating with oligo(dimethylsiloxane) segments. Both polymers can form hydrogen bonds with Nile Red so that before exposure to DMMP, the immobilized NR exhibits red shift relative to chloroform solution. DMMP sorption changes the environment of the dye and dramatically modifies its fluorescence spectrum and intensity, resulting in a strong fluorescence enhancement. The highest responses were obtained with the polymer containing fluorinated bisphenol groups. DMMP detection has been demonstrated at sub-ppm concentrations, with very low detection limits compared to previous NR-supported vapor sensors.

Micromolar solutions of mercurochrome in aqueous buffers at pH values ranging from 5.5 to 10.0 have proved to be highly fluorescent but no room temperature phosphorescence (RTP) was observed either in the presence of oxygen or in deoxygenated media. However, the presence of heavy atoms in the molecule provokes an efficient intersystem crossing and a high RTP emission signal can be observed when mercurochrome is immobilized on a solid support [4]. Solid-surface room temperature phosphorescence is especially well suited to *moisture* sensing because of the RTP quenching due to the weakening of the hydrogen bonds between the immobilized dye and the solid support by the incoming water molecules. The looser bonding leads to a decrease in the rigidity of the phosphor and favors oxygen access to the luminescent species. Different supports such as a Dowex 1X2-200 resin or a silica sol-gel matrix can impart a different rigidity to the indicator dye so that an analytically useful RTP signal is obtained with a detection limit of 0.11% relative humidity in air.

2.1.2 Linearity of the Response

If we take as an example sensors for *molecular oxygen* based on luminescent indicators, if the indicator is distributed uniformly throughout a homogeneous encapsulating medium (at least at the macroscopic level) then the quenching of its luminescence fits a simple linear Stern–Volmer plot (see Sect. 3.1). In practice, the majority of current optical oxygen sensors exhibit downward-curved Stern–Volmer plots in a degree that is dependent on the organic polymer dispersing medium and the indicator nature [5]. Downward-curved Stern–Volmer plots are usually attributed to a distribution of indicator species due to microheterogeneity within the sensing layer, each of them with its own Stern–Volmer constant [6]. When the indicator molecules are soluble in the polymer matrix and are homogeneously incorporated to a defined microdomain of the polymer support, the Stern–Volmer plots tend to be linear over the entire oxygen concentration range in the gas, aqueous or organic solvent phase, as demonstrated with Ru(II) polypyridyl complexes immobilized in poly(dimethylsiloxane) [7, 8], Nafion-117 ionomer [9] and with Pt(II) porphyrins supported on poly(isobutyl methacrylate-*co*-tetrafluoropropyl methacrylate) [10], and MF-4SK ion-exchanger [11].

2.2 Effect of the Polymer-Supported Indicator Concentration

Some important features concerning the indicator concentration in the sensing phase that can influence the sensor response, especially its sensitivity, are (i) the necessary amount of indicator molecules to give satisfactory signal-to-noise ratios and (ii) the possibility of self-quenching due to aggregation of the dye, when a high concentration of indicator molecules decreases the analytical signal in the absence of analyte.

Nanosensors or PEBBLE (probes encapsulated by biologically localized embedding)-based sensors can be prepared in the 0.5–1 μm size range and may work with as little as a single indicator dye per nanobead, provided that the number of nanobeads used is sufficient for generating a satisfactory signal-to-noise ratio. If only a single nanobead is used, it normally has to contain a sufficient number of dyes, in the order of 1000, to provide high enough signal-to-noise ratio [12].

Provided the indicator concentration is high enough, another factor that influences the noise in the measured signal is the polymer surface topography. A uniform surface that presents low reflectance and also minimal scattering always produces a more uniform signal [13].

To avoid the effects of non-fluorescent aggregates on the sensor performance, films are usually prepared from solutions containing a low dye concentration. For example, at concentrations as low as 2.5×10^{-4} M, additional quenching of Nile Red fluorescence in spin-cast fluorinated sensor films for DMMP vapors was observed [3]. Most likely, the reason for such quenching is aggregation of the polymer-supported dye leading to an exciton coupling effect, which is typical of dyes having planar conjugated chromophores.

An alternative to avoid self-quenching is to use the electrostatic layer-by-layer assembly (ELBL) technique. It eliminates this effect because the indicator is distributed in consecutive (nanometer-thick) polymer layers, each of them having opposite polarity to the previous one. In fact, this technique can be utilized to adsorb the optimum dye concentration with adequate separation in thin film optical sensors [14, 15].

2.3 Effect of the Support on the Analyte-Indicator Interaction

In general, when we consider sensors for molecular oxygen based on luminescence quenching, the higher the oxygen permeability of the encapsulating polymer, the larger the analyte sensitivity of the film. Permeability of a gas through a polymer is defined by the product of the gas solubility in the polymer, *S*, and the diffusion coefficient of the gas in the polymer, *D*. While oxygen solubility does not vary very much with the sensing phase considered, its diffusion coefficient does change. The slope of the dose-response plot of the oxygen sensing system, which determines its sensitivity, is equal to the product of the indicator emission lifetime in the absence of quencher (τ_0) and the bimolecular quenching constant by oxygen (k_q). The latter is very dependent on the permeability and, therefore, *D* for the encapsulating medium because the quenching reaction is very largely diffusion controlled. This dependency is so pronounced that by using the same oxygen quenchable indicator in a range of different polymer encapsulating media of different oxygen permeabilities (hydrophobic polymers such as silicone rubber, PVC, polystyrene, poly(methyl methacrylate) and cellulose), a variety of oxygen sensors of markedly different oxygen sensitivities can be generated. The high value of *D* for oxygen in silicone rubber, compared with most other polymers, is attributed to the opening and closing of substantial void volumes contained in the polymer through segmental movement [6].

The analyte sensitivity and response time of a PVC film can also be markedly improved with addition of plasticizers such as tributyl phosphate (TBP) or tripropyl phosphate (TPP), since it is well established that the presence of plasticizer in the polymer-encapsulating medium increases gas diffusion rate by increasing the degree of polymer segmental mobility. It is important to select compatible plasticizer and polymer combinations. In general, the plasticizer with the lowest viscosity produces films of the highest oxygen sensitivity. In addition, faster response times are obtained with higher plasticizer levels [16, 17].

In a similar manner, the presence of additives in the silicone rubber films for oxygen sensing can considerably alter the optode sensitivity [18]. Copolymer systems can be used to increase the solubility of indicator molecules in non-polar polymers and also to increase the polymer rigidity, which are important features of the sensing matrix, as discussed above. However, caution must be taken since it is possible to shut off quenching due to poor local oxygen diffusion or solubility around the indicator molecule. Additionally, severe heterogeneity (non-linear Stern–Volmer plots) may result even though a polymer matrix can be a good solvent of the indicator dye [5].

Engineering of the highly hydrophobic tris(4,4′-dinonyl-2,2′-bipyridine)-ruthenium(II) complex and its adsorption on a poly(tetrafluoroethylene) film has led to the development of a sensing layer for hydrocarbons in water [19]. The effect of the dye loading, excitation modulation frequency, sample temperature, dissolved oxygen, hydrocarbon type and potential interfering compounds on the luminescent material response has been investigated. Such a material enables another use of the modern emission lifetime-based field luminometers for dissolved oxygen monitoring. The sensitivity to hydrocarbons is a consequence of the charge-transfer nature of the metal complex (electronic) excited state and the affinity of both the polymer support and the indicator dye for the analytes.

Using a similar polymer support, a simple, inexpensive, robust optode for relative humidity monitoring in air has also been fabricated using the luminescent complex $[Ru(phen)_2(dppz)]^{2+}$, where phen and dppz stand for 1,10-phenanthroline and dipyrido[3,2-a : 2′,3′-c]phenazine, respectively [20]. The water molecules switch-off the indicator luminescence by hydrogen bonding to the dppz ligand and promote deactivation via a phenazine moiety-based excited state. In the absence of water, a phen moiety-based emissive excited state is the lowest lying level. Response to humidity is enhanced by static quenching of the supported indicator due to the hydrophobic character of the polymer thin film. The sensor layer performs particularly well under high humidity condensing conditions where conventional electrical devices fail to give accurate measurements.

A more sophisticated class of optical sensors with high selectivity towards ions are the ion-selective optodes (ISOs) [21], where the matrix (hydrophobic polymer such as PVC) contains a selective lipophilic ionophore (optically silent), a chromoionophore, a plasticizer and an anionic additive. The measurement principle is based on a thermodynamic equilibrium that controls the ion exchange (for sensing cations) or ion coextraction (for sensing anions) with the sample. The source of optode selectivity is a preferential interaction between the target ion and an ionophore. For a dye to act as a chromoionophore, it must be sufficiently lipophilic to remain in the organic membrane phase and must demonstrate fairly strong ion-induced changes in UV-vis absorption or fluorescence. To achieve sensor miniaturization, fluorescence rather than absorbance has been utilized. Therefore, the chromoionophore indicators that have been most widely employed are hydrophobic derivatives of the H^+-sensitive Nile Blue and fluorescein. PVC is the typical constituent of the reagent layer, a fact that can be attributed to its relatively cheap cost, good mechanical properties and ease of plasticization. Other polymers also employed are derivatized PVC, silicone rubbers, polyurethanes or polystyrene. Membranes typically contain a high percentage of plasticizer, which acts as the solvent for all membrane constituents and, hence, should be compatible with these species, specially the ionophore. The last component is normally an ionic additive that is lipophilic enough to remain in the organic membrane phase, even in contact with the aqueous solution. Two usual types of ionic additives are employed: quaternary ammonium salts (cationic sites needed) and tetraphenylborate salts (anionic sites needed). These ionic additives govern the overall ionic strength and electroneutrality conditions of the membrane. Therefore, care must be taken to balance ratios of ionophore, chromoionophore and other additives within the membrane [22].

For measuring a significant change of the optical signal from the optode membrane, there must be a change in the bulk composition of the membrane to influence the chromoionophore. Simultaneously, electroneutrality within the bulk membrane phase must be maintained. Therefore, the mechanism of optode response must be either an ion-exchange of two anions or two cations, or must be a coextraction of two differently charged ions where one ion of the pair alters the optical properties of the ionophore, as shown in Fig. 1.

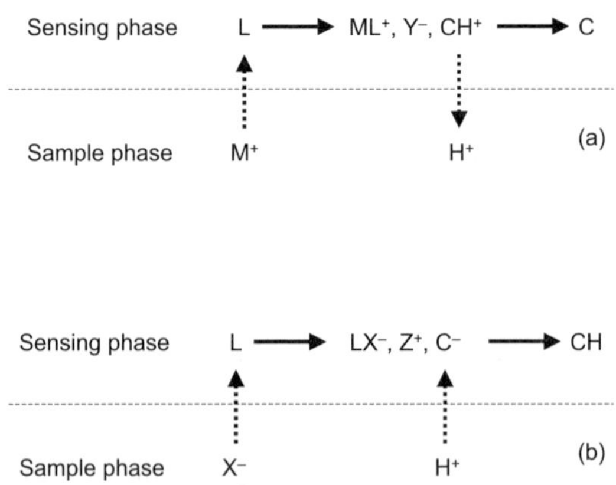

Fig. 1 Mechanisms of ion-selective optode response in aqueous samples containing **a** cationic analytes, M^+, or **b** anionic analytes, X^-. L represents the ionophore ligand, C the chromoionophore and Y^- and Z^+ the corresponding ionic additives (counterions)

2.4 Effect of the Support on the Indicator Photostability

We have chosen Ru(II) luminescent indicators containing polyazaheterocyclic ligands to discuss their photostability in oxygen-sensing layers since they are very photostable compared to typical organic indicators and also because their photochemistry illustrates not only the effect of direct irradiation (ligand loss), but also the effect of reaction with generated singlet molecular oxygen due to the indicator excited state quenching by oxygen.

Although Ru(II) complexes display low photodegradation quantum yields and have microsecond excited-state lifetimes (a photophysical parameter which, in principle, is independent of the photoprocesses that take place in the sensing phase provided the photodecomposition products do not emit), changes in the absorption spectrum and decay times of Ru(II) dyes immobilized on silicagel, polystyrene (PS) or PVC have been detected, even though rigid supports can largely inhibit the ligand loss photochemistry typical of Ru(II) complexes [23]. These changes are dependent on: (i) the nature of the dye (2,2′-bipyridine vs. 1,10-phenanthroline derivatives), (ii) the polymer type, (iii) the oxygen concentration, and (iv) the dye loading [24, 25]. While complexes with bipyridyl ligands adsorbed on silicagel did not show any decay time decrease, complexes having phenanthroline ligands exhibited significant photodegradation which is evidenced by a decrease in the luminescence intensity, a reduction in the emission lifetime and specific absorption spectral changes (absorption bands of photoproducts above 500 nm). The same trend was also observed in organic polymer matrices, where the photodegradation effects are more pronounced (especially in PVC compared to PS, where the singlet oxygen lifetime is relatively short). The effects depend strongly on the oxygen and dye concentrations present and increase with their level. Reaction with photosensitized singlet oxygen seems to be the dominant process. The absorption bands of the photoproducts of Ru(II)-phenanthroline complexes overlap with the emission spectra of the intact indicator molecules. Therefore, Förster resonance energy transfer from the indicator molecules to their own photoproducts, generated by singlet oxygen attack of the phenanthroline 5,6-bridge of the ligands (due to its higher π-electron density), is suggested to be responsible for the observed reduction in the measured decay times. Unlike in solution, the photogenerated singlet oxygen in a polymer is not able to diffuse away from the indicator, due to the higher viscosity, and has an enhanced opportunity to attack it [26].

3 Inorganic Supports

Inorganic polymer supports for the optical indicator dye are preferred when chemical inertia, dimensional stability (absence of swelling) and/or temperature endurance are a must. However, the number, nature and available formats of inorganic materials for optical sensing are more limited than those of their organic counterparts. Unlike organic polymers, some of them are difficult to manufacture in thin films, display significant fragility or do not show the required degree of porosity to the analyte or transmission of light once they are in non-particulate form. Moreover, the variety and surface density of functional groups of commercially available inorganic materials, for chemical attachment of indicator molecules, is also smaller than that of the carbon-based polymers [27, 28]. In order to combine the best features of both worlds, it is not surprising that organic–inorganic composites and nanostructured materials are increasingly used to develop optochemical sensors [29, 30].

3.1 Glass, Fused Silica and Silica Gels

Glass, fused silica and silica gels (silicon dioxide) have been the most traditional indicator supports of the inorganic polymer type. Sometimes, the fiber core itself has been chemically etched to bind indicator molecules. Even after proper etching, glass and fused silica are characterized by a small surface area so that their use is restricted to support fluorescent indicator molecules with high emission quantum yield (e.g., fluoresceins, cyanines, rhodamines, polycyclic aromatic hydrocarbons). This is certainly not the case for silica gel: for instance, pyrolytic ("fumed") silicas are produced in the form of a few nanometer particles with surface areas in excess of 300 $m^2\,g^{-1}$ by vapor-phase hydrolysis of silicon chlorides in a hydrogen-oxygen flame. Due to the surface silanol groups (~ 5.5 OH nm^{-2}), untreated silica gels are hydrophilic but treatment with silanes or siloxane oligomers or polymers renders them hydrophobic due to reduction of the number of surface silanols and the chemical nature of the capping molecules. These powders are extremely active in adsorbing the indicator dyes and the analytes, depending on the surface polarity of the silica gel used. Obviously, they also strongly adsorb many other species found in the sample that may interfere with the optical measurements, unless they are embedded in an analyte-permeable polymer film (typically organic, see below) to form the sensitive layer.

3.2 Porous Glasses

In order to benefit from both the mecano-chemical properties of glass-type indicator supports and the surface features of silica powders, porous glasses seem to be the materials of choice [31]. They can be produced in many geometric forms (beads, rods, fibers, hollow tubes, thin membranes), with pore sizes in the 0.3–1000 nm range and a very reactive surface. Porous glasses result from the acid or combined acid/alkaline treatment of phase-separated borosilicate glasses. Heating the latter to 500–700 °C causes separation of two phases: an alkali-rich borate phase that is soluble in hot mineral acids or water, and another that is made of almost pure silica. The chemical and structural properties of the porous glasses are determined by (i) the initial glass composition, (ii) the heat treatment applied (temperature, time) and (iii) the leaching out conditions. In the lower temperature range (< 580 °C), nanoporous glasses are formed after acid treatment (e.g., the commercial porous Vycor glass, with 7 nm average pore size). Heating beyond 580 °C yields mesoporous materials after removing the colloidal silica deposits that remain in the cavities of the main silica framework by treatment with dilute alkaline solutions (e.g., the commercial "controlled-pore glass", CPG, with selectable 50–1000 nm pores). Depending on the conditions of the thermal pre-treatment, the surface of porous glasses contains between 4 and 8 OH nm^{-2}, including both weakly-acidic Si – OH and B – OH groups. A number of surface-modified porous glasses are also available for hosting enzymes or indicator dyes. Functionalization procedures may lead to hydrophobic glasses, chemically-reactive surface groups (alkyl-amino, -thiol, -carboxyl, -hydroxyl) for covalent bonding, or porous glasses with enhanced, more uniform non-covalent binding capacity.

An illustrative example of the advantages of robust covalent bonding of indicator dyes to CPG for optical sensing in aggressive environments can be found in [32]. Ru(II) polypyridyls are the paradigm indicators in luminescence optosensing of O_2 either in the gas or aqueous phases. However, the (organic) polymer-sensitive layers used to that end do not withstand organic solvents at elevated pressure and temperature, operational conditions present in the petrochemical and aeronautics industries, among others. Formation of robust amide bonds between amino-derivatized CPGs and sulfonated phenanthroline Ru(II) complexes provides the required stability to the sensitive phase for measuring dissolved oxygen (down to at least 6 μM) in such media using the very same instrumentation available, e.g., as for environmental sensing.

3.3 Sol-Gel Materials

The attractive features of porous glass materials for optical sensing are only hampered by their significant cost and the possibility of incorporating the indicator dye only after the glass is formed, as the temperatures involved in the manufacturing process prevent addition of the indicator at an earlier step. The sol-gel process provides glass-like materials at room or moderate temperature in a huge variety of formats (powders, ultrathin films, spray coatings and bulk monoliths of virtually any shape and size). The most widespread method for preparing such materials involves polymerization of a metal alkoxide precursor through a sequence of hydrolysis and condensation reactions to form a highly cross-linked solid after sol and gel phase transition [33]. The feasibility of "low" (compared to usual glass) temperature preparation and processing allows for incorporation of almost any type of indicator molecules, biocatalysts and even whole cells, either by encapsulation or copolymerization [34]. Moreover, addition of organically modified metal alkoxides or organic polymer dopants to the sol phase yields organic–inorganic nanocomposite materials with a fine control of the polarity, homogeneity, surface nature, pore size, processability and other key features of the sol-gel product.

Although many elements of the periodic table can be used to obtain sol-gel materials from their $M(OR)_4$ precursors, M is commonly Si, Al or Ti, and R is an alkyl or aryl group. Organically modified silicates ("ormosils") or the more general term, organically modified ceramics ("ormocers"), contain also alkoxides of the general formula $R'_nM(OR)_{4-n}$ or $R'R''M(OR)_2$. If covalent incorporation of the indicator dye is sought, at least one ω-alkyltrialkoxisilane moiety is introduced in its structure and the indicator is copolymerized with the starting material(s) [35]. All the precursors are normally hydrolyzed, under acidic or basic catalysis, with a controlled amount of water and alcohols. The latter ensure homogeneity of the reaction mixture as most alkoxide precursors are not water-soluble. It is known that the sol-gel process is kinetically controlled, so that evolution continues long after gelation has taken place [29]. This fact very often provokes significant drift of the sensor response to the analyte as a consequence of either slow changes of the microenvironment around the indicator molecules, or variations in the permeability of the sol-gel to the analyte. Therefore, sol-gel materials are often annealed after preparation by rigorously controlled heating to avoid excessive pore contraction, which would hinder entrance of the analyte. Thermal annealing also helps to encapsulate indicator molecules that are not covalently bound to the gel network.

Optimization of a sol-gel matrix for optical sensing of a particular analyte is a rather complex procedure as the preparation and aging conditions will strongly influence the physicochemical features of the resulting solid material [29, 34, 36, 37]. Among the multiple factors that affect the sol-gel sensitive layer, the type and relative amounts of precursors, catalyst, water/M ratio, cosolvent(s), pH, temperature, indicator and modifier level, and casting procedure have to be considered [38, 39]. To further complicate the picture, some of these factors are mutually dependent. Small organic probe molecules such as 7-azaindole, dansyl, fluorescein, prodan, pyranine, pyrene, rhodamine 6G and Ru(II) complexes, among several others, have been used to investigate the structure of sol-gel nanocomposite materials [29] via environment-dependent variations in their luminescence (intensity, wavelength, anisotropy, decay kinetics, quenching, excimer formation or resonance energy transfer).

3.4 Zeolites

Zeolites are inorganic molecular sieves different from silica-based porous materials (including sol-gels) in that they possess structurally defined cages and channels of uniform size [40]. Optically active guest molecules of appropriate size can be incorporated within the cavities of these aluminosilicate materials by simple sorption from a solution or gas phase. Alternatively, exchange of ions originally present in the negatively charged molecular sieves is a useful method for loading cationic indicator dyes such as methylene blue, pyronine or thionine from their aqueous solutions (cation exchange from non-aqueous solvents is generally unsuccessful). Covalent grafting of dyes via chemical functionalization with silanes of the zeolite pore walls prevents indicator leaching. The microwave-assisted synthesis of zeolites is a relatively mild procedure that allows covalent tethering of dyes to the walls of the molecular sieves by previously attaching the chromophore to one of the building blocks of the host lattice (e.g., 3-aminotriethoxysilane) [40]. If an indicator dye, such as a metal polypyridyl coordination complex, exceeds the interconnecting channel diameter (but not the supercage size), it may still be incorporated by the "ship-in-the-bottle" strategy: the metal ion is exchanged in a first step and then the ligands are introduced to coordinate the former and form the target complex inside the zeolite [41].

3.5 Effects of the Supported Indicator Heterogeneity

Very often, the inorganic material in the form of micrometer- or nanometer-sized particles serves both as a support for the indicator dye and to reinforce the sensitive layer while the organic polymer provides a medium for analyte transport from the sample (liquid or gaseous). However, the latter not only performs as a passive matrix for embedding the indicator-loaded inorganic material and preventing its leaching into the sample medium but, more importantly, it confers the required selectivity to the chemical sensor. A case in point to illustrate this essential issue is the oxygen optosensor (Fig. 2) (see also Sect. 1). Many papers in the analytical chemistry literature describe highly O_2-sensitive layers made of cross-linked silicone films containing luminescent dyes (ruthenium complexes, porphyrins) adsorbed onto silica gel, CPG or sol-gel beads ([30] and references therein). Gas-phase O_2 measurements are not subject to many chemical interferences; however, oxygen optosensing in aqueous phase would be impossible in the real world without the organic polymer layer due to efficient luminescence (both fluorescence and phosphorescence) quenching of the indicator dye by various species present in the water (cations, anions), and fast biofilm formation [42].

The presence of inorganic particles in the sensitive layer complicates the quenching processes of luminescence-based sensors and provokes departure from the simple Stern–Volmer kinetics observed for homogeneous solutions of a luminophore (Eq. 1) [43]:

$$\frac{I_0}{I} = \frac{\tau_0}{\tau} = 1 + k_q \tau_0 [Q] \qquad (1)$$

where I is the luminescence intensity and τ is the luminescence lifetime in the presence of a quencher Q. The subscript 0 denotes absence of quencher and k_q is the bimolecular quenching rate constant. The product $k_q\tau_0$ is usually called the Stern–Volmer constant or K_{SV}. Equation 1 holds for purely *dynamic* quenching processes, i.e., the luminophore and the quencher are not associated before photoexcitation of the former.

The reasons behind a complex behavior in the relative emission intensity or lifetime vs quencher concentration plots (typically the quencher is the analyte species or a third party, the concentration of which depends on the analyte level [44]) are manifold and may be dependent on the nature of the inorganic support surface, its interaction with the organic polymer matrix, the

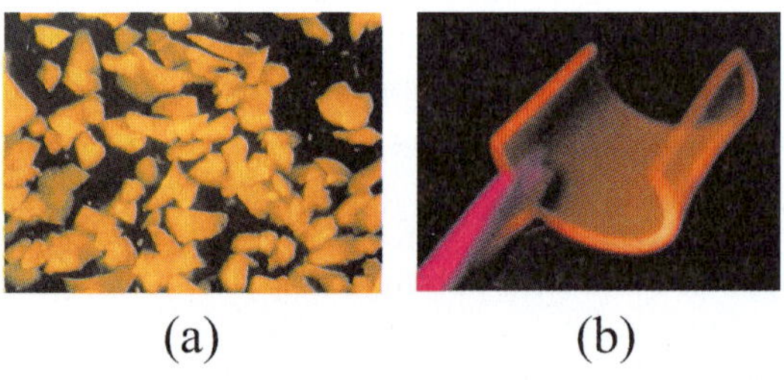

Fig. 2 Different orange-red luminescent layers for O_2 sensing. **a** Indicator layer containing the tris(4,7-diphenyl-1,10-phenanthroline)ruthenium(II) dye adsorbed on CPG particles embedded in a poly(dimethylsiloxane) film [8]. **b** Perfluorinated (Nafion) ionomer membrane doped with the same cationic dye. Although the former appears as a homogeneous material, even under the optical microscope, the luminescent indicator is actually sitting down in various domains as evidenced by its emission kinetics profile [9]

chemical determinand species, the aggregation of the dye and the particular quenching mechanism (electron, energy or proton transfer [44]).

A detailed discussion of the various possible situations is beyond the scope of this chapter; however, the optosensor scientist must take into account the following effects in order to rationalize the device response (where applicable):

(i) The intrinsic *heterogeneity* of the dyed particles at the microscopic or nanoscopic level [30]. Immobilization of the fluorophore onto a solid support yields species located in different domains, each of them displaying a different Stern–Volmer constant (possibly including also $K_{SV} = 0$) due to restricted accessibility of the quencher species (Fig. 3). Aggregation of the indicator molecules and variation in their radiative or non-radiative deactivation constants are also possible, depending on the dye concentration and nature (e.g., polycyclic aromatic hydrocarbon molecules tend to aggregate upon adsorption). Partition of the adsorbed indicator into the organic polymer film also needs to be considered. As a case in point, different multi-site models have been developed to account for the typical downward curved Stern–Volmer plots observed for the oxygen quenching of inorganic solid-supported luminophores embedded into organic polymer films [7]. The (adsorbed) photoexcited dye can be quenched by direct collision with an oxygen molecule diffusing in the gas phase or, for dyed particles dispersed into an organic polymer, diffusing through the latter. In addition to or alternatively, the oxygen molecules may (reversibly) adsorb onto the surface of the dyed particle and then diffuse in a 2-D space to find the photoexcited dye [30]. The direct quenching mechanism is expected to occur at a lower rate than

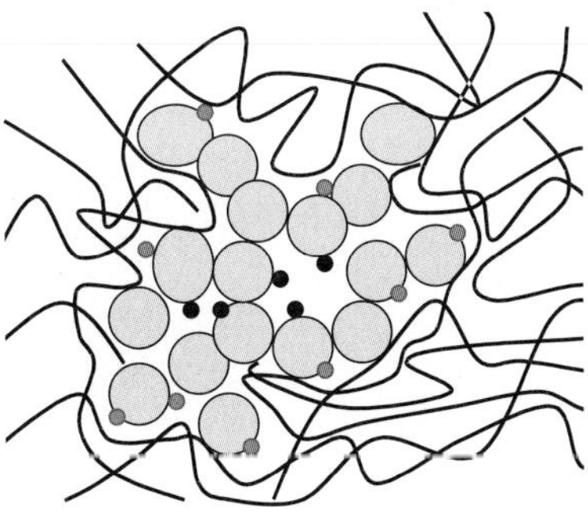

Fig. 3 Example of the microdomains created by embedding luminescent indicator-loaded silica gel nanoparticles into poly(dimethylsiloxane) before cross-linking to fabricate films for oxygen optosensing. The *largest circles* represent the silica particles; the *black lines* represent the PDMS polymer chains, while the *smallest circles* represent the indicator dye molecules (the *black circles* depict those located in the organic polymer-free silica regions and the *gray circles* stand for those adsorbed on silica regions in contact with the PDMS)

the same process in homogeneous solution due to protection of the dye by the solid surface (leading to a decrease in the collision frequency factor). However, the adsorption-mediated O_2 quenching of the immobilized dye may be faster than the rate of the corresponding 3D process due to both a concentration effect and the need to explore a smaller space by the mobile quencher before collision occurs (assuming that the photoexcited dye does not move significantly during its lifetime). Such an enhancement or decrease of the quenching rates have been well documented for luminophores bound ionically to polyelectrolytes (e.g., DNA) and spheroidal oligomers (see, for instance, [45] and [46]), depending on whether the quencher also binds or not to the polymer surface.

(ii) The contribution of *scattered light* that reaches the detector (at the monitored emission wavelength) from the powerful excitation source, even when using wavelength selectors. This effect is particularly dramatic for analyte-permeable sensitive layers containing the fluorophore immobilized on highly reflective inorganic particles (e.g., SiO_2, TiO_2 or Al_2O_3) and provides a constant signal background on top of the luminescence quenching phenomenon that is not considered in Eq. 1.

(iii) The feasibility of *distance-* or *time-dependent* quenching rate "constants" [47]. Equation 1 assumes that every photoexcited indicator molecule reacts with the analyte (or an analyte-dependent relay species) at exactly the same rate. This is the natural situation for excited state quenching in fluid solution but it may not be the case for a surface-bound indicator dye [48]. The coverage fraction (indicator loading) is expected to play a key role in the sensor performance when such situations take place [49].

(iv) The effect of reduced or fractal *dimensionality* (i.e., $< 3D$) on the quenching reaction space. Fluorescence deactivation upon adsorption of the quencher onto a polymer material may take place by diffusion of the species within a 2D surface, or even a confined monodimensional space (e.g., the narrow channels within a zeolite [50]). Sometimes semiempirical fractal models are used to fit the excited state quenching kinetics suggesting that the reaction is affected by the geometrical features of the solid [51].

(v) The possibility of acid/base reactions in the electronic excited state that yield luminophore species with different degree of protonation, each of them displaying distinct emission features. The complexity of the decay kinetics will be dependent on the excited state lifetimes of the different species compared to their proton transfer rates [52] and the relative concentration of the various excited species. The latter is a function of the acidity of the indicator microenvironment and, when non-reversible proton transfer in the excited state occurs, it is also a function of the pH regulator nature and concentration. If the solid support contains acidic or basic surface chemical groups, they may also influence the proton transfer to the bound indicator dye (e.g., the intrinsic surface silanol groups of silica or glass, or the carboxylic or amino moieties of functionalized silica or glass).

3.6 Some Examples

The above-mentioned issues have been discussed in numerous papers; a comprehensive collection of all of them would take too much space, so only a few illustrative (recent) examples are outlined.

The luminescence of tris(2,2′-bipyridine)ruthenium(II) cation, RB_3^{2+}, adsorbed on mesoporous silica was investigated by Ogawa et al. as a function of the dye loading and hydration degree of the solid support [53]. Upon dehydration of the dyed silica the adsorbed RB_3^{2+} ions aggregate and a strong luminescence self-quenching occurs. Re-hydration of the support restores the luminescence intensity because the dye molecules redistribute themselves into the pores. Obviously, the sensitivity of the indicator dye towards an external analyte (quencher) will be strongly dependent on the loading level of the inorganic particles.

A combined dynamic and static quenching mechanism has been proved in an oxygen sensor fabricated by encapsulating erythrosin B in a silica sol-gel/fluoropolymer thin film composite [54]. The phosphorescence intensity Stern–Volmer plot is linear at low dissolved oxygen concentrations ($< 10\ \mathrm{mg\,L^{-1}}$, $I_0/I = 2$) but shows a pronounced upwards bending at higher O_2 levels ($I_0/I = 80$ at $40\ \mathrm{mg\,L^{-1}}$). As the formation of a ground state complex between the dye and the oxygen molecule is highly improbable, the authors propose a more likely model based on the sphere-of-action quenching mechanism. According to these mechanism, the photoexcited dye is surrounded by so many quencher molecules that the probability of deactivation is unity within a certain distance [55]. The authors postulate that such a situation originates from microdomains in the composite material: the polar dye molecules reside in the sol-gel regions while the (hydrophobic) O_2 tends to accumulate at the interface of the perfluorinated polymer domains that separate the inorganic clusters. Nevertheless, by using hybrid xerogels that are composed of alkyl and perfluoroalkyl organically modified silicates doped with tris(4,7-diphenyl-1,10-phenanthroline)ruthenium(II), RDP_3^{2+}, Bukowski et al. have manufactured very sensitive O_2 sensors ($I_0/I = 35$ under pure oxygen) displaying linear Stern–Volmer plots and exponential luminescence decay profiles [56]. Due to the high diffusivity of O_2 within these hybrid xerogels, the steep slope of the plot is postulated to arise from an enhancement in k_q and not from a significant change in τ_0.

By preparing polycyclic aromatic hydrocarbon nanocrystals grown in porous silicate films, Botzung-Appert et al. [57] have prepared potential fluorescent sensors based on electronic energy transfer quenching to an acceptor adsorbed on the surface of the nanocomposite. Their first report is based on rubrene or tetracene nanocrystals in a sol-gel matrix of mixed alkoxy- and alkylalkoxysilanes, and methylene blue (MB) as acceptor. The authors suggest that the observed monoexponential decay profiles are a consequence of the dye aggregation in the nanocrystals and the week interactions between the polyaromatic molecules located at the surface of the crystal and the silicate network. The color change of MB in the presence of reducing sugars (redox indicator) has been used to make a proof-of-principle fluorescent sensor for glucose. Analyte-sensitive energy acceptors might replace methylene blue to fabricate different optical sensors based on such principle.

Aminofluorescein has been immobilized in various organically modified silicates, obtained by copolymerization of methyltrimethoxysilane, phenyltrimethoxysilane, dimethyldimethoxysilane and diphenyldimethoxysilane with tetramethoxysilane at different ratios, to fabricate ammonia-sensing films [58]. The micropolarity of the films was probed with the $E_T(30)$ Reichardt's solvatochromic dye. The dynamic range and response time for ammonia were found to be strongly dependent on the nature and contents of organosilicon precursors. Unlike fluorescein-based pH and ammonia optosensors, the authors propose that their sensor fluorescence enhancement in the presence of ammonia is due to reduction of the aminofluorescein photoinduced electron transfer self-quenching via hydrogen bonding of the analyte to the indicator molecules within the modified silicates.

Eriochrome Cyanine R entrapped in methyltriethoxysilane-modified silica sol-gel materials serves as a pH/Cu^{2+} optical sensor [59]. Incorporation into polyisobutylene of the alkyl-modified silica ground to fine particles decreases 300-fold the response time to Cu^{2+} of the corresponding bulk (monolith) inorganic material and eliminates cracking during the gelation process. Optimum sensitivity to this cation is achieved for gels containing 10% methyltriethoxysilane. The sensor features are rationalized in terms of the pore size distribution and the structural effects provoked by the organic modifier.

A smart way to avoid leaching of indicator dyes encapsulated into porous sol-gel supports is to covalently bind them to polyamidoamine (PAMAM) dendrimers [60]. The molecular weight of this spheroidal oligomer (14 215 Da) suppresses all detectable washing from water-dip monoliths prepared from tetramethoxysilane and PAMAM-erythrosin isothiocyanate conjugate. Interestingly, a prior study demonstrated that significant leaching occurs when the when the same dye is conjugated to 70 kDa dextran. It seems that the successful strategy requires a macromolecule having an *organized* secondary structure to eliminate indicator leaching.

4 Biosensing Layers

According to a recently proposed IUPAC definition [61]:"A biosensor is a self-contained integrated device which is capable of providing specific quantitative or semi-quantitative analytical information using a biological recognition element (biochemical receptor) which is in direct spatial contact with a transducer element. A biosensor should be clearly distinguished from a *bioanalytical system*, which requires additional processing steps, such as reagent addition. Furthermore, a biosensor should be distinguished from a *bioprobe* which is either disposable after one measurement, i.e. single use, or unable to continuously monitor the analyte concentration." Biosensors that include transducers based on integrated circuit microchips are known as biochips [62].

The recognition process may be based on either a *catalytic* reaction (metabolism biosensors) or a *binding* process (affinity ligand-based biosensors). The former are kinetic devices based on the achievement of a steady state concentration of a transducer-detectable species. The biocatalyst can be an isolated enzyme, a microorganism, a subcellular organelle or a tissue slice. In the second group, the bioreceptor molecule binds the analyte "irreversibly" and non-catalytically, and the binding event originates a physicochemical change that will be measured by the transducer. An antibody, a nucleic acid or a hormonal receptor are the recognition elements included in this group.

It is well known that the choice of an appropriate support and immobilization procedure is a key step in the fabrication of optical biosensors and will largely determine the characteristics of these devices. Although a great number of immobilization techniques for biological elements have been developed, there is not one general method recommended for all cases. Many aspects have to be considered, such as the retention of a significant portion of the activity of the biomolecules after immobilization, the stability of the immobilized biological component, the ease of preparation and regeneration of the biosensing layer and the cost of the process, among others. Ideally, the immobilization procedure should led to: (i) an increase in the affinity of the substrate by the biological receptor, (ii) a broadening of the working pH range, and (iii) a reduction of the microbial contamination. Methods for immobilization of biomolecules are traditionally classified into two main classes [63–66]:

- *Physical* methods that involve the supporting of a biological element in any way that is not depending on covalent bond formation (e.g., adsorption, entrapment, microencapsulation). These procedures are simple and in most cases the biocatalyst remains unchanged.
- *Chemical* methods that involve the formation of at least one covalent bond (attachment of enzyme to water-insoluble functionalized polymers, intermolecular cross-linking of enzyme molecules using multifunctional

reagents). This immobilization is usually irreversible and the biocatalyst is not readily desorbed.

Some of the advantages and disadvantages of the immobilization methods described above are summarized in Table 1. There is a great variety of supports available for the immobilization of the biomolecules that differ in size, density, porosity or shape. These materials can be classified in two categories:

- *Inorganic supports* such as silica gel, alumina, controlled pore glass, sol-gel glass, etc.
- *Organic supports*. These can be further divided into:
  - Natural polymers: polysaccharides (celullose, starch, pectins, dextrans, agar, agarose, alginate, chitine, chitosan, etc.) and fibrous proteins (collagen, keratine, etc.).
  - Synthetic polymers: polyolefines (polystyrene, polystyrenedivinylbenzene), acrylic polymers (polyacrylate, polyacrylamide, polymethacrylate, etc.) and others such as polyvinyl alcohols, polyvinyl chloride, polytetrafluoroethylene, polyamides, polyurethane, silicone, etc.

Table 1 Advantages and disadvantages of the main immobilization methods of the recognition element of biosensors

Immobilization method	Advantages	Disadvantages
Adsorption (physical, ionic binding, chelation)	Simple Mild binding conditions Low biomolecule changes	Weak binding force Binding stability highly dependent on pH, solvent and temperature
Entrapment (gel entrapment, microencapsulation)	Mild procedure Intermediate binding forces Possible protection from microbiological attack High stability	Possible loss of biomolecule activity by leakage Denaturalization by free radicals formed in the process
Cross-linking	High stability Strong binding force Used for stabilization of proteins covalently bound to a solid support Possible protection from microbiological attack	Reaction difficult to control High amounts of biomolecule are required Possible loss of protein activity
Covalent binding	Stable binding Low leakage High stability	Time consuming and laborious reactions Activity losses if the groups required for the biological activity are involved in the binding

4.1 Enzyme-Based Biosensors

4.1.1 Physical Entrapment

From the experimental point of view, physical entrapment is an easy way to immobilize enzymes and basically does not affect the quaternary structure of the protein or the active site of the enzyme. Furthermore, it requires relatively small amounts of biomolecule to produce active enough sensitive layers, and can shield biomaterials from outer destructive factors (e.g., microorganisms or proteases). However, low molecular weight substances are not efficiently prevented from diffusing out of the gel and it has been shown that adsorption does not always give a density of proteins as high as covalent attachment [67]. In earlier models, an enzyme solution was deposited on a solid support and kept at the sensing tip of the biosensor by means of a dialysis or ultrafiltration membrane. This is a simple approach although the robustness and response time of the devices can be a limitation for some applications.

Gel entrapment is another method for physically retaining biomolecules in a solid support [68–73]. The method is not only suitable for enzymes but also for macromolecules, organelles and cells [74]. The biocatalyst is suspended in a monomer solution, which is then polymerized, and the enzymes are entrapped within the polymer lattice during the cross-linking process. In this case, unlike covalent binding and cross-linking, the enzyme itself does not bind to the gel matrix. The polymer matrix can be formed on the surface of the fiber itself or polymerized onto a transparent support (for instance, glass) that is then coupled to the fiber. Polyacrylamide, silicone rubber, polyvinyl alcohol or polyurethanes have been applied to this aim, and the sol-gel method for the synthesis of inorganic materials is also widespread.

The sol-gel procedure enables encapsulation of enzymes in optically transparent, porous silicate matrices, under mild room-temperature conditions. The small pores prevent microbial degradation and, due to the biomolecule size, they will not diffuse out of the polymer network. The physical encapsulation avoids self-aggregation effects as well as protein unfolding and denaturalization. At the same time, the catalytic activity is maintained as the enzymes are able to react with small substrates that can transfer across or within the support, assuring continuous transformations [75].

The enzyme, normally in solution, is added and after aging either a thin film or a gel can be formed with the encapsulated molecules. Depending upon the acid used, the solution pH and other conditions, gelation can take from 1 min up to several days. The sol-gel matrix shows many advantages: (i) the ability to entrap a large amount of enzyme, (ii) retention of the enzymatic activity due to the sufficiently mild conditions of the sol-gel process,

(iii) chemical inertness and optical transparency, and (iv) the possibility of preparing films or monoliths [76]. Unfortunately, it has also been shown that some enzymes lose their activity upon encapsulation in the matrix and sometimes leach out from the sol-gel matrix [76]. Several sol-gel enzyme-based optical biosensors have been described in the literature for different analytes (e.g., [77–79]).

Enzyme micro-encapsulation is another alternative for sensor development, although in most cases preparation of the microcapsules may require extremely well-controlled conditions. Two procedures have usually been applied to microcapsule preparation, namely interfacial polymerization and liquid drying [80]. Polyamide, collodion (cellulose nitrate), ethylcellulose, cellulose acetate butyrate or silicone polymers have been employed for preparation of permanent microcapsules. One advantage of this method is the double specificity attributed to the presence of both the enzyme and the semipermeable membrane. It also allows the simultaneous immobilization of many enzymes in a single step, and the contact area between the substrate and the catalyst is large. However, the need for high protein concentration and the restriction to low molecular weight substrates are the important limitations to this approach.

4.1.2 Covalent Binding

Enzyme molecules may be cross-linked either with themselves or co-cross-linked with an inactive protein (such as albumin, either bovine or human) using multifunctional low molecular weight reagents. Success of this procedure requires a strict control of the experimental factors, such as the enzyme and multifunctional agent concentration, pH, ionic strength, temperature and reaction time. The mechanical stability of these layers is usually limited for sensors and, in general, it is preferred to combine cross-linking with other methods such as enzyme adsorption or entrapment in different supports such as silica, cellophane or active carbon. The most common reagent used for cross-linking is glutaraldehyde.

The covalent attachment of enzymes to water-insoluble carriers is usually the preferred immobilization method for sensor manufacturing. Obviously, the selected procedure should avoid the loss of enzymatic activity and keep the accessibility of the binding site to the substrate molecules. Unfortunately, this is usually not the case and due to the severe conditions of many of these procedures, major activity losses and/or changes on the substrate selectivity are produced during immobilization. Some authors have pointed out that the enzyme activity decreases approximately one fifth per formed bond [66].

The methods for covalent enzyme immobilization have been reviewed extensively in the literature [64, 65, 80, 81]. The functional groups of the enzyme involved in the chemical bonding can be the *N*-terminal and ε-amino groups (from lysine) as well as the carboxyl (aspartic or glutamic acid), sulfhydryl (cysteine), hydroxyl (serine, threonine), indole (tryptophan), imidazole (hystidine) and phenolic (tyrosine) functions.

Natural supports (agarose, dextran, cellulose, porous glass, silica, the optical fiber itself or alumina) and synthetic resins (acrylamide-based polymers, methacrylic acid-based polymers, maleic anhydride-based polymers, styrene-based polymers or nylon, to name a few) have been applied for covalent attachment of enzymes. These materials must display a high biocatalyst binding capacity (as the linearity and the limit of detection of the sensing layers will be influenced by this value), good mechanical and chemical stability, low cost, and ease of preparation.

4.2
Whole-Cell Biosensors

The main advantages of immobilizing whole cells over the isolated enzymes are: (i) cells are more resistant to changes in pH or temperature and (ii) a single cell can contain all the enzymes and cofactors needed for bioconversion of the analyte [82]. Most of the whole cell-based optical biosensors used in toxicity or bioremediation studies [83, 84] employ microorganisms [85–87], animal [88], plant tissues [89, 90] or cell receptors [91] as biorecognition elements.

The two most popular methods for immobilization of whole-cells are entrapment in either synthetic or naturally occurring polymer matrices (e.g., alginate, agarose, gelatin, polyacrylamide, polyvinyl alcohol) [85, 86, 92, 93] and adsorption onto the surface of dialysis or filtration membranes (e.g., cellulose, glass fiber, etc.) [89–91]. Temperature, pH and ionic strength play an important role in maintaining the attachment, thus sometimes cell loss may occur during the assay runs. Moreover, the low physical and chemical stability of some of the aforementioned immobilization materials avoid their use in continuous measurements.

Covalent attachment, through the binding of a reactive moiety on the cell such as an amine or carboxyl group, to a reactive group on the solid support, via bifunctional cross-linkers, can be an alternative to avoid some of these problems. In this regard, organosilane-derivatized glass has been examined for its ability to covalently immobilize bacterial cells [94]. Siran porous glass beads have a large surface area, are chemically and biologically inert, can be sterilized, and have superior physical properties compared to other potentially useful immobilization materials. Covalent attachment chemistries using aminosilane-Siran/carbodiimide and mercaptosilane-Siran/*N*-succinimidyl 4-maleimidobutyrate (GMBS) have provided a threefold higher density of cells than that observed with the aminosilane-Siran/glutaraldehyde method. Although glutaraldehyde is a universal cross-linker, there are concerns about its effects on the cells and proteins, as well as with batch-to-batch reproducibility and polymerization of the stock glutaraldehyde.

4.3 Affinity Ligand-Based Biosensors

4.3.1 Immunosensors

Firm and specific interactions between antigens and antibodies allow their application in biosensor technology by employing either immobilized antibodies or antigens. The selection of the solid support for immobilization is also a key parameter in the immunosensor design. The solid support must be easily activated to allow antibody attachment, the pore sizes should be large considering the size of the antibodies and it should be hydrophilic in order to avoid any non-specific interactions [95]. Traditional supports used in immunosensing include silica, agarose, glass beads, cellulose and synthetic polymers. In many cases, the surface of a properly modified fused-silica fiber is used as a support in evanescent-wave immunosensors [96–98].

Ideally, the immobilization procedure should retain the biospecific activity of antibodies. Numerous coupling strategies have been developed for supporting antibodies on different solid surfaces through the formation of defined linkages, in which glutaraldehyde, carbodiimide and other reagents such as succinimide ester, maleimide and periodate are widely used. However, problems associated with the loss of biological activity of the antibodies upon immobilization have been noticed in many cases [99]. Antibodies can be directly immobilized onto a solid support via the $F(ab')_2$ or Fc fragments. Usual methods of covalent immobilization result in binding of all three fragments in a random way (Fig. 4). The result of such random orientation is a decrease or complete loss of their antigen-binding properties. Nevertheless, some approaches can be used to increase the yield of active antibodies. Methods for antibody immobilization can be divided into three main categories [95–100].

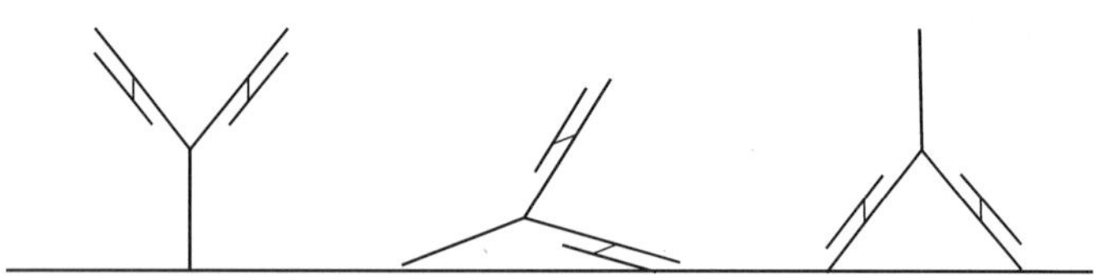

(a) (b) (c)

Fig. 4 Scheme of the different situations that can occur during IgG antibody immobilization by random coupling procedures: **a** fully active antibody, **b** partially active antibody and **c** inactive antibody

4.3.1.1
Entrapment

The sol-gel technique has recently been used successfully for entrapping antibodies against PAHs, 1-nitropyrene, *s*-triazines and TNT in the pores of a hydrophilic glass matrix [96]. Activation and bonding are no longer necessary and the immobilization is carried out under mild conditions, so that one can expect that the antibodies retain their affinity and specificity. The entrapment is a simple and rapid two-step procedure in which hydrolysis is followed by polymerization of tetramethoxysilane after addition of the antibody solution. Leaching of antibodies can be a problem in sol-gel techniques because of the high porosity of the matrix and the fact that antibodies are not covalently bound to it. In some cases, aging of the material has also been shown to affect the response of these devices [101].

4.3.1.2
Covalent, Non-Site-Directed Attachment

A variety of reagents have been used to covalently attach antibodies onto a solid phase, upon derivatization of the solid support and/or the antibody:

- *Soluble activators*, such as carbodiimide and succinimide, that allow direct chemical attachment of an amino group in one element to a carboxylic group in the other
- *Soluble bifunctional cross-linkers*, e.g., glutaraldehyde and maleic anhydride, that bridge between two amine groups. Alternatively, heterobifunctional agents, such as *N*-succinimidyl 4-maleimidobutyrate (GMBS) can bind thiol groups on the support and amine groups in the protein
- *Solid-phase-bound activators* include polymers carrying active groups, such as epoxy derivatives, that allow covalent attachment to $-NH_2$, $-OH$ or $-SH$ groups by gentle mixing with the protein

As mentioned previously, the main disadvantage of these immobilization methods is the non-specific nature of the binding reaction, which often results in multiple site attachments and multiple orientations. As a consequence, the immobilized antibodies show a decrease or complete loss of their antigen-binding capacity. To eliminate these drawbacks, several approaches leading to oriented antibody coupling have been developed over the last few years [99].

4.3.1.3
Covalent, Site-Directed Attachment

In this case, immobilization is made in such a way that the antigen-combining sites of the antibody are not involved, thereby increasing the fraction of fully functional receptor. Several procedures for covalent, site-directed binding have been described:

- *Direct attachment to solid phase.* The most successful method relies on chemical oxidation with periodate of the carbohydrate moieties linked mainly to the CH_2 domain of the Fc fragments (Fig. 5a). Under mild conditions, the hydroxyl groups of the carbohydrates can be oxidized to aldehyde, which are then chemically bound to amino- or hydrazide-activated solid supports such as agarose, glass, silica, acrylic-based copolymers or cellulose by forming covalent imine or hydrazone bonds, respectively. The solid phase produced in this way is highly functional, because the antibody-combining sites remain free to interact with the antigen.
- *Attachment via affinity interactions.* These methods are based on the high affinity of Fc receptors, such as protein A (PrA), protein G (PrG) or recombinant protein A/G, for the amino groups in the Fc region of antibodies. In these procedures, the Fc receptor is immobilized on a solid support and is the key element for the oriented immobilization of the antibody (Fig. 5b). However, not all Fc receptors have the same affinity to all IgG isotypes. PrA has been successfully used to bind the Fc portion of antibodies from many mammalian species, except for goat, sheep, cow and horse. PrG has some advantages over protein A, as it reacts with more IgG isotypes and to a lower extent with other immunoglobulins, such as human IgM and IgA. How-

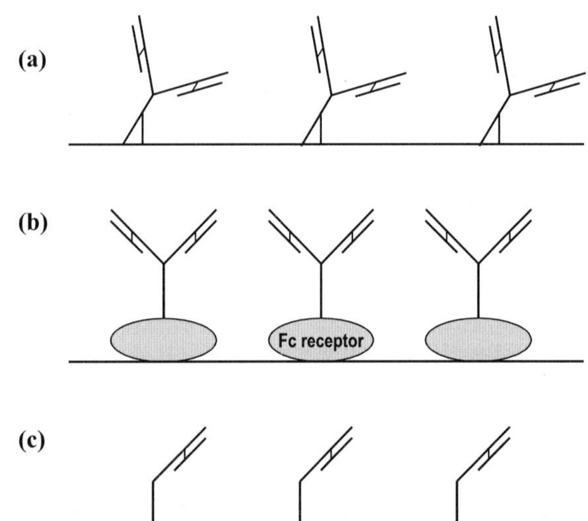

Fig. 5 Summary of the different oriented IgG antibody immobilization methods: **a** antibody is coupled to the solid support via an oxidized carbohydrate moiety in the CH_2 domain of its Fc fragment; **b** antibody binds to Fc receptors on solid supports; **c** monovalent Fab′ fragment is bound to an insoluble support via a sulfhydryl group in the *C*-terminal region of the fragment

ever, it does not bind strongly to several IgGs with which protein A reacts well. The third generation of IgG-binding Fc receptor is a recombinant protein A/G that contains four Fc binding domains from protein A and two from protein G, combining the IgG binding profiles of both proteins. All three Fc receptors have been successfully immobilized on different affinity gels to bind antibodies, and several commercial supports are now available on the market [102]. A disadvantage of these approaches is the possibility of a cleavage of the Fc receptor-antibody complex during elution of the antigens. This can be overcome with the use of a bifunctional cross-linker reagent (e.g., dimethylsuberimidate), which ensures the stability of the immobilized antibody.

The avidin-biotin interaction has also been used to immobilize antibodies and proteins, especially in commercial systems based on surface plasmon resonance (SPR) measurements (e.g., the BIAcore). The extraordinary affinity ($K_d \sim 10^{-15}$ M) of avidin (or its bacterial relative, streptavidin) for the vitamin biotin is the basis of this immobilization procedure. A solid support (e.g., glass beads, sensor chip, optical fiber) covered with avidin can be used as an activated carrier for a very sturdy immobilization of previously biotinylated antibodies. In spite of the many methods for biotinylating proteins described in the literature, the use of biotinyl *N*-hydroxysuccinimide ester (BNHS) and similar derivatives, remains the most useful [65].

- *Immobilization of antibody fragments via sulfhydryl groups.* This can be achieved by digesting the $F(ab')_2$ fragments with pepsin, followed by reduction of the disulfide bond between the monovalent Fab′ fragments. Alternatively, Fab′ fragments produced by molecular genetic techniques can be immobilized directly. The sulfhydryl group formed in the *C*-terminal region of the fragment can be coupled to a derivatized solid support in such a way that its antigen binding site remains available for interacting with the antigen (Fig. 5c).

Those procedures that allow an oriented immobilization of intact antibodies, or antibody fragments, on solid supports have been the most applied in the fabrication of different immunosensors.

4.3.2
Protein Microarrays

Microarrays can be divided in two general types: (i) biochips, generated from biochemical or biological components (such as enzymes, antibodies, DNA, cell or tissue) and (ii) chemical microarrays that consist of arrays of small compounds (sugars, organic molecules or peptides) [103]. Preparation of these arrays can be tackled by in situ synthesis or by spotting. The second approach is more versatile and can be used to immobilize almost any molecule. Photolithography, mechanical spotting and ink-jet applications are the main methods for microarray spotting. Electrospraying has also been used for protein deposition in a highly reproducible way [104].

Although still fairly new, peptide and protein-based microarrays hold a great potential for the future of proteomics, enabling assay of a wide range of biochemical interactions (e.g., protein–protein, protein–lipid, protein–nucleic acid and enzyme–substrate), as well as drug and drug target identification [105, 106]. One of the current research areas in this field is the development of novel surface and linkage chemistry for immobilization of peptides and proteins. While protein microarray technology borrows extensively from DNA microarray analysis, it is still in its infancy due to fundamental differences between proteins and DNA. Unlike DNA, proteins are chemically and physically heterogeneous, have a three-dimensional structure that is critical to their function and most of them will absorb non-specifically to commonly used substrate materials.

Surface chemistry is a key technology for protein microarray development. The supports used for protein immobilization have to fulfil some important requirements: they must provide good quality spots, low background, simplicity of manipulation and compatibility with detection systems. An ideal surface or immobilization procedure for all proteins and applications does not exist; however current methods are more than adequate for many applications. Basic strategies for protein immobilization consider covalent versus non-covalent and oriented versus random attachment, as well as the nature of the surface itself [106]. It has been demonstrated that the specific orientation of immobilized antibody ("capture agents") consistently increases the analyte-binding capacity of the surfaces, with up to 10-fold improvement over surfaces with randomly oriented capture agents [107].

Most supports used for the immobilization of microarrays are made of glass, but plastics and polymer membranes can also be used [108]. The immobilization of proteins is often carried out using non-covalent protein surface interactions with hydrophobic (nitrocellulose, polystyrene) or positively charged (polylysine, aminosilane) surfaces. Covalent attachment using a variety of chemically activated surfaces can also be used [109]. Several groups have developed reactive surfaces on glass that can be covalently cross-linked to proteins [105]. In general, a bifunctional silane linker is used to form a self-assembled monolayer (SAM), which has one functional group that reacts with the hydroxyl groups on the glass surface, and another one free that can either react directly with primary amine groups of proteins (e.g., aldehyde or epoxy groups) or can be further chemically modified to reach maximum specificity. Gold-coated glass surfaces are another variation. To form a SAM on a gold surface, a bifunctional thio-alkylene is usually used; its SH-group reacts with gold and another free one reacts with capture molecules.

The glass slides can be covered with a polymer layer, such as polyacrylamide, agarose or gelatin, that provides a porous structure and increases the efficiency of protein immobilization [110].

Probably the best means of protein attachment is through highly specific affinity interactions (e.g., streptavidin-biotin or His-tag-nickel-chelates) [111]. Proteins fused with a high-affinity tag, at their amino or carboxyl terminus, are linked to the surface of the chip via this tag, and hence all the attached proteins should orient uniformly away from the surface. Using this method, immobilized proteins are more likely to remain in their native conformation, while the analytes have easier access to the active sites of the proteins.

Polycarbonate compact discs have also been used as supports for microarray development [112]. The microarray is generated using an inkjet applicator that employs an electric current to dispense monodisperse droplets containing antibodies onto the disk. The sensing reaction is based on competitive inhibition immunoassays using fluorescent antibodies and final readout is accomplished using a fluorescence scanner.

Fiber-optic microarray biosensors have been prepared by patterning an array of wells at the tip of an optical imaging fiber bundle [113]. Micro-sized sensors can be prepared by covalently immobilizing indicating chemistries on the surface of microbeads (3 μm). Oligonucleotides, proteins or chemical indicators can be bound to the beads. The microwells are filled just with one bead and a unique encoding system has been developed using dyes in different ratios. The optical signal is registered on a charge-coupled device (CCD) detector. In this approach the packing density of the array is very high although the decoding step may be an important drawback.

Radioactivity, chemiluminescence or fluorescence have traditionally been used to monitor signals with protein arrays. The use of fluorescent dyes allows for increased sensitivity of detection, the ability to multiplex detection of several targets, and easy interfacing with detection instrumentation and automation. Surface plasmon resonance spectroscopy is another alternative, but successful development and commercialization has not yet been achieved.

4.3.3 DNA Biosensors

During the last few years a substantial amount of work has been devoted to immobilization of DNA oligonucleotides on the surface of an optical substrate (e.g., optical fibers and glass slides) for the development of DNA biosensors and microarrays. Nucleic acids are a natural choice as recognition elements, due to their intrinsic molecular base pairing ability, high degree of selectivity and stability. Particular aspects of immobilization that have received significant attention include the efficiency of DNA coupling, the stability of the attachment and the steric accessibility of oligonucleotides for hybridization. Another factor to take into account is the density of immobilized oligonucleotides onto the solid surface. Thus, a low surface loading originates poor analytical signals, whereas a high surface loading may prevent the interaction of immobilized single-stranded DNA with complementary strands by steric hindrance, and thereby also decrease the analytical signal.

DNA oligonucleotides can be immobilized on sensor surfaces using methods of attachment similar to those used for enzyme-based biosensors, including adsorption, cross-linking, encapsulation, avidin-biotin interactions and covalent binding [81, 114]. Optical fibers provide an ideal platform for single target oligonucleotide detection, combining the selectivity of nucleic acid recognition with fiber-optic transduction mechanisms. The oligonucleotides can be either immobilized on the distal end of the fiber core or coupled to portions of the fiber core where the clad has been removed. Single-stranded DNA probes are covalently bound to the surface via hydrophilic linkers (e.g hexaethylene or pentaethylene glycol) [115, 116]. Another possibility is the fabrication of fiber-optic DNA microarrays, by selectively etching the fiber core and forming an array of microwells that can be filled with oligonucleotide-functionalized silica microspheres [116]. This microarray configuration provide certain advantages, such as the reduced array size that enables smaller volumes to be interrogated, and the faster responses compared to most array-based systems. Moreover, fiber-optic microsphere-based microarrays can be regenerated for more than 100 hybridizations using high temperatures or organic denaturants without significant signal degradation [117].

Another type of microsphere-based array employs a different approach by changing the nucleic acid probe design. Detection using molecular beacons (MBe) [118] and aptamers [119] provide another level of nucleic acid interactions not established with other platforms. MBe probes have been adapted to microarray analysis in order to avoid some limitations of optical DNA biosensors, such as the need for labeled targets, intercalation reagents or competitive assays. A MBe is a single-stranded oligonucleotide containing a reporter fluorophore that is quenched by an adjoining quenching molecule via fluorescence resonance energy transfer (FRET). The quenching molecule can be either a non-fluorescent species that non-radiatively captures the reporter fluorophore energy, or another fluorophore with an excitation wavelength overlapping the emission wavelength of the reporter fluorophore. In the absence of target molecules, the quencher and the reporter fluorophore are close to one another due to the self-complementary DNA sequences or stem structures present in the molecular beacon structure. When proximal, the fluorescence energy is transferred from the MBe reporter fluorophore and a low signal is measured. The probe sequence connecting the two self-complementary stems is termed a loop sequence. Interactions between the MBe loop sequence and a complementary DNA target cause a conformational change that forces the stem apart leading to restoration of fluorescence. Fluorescein and 4-(4-dimethylaminophenylazo) benzoic acid can be used as universal fluorophore and quencher, respectively.

Biotinylated MBes have been designed and immobilized onto the core surface of an optical fiber by means of biotin-streptavidin interactions [120]. However, the adsorption method for streptavidin immobilization onto the optical fibers has shown to be not very efficient, so that new strategies to immobilize biotinylated MBes through bridge structures have been developed [118]. Thus, MBes were biotinylated at the quencher side of the stem and then linked to a biotinylated glass through streptavidin, which acts as a bridge between the MBe and the glass matrix. The bridge immobilization method originates faster and more stable responses, as compared to those cases in which direct immobilization of streptavidin onto the glass support is made. In fact, the bridge structure reduces steric hindrances allowing the target DNA to easily bind the immobilized MBe after reaching the sensor surface. More detailed information on DNA biosensors can be found in Massey and Krull, this volume.

5
Conclusions and Outlook

The variety of indicator and biochemical receptor supports represents both a problem to select the optimum one and an advantage to improve the sensitivity and selectivity of optical chemical sensors. Whether colorimetric or luminescent, the very best indicator dye is useless if a proper polymer material is not found for sensing the target analyte. Although, unfortunately, many reports and publications are based on a few days of continuous measurements or a few weeks of intermittent determinations, indicator leaching and photobleaching are the most important issues in long-term monitoring. The former problem is minimized or suppressed by covalent attachment to the support surface. Given the diversity of functionalized organic and inorganic polymer materials currently available in the market, it should be possible to tether virtually any indicator molecule or biochemical receptor to a solid support. The only obstacle for the synthetic chemist might be the lack of enough mechanistic studies dealing with chemical reactivity between solution and supported reagents as, partially out of difficulty and partially from the rapid speed of progress, few basic studies on the role of the physical structure of the solid support have been undertaken. Photostabilization of the indicator dye is also a polymer business. If the solid support prevents aggregation and is chemically inert towards the photoexcited dye, long-lasting monitoring with the optode is assured. Nanocomposites, particularly those involving sol-gel and organically-modified sol-gel materials embedded in permeable organic polymers have a promising future that is already opening up. Last, but not least, biofoolproof coatings and materials are currently available to provide a long life for optodes in environmental and clinical measurements. Many aces are at our disposal to win the interplay of indicator, support and analyte species!

Acknowledgements The authors express their gratitude to the many collaborators, funding institutions and companies that have made possible their research in the optical chemical sensor field over the last 15 years. Current grants from the Spanish Ministry of Education and Science (BQU-2002-04515-C02) are particularly acknowledged.

References

1. Steinberg IM, Lobnik A, Wolfbeis OS (2003) Sensors Actuators B Chem 90:230
2. Ertekin K, Alp S, Karapire C, Yenigül B, Henden E, Içli S (2000) J Photochem Photobiol A 137:155
3. Levitsky I, Krivoshlykov SG, Grate JW (2001) Anal Chem 73:3441
4. Costa JM, Sanz-Medel A (2000) Anal Chim Acta 407:61
5. Kneas KA, Xu W, Demas JN, DeGraff BA (1997) Appl Spectrosc 51:1346
6. Mills A (1998) Sensors Actuators B Chem 51:60
7. Orellana G, García-Fresnadillo D (2004) In: Narayanaswamy R, Wolfbeis OS (eds) Optical sensors: industrial, environmental and diagnostic applications. Springer, Berlin Heidelberg New York, p 338
8. Mingoarranz FJ, Moreno-Bondi MC, García Fresnadillo D, de Dios C, Orellana G (1995) Mikrochim Acta 121:107
9. García-Fresnadillo D, Marazuela MD, Moreno-Bondi MC, Orellana G (1999) Langmuir 15:6451
10. Amao Y, Miyashita T, Okura I (2001) React Funct Polym 47:49
11. Borisov SM, Vasil'ev VV (2004) J Anal Chem 59:155
12. Brassuel M, Kopelman R, Miller TJ, Tjalkens R, Philbert MA (2001) Anal Chem 73:2221
13. Brook TE, Narayanaswamy R (1998) Sensors Actuators B Chem 51:77
14. Grant PS, McShane MJ (2003) IEEE Sensor J 3:139
15. Lee SH, Kumar J, Tripathy SK (2000) Langmuir 16:10482
16. Mills A, Thomas MD (1998) Analyst 123:1135
17. Mills A, Williams FC (1997) Thin Solid Films 306:163
18. Xu W, McDonough RC, Langsdorf B, Demas JN, DeGraff BA (1994) Anal Chem 66:4133
19. Castro AM, Delgado J, Orellana G (2005) J Mater Chem 15:2952
20. Bedoya M, Díez MT, Moreno-Bondi MC, Orellana G (2005) Sensors Actuators B Chem (DOI: 10.1016/j.snb.2005.07.006)
21. Johnson RD, Bachas LG (2003) Anal Bioanal Chem 376:328
22. Seiler K, Simon W (1992) Sensors Actuators B 6:295
23. Adelt M, Devenney M, Meyer TJ, Thompson DW, Threadway JA (1998) Inorg Chem 37:2616
24. Hartmann P (2000) Anal Chem 72:2828
25. Hartmann P, Leiner MJP, Kohlbacher P (1998) Sensors Actuators B Chem 51:196
26. Fuller ZJ, Bare WD, Kneas KA, Xu WY, Demas JN, DeGraff BA (2003) Anal Chem 75:2670
27. Hodge P, Sherrington DC (1980) Polymer-supported reactions in organic synthesis. Wiley, Chichester
28. Zaragoza Dorwald F (2002) Organic synthesis on solid phase, 2nd edn. Wiley-VCH, Weinheim, p 17
29. Keeling-Tucker T, Brennan JD (2001) Chem Mater 13:3331

30. Lu X, Winnik MA (2001) Chem Mater 13:3449
31. Enke D, Janowski F, Schwieger W (2003) Microporous Mesoporous Mater 60:19
32. Xavier MP, García-Fresnadillo D, Moreno-Bondi MC, Orellana G (1998) Anal Chem 70:5184
33. Husing N, Schubert U (1998) Angew Chem Int Ed Eng 37:23
34. Wolfbeis OS, Reisfeld R, Oehme I (1996) Struct Bonding 85:51
35. Schottner G (2001) Chem Mater 13:3422
36. Lin J, Brown CW (1997) Trends Anal Chem 16:200
37. Lechna M, Holowacz I, Ulatowska A, Podbielska H (2002) Surface Coatings Technol 151–152:299
38. MacCraith BD, McDonagh C (2002) J Fluorescence 12:333
39. Lavin P, McDonagh CM, Maccraith BD (1998) J Sol-Gel Sci Technol 13:641
40. Schulz-Ekloff G, Wohrle D, van Duffel B, Schoonheydt RA (2002) Microporous Mesoporous Mater 51:91
41. Meier B, Werner T, Klimant I, Wolfbeis OS (1995) Sensors Actuators B Chem 29:240
42. Navarro-Villoslada F, Orellana G, Moreno-Bondi MC, Vick T, Driver M, Hildebrand G, Liefeith K (2001) Anal Chem 73:5150
43. Gilbert A, Baggott J (1991) Essentials of molecular photochemistry. Blackwell, Oxford
44. Orellana G (2005) Fluorescence-based sensors. In: Martellucci S, Baldini F (eds) Optical chemical sensors. Springer-Kluwer, New York (in press)
45. (a) Orellana G, Kirsch-De Mesmaeker A, Barton JK, Turro NJ (1991) Photochem Photobiol 54:499; (b) Lecomte JP, Kirsch-De Mesmaeker A, Orellana G (1994) J Phys Chem 98:5382
46. Moreno-Bondi MC, Orellana G, Turro NJ, Tomalia DA (1990) Macromolecules 23:910
47. Lianos P (1996) Heterogeneous Chem Rev 3:53
48. Bossi ML, Daraio ME, Aramendia PF (1999) J Photochem Photobiol A 120:15
49. Trammell SA, Yang P, Sykora M, Fleming CN, Odobel F, Meyer TJ (2001) J Phys Chem B 105:8895
50. Klafter J, Drake JM (1989) (eds) Molecular dynamics in restricted geometries. Wiley, New York
51. Charlesworth JM, Gan TH (1997) Langmuir 13:2699
52. Demas JN (1983) Excited state lifetime measurements. Academic, New York, p 64
53. Ogawa M, Nakamura T, Mori J, Kuroda K (2000) J Phys Chem B 104:8554
54. Gillanders RN, Tedford MC, Crilly PJ, Bailey RT (2004) J Photochem Photobiol A Chem 163:193
55. Lakowicz JR (1999) Principles of fluorescence spectroscopy, 2nd edn. Kluwer/Plenum, New York, p 244
56. Bukowski RM, Ciriminna R, Pagliaro M, Bright FV (2005) Anal Chem 77:2670
57. Botzung-Appert E, Monnier V, Ha Duong T, Pansu R, Ibanez A (2004) Chem Mater 16:1609
58. Chen X, Lin L, Li PW, Dai YJ, Wang XR (2004) Anal Chim Acta 506:9
59. Wright JD, Higginson NAC (2004) J Mater Chem 14:201
60. Senarath-Yapa MD, Saavedra SS (2001) Anal Chim Acta 432:89
61. Thevenot DR, Toth K, Durst RA, Wilson GS (1999) Pure Appl Chem 71:2333
62. Vo-Dihn T, Cullum B (2000) Fresenius J Anal Chem 366:540
63. Zaborsky O (1972) Immobilized enzymes. CRC, Cleveland
64. Carr PW, Bowers LD (1980) Immobilized enzymes in analytical and clinical chemistry. Fundamentals and applications. Wiley-Interscience, New York

65. Cass T, Ligler FS (1998) (eds) Immobilized biomolecules in analysis. A practical approach. Oxford University Press, Oxford
66. Hermanson GT, Krishna Mallia A, Smith PK (1992) Immobilized affinity ligand techniques. Academic, San Diego
67. Ulbrich R, Golfik R, Schellenberger A (1991) Biotech Bioeng 37:280
68. Li X, Rosenzweig Z (1997) Anal Chim Acta 353:263
69. De Marcos S, Hortigüela R, Galban J, Castillo JR, Wolfbeis OS (1999) Mikrochim Acta 130:267
70. Tsafack VC, Marquette CA, Lecca B, Blum LJ (2000) Analyst 125:151
71. Choi JW, Kim YK, Song SY, Lee IH, Lee WH (2003) Biosens Bioelectron 18:1461
72. Paranjpe P, Dutta S, Karve M, Padhye S, Narayanaswamy R (2001) Anal Biochem 294:102
73. Issberner JP, Schauer CL, Trimmer BA, Walt DR (2002) J Neurosci Methods 120:1
74. Trau D, Renneberg R (2003) Biosens Bioelectron 18:1491
75. Dave BC, Dunn B, Valentine JS, Zinc JI (1998) Sol-gel matrices for protein entrapment. In: Immobilized biomolecules in analysis. A practical approach. Oxford University Press, Oxford
76. Li CI, Lin YH, Shih CL, Tsaur JP, Chau LK (2002) Biosens Bioelectron 17:323
77. Doong RA, Tsai HC (2001) Anal Chim Acta 434:239
78. Ramos MC, Torijas MC, Diaz AN (2001) Sensors Actuators B: Chem 73:71
79. Zhu L, Li Y, Zhu G (2002) Sensors Actuators B: Chem 84:265
80. Moreno-Bondi MC, Benito-Peña E (2005) Fundamentals of enzyme-based sensors. In: Martellucci S, Baldini F (eds) Optical chemical sensors. Springer-Kluwer, New York (in press)
81. Choi MMF (2004) Mikrochim Acta 148:107
82. Marazuela MD, Moreno-Bondi MC (2002) Anal Bioanal Chem 372:664
83. Bentley A, Atkinson A, Jezek J, Rawson DM (2001) Toxicol in Vitro 15:469
84. Ramanathan S, Ensor M, Daunert S (1997) TIBTECH 15:500
85. Leth S, Maltoni S, Simkus R, Mattiasson B, Corbisier P, Klimant I, Wolfbeis OS, Csoregi E (2002) Electroanal 14:35
86. Gil GC, Kim YJ, Gu MB (2002) Biosens Bioelectron 17:427
87. Biran I, Rissin DM, Ron EZ, Walt DR (2003) Anal Biochem 315:106
88. Biran I, Walt DR (2002) Anal Chem 74:3046
89. Naessens M, Leclerc JC, Tran-Minh C (2000) Ecotox Environ Safe 46:181
90. Vedrine C, Leclerc JC, Durrieu C, Tran-Minh C (2003) Biosens Bioelectron 18:457
91. Ignatov SG, Ferguson JA, Walt DR (2001) Biosens Bioelectron 16:109
92. Polyak B, Bassis E, Novodvorets A, Belkin S, Marks RS (2001) Sensors Actuators B: Chem 74:18
93. Pattanapipitpaisal P, Brown NL, Macaskie LE (2001) Biotechnol Lett 23:61
94. Shriver-Lake LC, Gammeter WB, Bang SS, Pazirandeh M (2002) Anal Chim Acta 470:71
95. Hennion MC, Pichon V (2003) J Chromatography A 1000:29
96. Preininger C, Mencaglia A, Baldini F (2000) Anal Chim Acta 403:67
97. Sharma SK, Sehgal N, Kumar A (2003) Current Appl Phys 3:307
98. Vijayendran RA, Leckband DE (2001) Anal Chem 73:471
99. Lu B, Smyth MR, O'Kennedy R (1996) Analyst 121:29R
100. Nisnevitch M, Firer MA (2001) J Biochem Biophys Methods 49:467
101. Pulido-Tofiño P, Barrero-Moreno JM, Pérez-Conde MC (2001) Anal Chim Acta 429:337
102. http://www.millipore.com, http://www.piercenet.com, http://www.bio-rad.com

103. Xu Q, Lam KS (2003) J Biomed Biotech 5:257
104. Avseenko NV, Morozova TY, Ataullakhanov FI, Morozov VN (2002) Anal Chem 74:927
105. Zhu H, Snyder M (2003) Curr Opin Chem Biol 7:55
106. Predki PF (2004) Curr Opin Chem Biol 8:8
107. Peluso P, Wilson DS, Do D, Tran H, Venkatasubbaiah M, Quincy D, Heidecker B, Poindexter K, Tolani N, Phelan M, Witte K, Jung LS, Wagner P, Nock S (2003) Anal Biochem 312:113
108. Askari M, Alarie JP, Moreno-Bondi MC, Vo-Dinh T (2001) Biotech Prog 17:543
109. Glöcker J, Angenendt P (2003) J Chromatography B 797:229
110. Sakanyan V (2005) J Chromatography B 815:77
111. Templin MF, Stoll D, Schrenk M, Traub PC, Vöhringer CF, Joos TO (2002) Trends Biotechnol 20:160
112. Kido H, Maquieira A, Hammock BD (2000) Anal Chim Acta 411:1
113. Walt DR (2000) Science 287:792
114. Pearson JE, Gill A, Vadgama P (2000) Ann Clin Biochem 37:119
115. Almadidy A, Watterson J, Piunno AE, Raha S, Foulds IV, Horgen PA, Castle A, Krull U (2002) Anal Chim Acta 461:37
116. Epstein JR, Leung APK, Lee KH, Walt DR (2003) Biosens Bioelectron 18:541
117. Epstein JR, Biran I, Walt DR (2002) Anal Chim Acta 469:3
118. Li J, Tan W, Wang K, Xiao D, Yang X, He X, Tang Z (2001) Anal Sci 17:1149
119. Lee M, Walt DR (2000) Anal Biochem 282:142
120. Liu X, Farmerie W, Schuster S, Tan W (2000) Anal Biochem 283:56

Springer Ser Chem Sens Biosens (2005) 3: 227–260
DOI 10.1007/5346_007

Published online: 29 September 2005

Challenges in the Design of Optical DNA Biosensors

Melissa Massey · Paul AE Piunno · Ulrich J Krull (✉)

Department of Chemical and Physical Sciences, University of Toronto at Mississauga,
3359 Mississauga Rd. N., Mississauga, ON, L5L 1C6, Canada
ukrull@utm.utoronto.ca

Abstract The field of biosensors and biochips for nucleic acid diagnostics has developed significantly over the last decade. High-throughput techniques offering the advantages of sensitivity and selectivity combined with rapid analysis to provide reproducible and accurate results are highly sought after in the areas of medical diagnostics, forensics, environmental monitoring, and bioterrorism. This chapter gives a short review of the necessary considerations for the preparation of immobilized nucleic acid films on a solid sensor substrate and the development of techniques utilized for the detection of selective hybridization of target binding materials. The fundamentals of fibre optic and surface plasmon resonance optical sensor platforms are outlined, followed by key developments in the area of fluorescent particle labels and dyes used for the detection of nucleic acid hybridization. Recent advances in hybridization assays include fluorescent cationic polymer, molecular beacon, and duplex probe technologies. Finally, current methods used for the detection of interfacial DNA hybridization are described, including a discussion of limitations and possible strategies to enhance the key design priorities of sensitivity and selectivity.

Abbreviations

A	Adenine
ATR	Attenuated total reflectance
C	Cytosine
DABCYL	4-((4′-Dimethylamino)phenyl)azo)benzoic acid
DAPI	4′,6-Diamidino-2-phenylindole
DNA	Deoxyribonucleic acid
dsDNA	Double-stranded deoxyribonucleic acid
ssDNA	Single-stranded deoxyribonucleic acid
EH	*Erwina Herbicola*
EtBr	Ethidium bromide
FAM	Carboxyfluorescein
FRET	Fluorescence resonance energy transfer
G	Guanine
HEX	6-Carboxy-2′,4,4′,5,7,7′-hexachlorofluorescein
JOE	6-Carboxy-4′,5′-dichloro-2′,7′-dimethoxyfluorescein
K_b	Binding constant
MB	Molecular beacon
PCR	Polymerase chain reaction
PNA	Peptide nucleic acid
PRO	Propidium iodide
RNA	Ribonucleic acid
ROX	Carboxy-X-rhodamine
SMN	Survival motor neuron
SNP	Single nucleotide polymorphism
SPR	Surface plasmon resonance
SRB	Sulfurhodamine B
STS	Synthetic target sequence
T	Thymine
TAMRA	Carboxytetramethylrhodamine
T_m	Thermal denaturation temperature
TE	*Tris*-EDTA
TET	6-Carboxy-2,4,7,7′-tetrachlorofluorescein
TIRF	Total internal reflection fluorescence
TMR	Tetramethylrhodamine
TO	Thiazole orange
TOTO	Thiazole orange homodimer
YO	Oxazole yellow
YOYO	Oxazole yellow homodimer

1 Introduction

There is tremendous demand for high-throughput analytical technologies for determination of nucleic acids that may be used in applications such as the elucidation of genetic factors that are associated with disease, the development of new drugs and therapies and the detection of pathogens. Current protocols for genotyping and expression analysis generally involve cell lysis, nucleic acid isolation, perhaps followed by one or a series of amplification steps, and finally some electrophoretic separation and analysis steps. Alternatives to these methods include a variety of analytical techniques based on solid-phase hybridization, such as membrane hybridization methods (e.g. Southern blot) or use of systems that are based on microarrays (e.g. combinatorial screening, mutation analysis, sequencing by hybridization, or allele-specific primer extension). Significant advancements in the development of technologies for nucleic acid analysis would include reduction of the time for analysis, reduction of the complexity of sample handling, and increased confidence in analyses.

Microarrays address some of these issues. These technologies permit for large-scale screening of nucleic acid targets by monitoring interfacial hybridization of a sample to an array of immobilized probe DNA that may contain many thousands of different sequences, each in a discrete and spatially addressable element of the array. However, the large volume of information that may be gleaned requires significant time and effort for bioinformatic processing and signal deconvolution. Furthermore, microarrays may not be the preferred method of analysis when rapid, quantitative and economical analysis is required. Another advancement towards rapid nucleic acid analysis has been made in the development of real-time PCR methods, where quantification can be provided simultaneously with amplification [1]. A concern here is that specific and non-specific amplification may not always be discernable by the detection system and that the time frame from the presentation of the sample to the final analytical results can be lengthy (> 1 h) and may not be suitable in certain rapid response situations, such as for the monitoring of biological warfare agents. Some discrimination between specific and non-specific products may be achieved by post-amplification thermal denaturation analysis methods, however, these are most sensitive to products that differ significantly from the target amplicon (e.g. primer-dimers) and may substantially increase the analysis time.

2 Considerations for the Preparation of Immobilized Nucleic Acid Films

In a diagnostic application, it may be desirable to rapidly screen a large number of samples for a specific mutation, or a small set of mutations. Optical nucleic acid biosensor technology may be well suited to such an application. Ideally, biosensor technologies offer significant improvements in time of analysis (on a per-sample basis), reusability, and provide the potential for automation and miniaturization of the analysis so that point-of-care analysis may ultimately be achieved. A feature that is common to all biosensor technologies is the use of probe oligonucleotides that are immobilized to the surface of a solid substrate. Some optimization of the analytical figures of merit of a particular device may be achieved through control of the chemistry applied for immobilization of the probe nucleic acids. The nature of the immobilized film in terms of synthetic quality, density and homogeneity of distribution of probe molecules and the free energy of the substrate surface are of significance in terms of the behaviour of the resultant device as these determine the type and degree of nearest-neighbour interactions amongst immobilized probes and interactions between immobilized oligonucleotides and the substrate surface. These interactions have been shown to affect the efficiency [2], rate [3], and selectivity of hybridization [4, 5].

Regardless of the type of immobilization, it is now clear that the local environment of the immobilized nucleic acids is a key determinant of their binding activity [5, 6]. Optimization of nucleic acid immobilization chemistry can concurrently alter the sensitivity and speed of the particular sensor methodology. Factors such as immobilization density [5, 6], interaction with the substrate [7, 8] probe nature [2, 4, 5, 9] and sample processing have been demonstrated to have direct effects on the analytical figures of merit.

Shchepinov et al. [2] showed that the synthetic yield of immobilization for systems of oligonucleotides immobilized on the surface of polypropylene substrates was reduced on surfaces that had not been functionalized with molecular spacer molecules. Such spacers reduce steric hindrance by imparting motional freedom to the tethered end of the immobilized oligonucleotide, which improves hybridization yield and binding kinetics. Further, it was observed that hybridization efficiency was restricted at immobilization densities where neighbouring oligonucleotides were close enough in proximity to interact with one another. Similarly, Herne and Tarlov [7] also observed a relationship between the density of immobilization of thiol-derivatized oligonucleotides and hybridization yield for a system in which mixed monolayers comprised of 5′-mercaptohexane functionalized oligonucleotides and mercaptohexanol diluents were immobilized on gold substrates. The reduction in hybridization yield at higher densities of immobilized oligonucleotides, as measured by use of ^{32}P-radiolabelling, was attributed to steric hindrance and electrostatic repulsion of the complementary oligonucleotides within the close-packed DNA monolayer.

In addition to the potential for affecting sensitivity, the density of immobilized oligonucleotides has also been shown to affect the selectivity, the dynamic range and the kinetics of hybridization assays [5, 6]. The immobilization of oligonucleotides has been shown to have significant ramifications on the thermodynamics of hybridization, where the enthalpy change associated with interfacial hybridization may be substantially lower in magnitude than that observed in bulk solution [10], and may also be accompanied by reductions in the thermal denaturation temperature (T_m). Conventionally, the selectivity of hybridization assays is optimized by controlling experimental conditions such as temperature and ionic strength so that the difference in hybrid stability (as indicated by the T_m) between a fully complementary target and a partially complementary nucleic acid interferent is maximized. In so doing, T_m data are used as indicators of stability of binding and the effects of surface oligonucleotide immobilization on hybrid stability can be evaluated.

3 Development of the Selectivity of Optical Biosensors

Control of immobilization chemistry (immobilization density, surface capping and minimization of side product formation) leads to organization of the probe oligonucleotides within immobilized films that promotes selective hybridization and reduces non-selective adsorption. Deviations in thermal denaturation temperature brought about by single base mismatches can be on the order of 8–11 °C in low stringency conditions (1 M NaCl) for immobilized probe sequences of about 20-mer length under conditions of high immobilization density [10, 11]. Previous research has indicated that non-selective adsorptive interactions between nucleic acids in solution and the sensor surface occur independently of selective hybridization events in certain concentration regimes [12]. This implies that through the use of the appropriate reference sensor(s), the contributions to signal due to non-selective adsorption of labelled materials may be directly subtracted, providing a method for automation of data analysis and a reduction in the amount of sample clean-up required. For example, rapid (< 2 min) and quantitative genotyping for single nucleotide polymorphisms associated with spinal muscular atrophy was done using a reusable ($\sim$ 80 cycles of application) fibre-optic biosensor [13]. Sensors were functionalized with oligonucleotide probes covalently immobilized at high density ($\sim$ 7 pmol cm^{-2}) to impart selectivity enhancement and used in a total internal reflection fluorescence (TIRF) detection motif to detect 202 base-pair PCR amplicons. Significantly, detection was done over a broad range of ionic strength conditions (0.1 – 1.0 M) without stringency rinsing and with maintenance of high selectivity, thereby facilitating sample preparation.

High immobilization density may be defined as the case where there is a high likelihood of significant interaction between neighbouring oligonucleotides (e.g. mean centre-to-centre strand separation distances of less than 50 Å) and tends to provide the best results in terms of selectivity. Hybridization may be considered as a process that injects "extra" electrostatic charge into the surface region and can alter the charge and steric packing in immobilized oligonucleotide films that are already dense. To achieve control of selectivity of binding and the dynamic range that could be achieved via control of the concentration of oligonucleotide sequences at an interface is complex. In practical applications, control of selectivity may be impossible if there is only reliance on the density of immobilized probe molecules as the density is a dynamic function of the degree of hybridization. While consideration of T_m values can be used to help evaluate the relative selectivity of a given hybridization experiment, consideration must also be given to the shape of each thermal denaturation transition. The shape provides information as to the relative population of fully complementary and partially complementary hybrids at a given temperature and ionic strength. The conventional treatment of thermal denaturation data considers denaturation to be a two-step process. This assumption is not valid at a surface. As the number of potential contributing interactions increases within the system (e.g. strand–surface interactions and interactions between immobilized strands) further deviations from the two-state model are observed. The increased interactions beyond that assumed by the two-state, hybridization-denaturation model tends to result in a reduction of the slope of the thermal denaturation transition, and a lengthening of the temperature window over which the denaturation occurs. Broad temperature transitions are particularly limiting in terms of control of selectivity, and poor control of immobilization chemistry can result in such broad curves.

Preliminary experiments have been done to investigate the challenge of creation of an environment for immobilized oligonucleotide probes that offers good structural regularity and reproducibility, where nearest neighbour interactions provide for control of selectivity, yet where the degree of hybridization does not alter nearest neighbour interactions. The hypothesis was that a "matrix isolation" method will produce the desired environment for the probe molecules (Fig. 1). To achieve this, polyelectrolytes with charged backbones and significant flexibility were co-immobilized with oligonucleotides as a mixed film. The intention was that the probe molecules would be isolated by surrounding each on average with a sheath of immobilized polyelectrolyte which would not participate in binding to complementary nucleic acids and so provide a method to tune and reduce drift in the selectivity of the sensor. A mixture of dT_{20} as the probe oligonuclotide, and ethylene glycol phosphate oligomer as a matrix isolation material in a 1 : 20 mole ratio, respectively, has been investigated.

Melt temperature measurements indicated that the thermodynamic stability of the probe molecules can be adjusted using the oligomer matrix to achieve lower T_m values by as much as 5 °C, with retention of selectivity for discrimination of single base pair mismatches [14]. The enhanced temperature sensitivity may ultimately provide for the development of analytical devices of extremely high selectivity. Devices containing immobilized

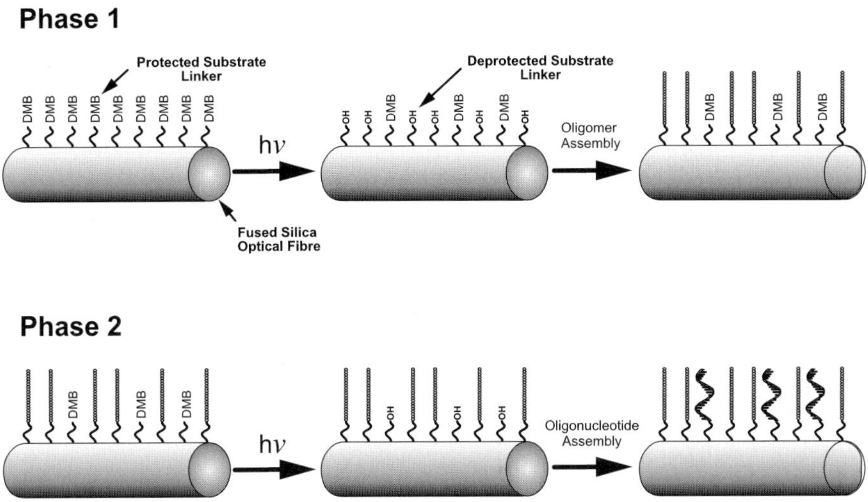

Fig. 1 Matrix isolation method of surface immobilization of probe oligonucleotide/polyelectrolyte mixed film for enhanced selectivity. *Phase 1* Photolabile dimethoxybenzoin (DMB) protecting groups are selectively exposed to electromagnetic radiation of appropriate wavelength to provide reactive sites in which polyelectrolyte "spacers" can be immobilized. *Phase 2* The remaining DMB-protected sites are photo-deprotected to expose sites for probe oligonucleotide immobilization onto the solid surface

films of high density of oligonucleotides, or mixed films of oligonucleotides and oligomers, may be created to operate at temperature and solution ionic strength conditions such that only hybrids with fully complementary nucleic acids or nucleic acid analogues may be detected. By designing an array of probe sequences, each with carefully designed immobilization density and chemistry, it may be possible to generate an array of probe sequences with similar T_m values, regardless of the G-C content of the hybrids formed. This would allow the simultaneous assay of a number of different sequences at the same temperature with reduced loss of calibration.

4 Optical Techniques for the Determination of Interfacial Nucleic Acid Hybridization

Nucleic acid biosensors based on optical modes of detection represent another common approach for generating analytical signals based on nucleic acid hybridization. The methods discussed herein are based on methods that are suitable for the study of materials on surfaces. There are a number of different optical methods that have been described, with the most common being attenuated total reflectance (ATR), total internal reflection fluorescence (TIRF) and surface plasmon resonance (SPR) [15]. All of these methods work on the principle that when light strikes an interface between two different materials of differing refractive indices (η_1 and η_2), total internal reflection results at the critical angle θ, where:

$$\sin\theta_c = \eta_2/\eta_1 \quad \text{where} \quad \eta_1 > \eta_2 \tag{1}$$

4.1 Fibre Optic Sensors

Optical fibres are routinely used for the transmission of electromagnetic radiation over long distances by confinement of the photons within guided modes of the cylindrical waveguide [15]. Optical radiation propagates through the permissible modes of optical fibres by total internal reflection (TIR) and at the points of reflection, at the interface between the waveguide core and cladding, localized zones of constructive interference are created. This serves to create a standing electric field intensity, known as an evanescent wave, that extends beyond the waveguide (i.e. in the medium of lower refractive index) and decays exponentially with increasing distance from the interface. The intensity of the evanescent wave can be determined by use of the following relationship:

$$E_T = E_{oT}\exp\left(\frac{-z}{\delta}\right) \tag{2}$$

where E_T is the magnitude of the electric field vector, E_{oT} is the magnitude of the electric field vector at the interface (where it is most intense), z is the distance along the normal to the plane defined by the interface and δ is the characteristic decay length. The characteristic decay length is dependent on the wavelength (λ) of light under consideration, the refractive index of the waveguide core (η_1) and outer medium (η_2) and the propagation angle (θ) of the guided photon(s) relative to the normal to the interface (i.e. TIR incidence angle) as given by:

$$\delta = \frac{\lambda}{2\pi\eta_1}\sqrt{\sin^2\theta - \left(\frac{\eta_2}{\eta_1}\right)^2} \tag{3}$$

The effective sampling depth or penetration depth (d_p) is representative of the characteristic decay length and represents the distance at which the intensity of the evanescent field has decayed to $1/e$ with respect to the maximum (at the interface), as defined by:

$$d_p = \frac{\lambda}{4\pi\eta_1}\sqrt{\sin^2\theta - \left(\frac{\eta_2}{\eta_1}\right)^2} \tag{4}$$

Excitation of a fluorophore that is close to the surface of the waveguide can be achieved via the evanescent field. The resulting fluorescence emission is isotropically distributed, however, some small component of the emitted light can be captured by the fibre in a guided mode and transmitted by TIR to filtering optics and a detector [16]. If an oligonucleotide probe is immobilized on the surface of an optical fibre and participates in the formation of a hybridized complex with a fluorophore-labelled target, then a fluorescent signal can be generated and transmitted through the fibre to a detector, providing an interfacially-sensitive method for detection of nucleic acid hybridization. Optical fibres provide the advantages of small size, flexible geometry, and potential for remote sensing applications [17–23].

4.2 Surface Plasmon Resonance

A commonly employed method for determination of nucleic acid hybridization on probe functionalized substrates is surface plasmon resonance (SPR). Plasmon excitation is a method of delivering radiant source energy to a conductive surface. A plasmon may be considered as a collection of electrons that undergo oscillation with a defined coherence length (in the conduction band of a thin metal film). The oscillations of the electrons create an oscillating electric field that takes the form of a wave along the surface, with a perpendicular electric field component that extends from the surface into the medium of lower optical density. SPR experiments are often done in the Kretschmann configuration, as shown in Fig. 2, where one face of a prism is coated with a thin metal film (usually a few tens of nanometers of Au or Ag) to which a layer of sensing material is immobilized.

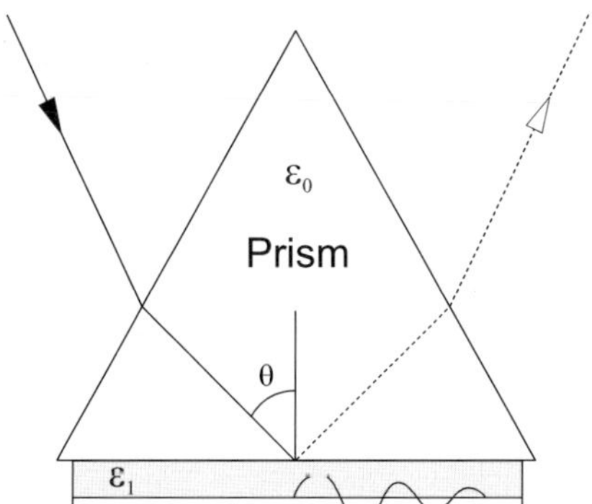

Fig. 2 Surface plasmon resonance device in the Kretschmann configuration. ε_0 is the refractive index of prism, ε_1 is the refractive index of thin metal film (usually Au or Ag), ε_2 is the refractive index of air, and θ represents the critical angle

When plane-polarized radiation is directed onto the metal film via the prism coupler, energy may be selectively deposited into the thin surface layer of the metal film. This phenomenon will occur at one critical angle where the photon wavevector matches that of the surface plasmon [15] and the intensity of reflected radiation will be dramatically reduced. In order to achieve photon-plasmon coupling, certain conditions of incidence angle, metal film properties, and the dielectric properties of any additional interfacial coating on the metal film surface must be satisfied [15]. The angle at which surface plasmon resonance is achieved is often used as the analytical parameter to investigate changes in the physical properties of the surface layer. Alternatively, the photon incidence angle can be kept constant and the wavelength of light can be altered to achieve resonance generation of surface plasmons. When nucleic acid probe molecules are immobilized onto the metal film, hybridization events with target nucleic acid will cause an increase in the refractive index of the biomolecular sensing layer. This will alter the wave propagation properties and will cause perturbations in the resonance conditions, which provides for analytical information regarding the concentration of hybridized target molecules at the surface. Other characteristics that can be measured include phase shifts, amplitude, polarization and spectral distribution [24].

Corradini et al. [25] have recently used SPR to develop an optical method for recognition of cystic fibrosis point mutations in DNA. Biotinylated 9-mer oligonucleotide DNA probe molecules were immobilized onto the surface of streptavidin functionalized gold coated sensor chips. Target DNA oligomers, peptide nucleic acid (PNA) oligomers and chiral PNA oligomers were examined. It was demonstrated that PNAs containing different chiral monomers may be used to alter affinity and selectivity for their DNA complements [26]. The chiral PNA oligomers were selected to contain a "chiral box" where three of the nucleotides within the box contained chiral monomers based on D-lysine (which has a positive charge associated with its chain) that were used to target a point mutation. Both wild-type and mutated oligomers were immobilized on the sensor surfaces to see if the DNA or the PNA targets exhibited selectivity to either sequence and also to determine if the PNA targets (specifically the PNA chiral box targets) and DNA targets provided differences in signal intensities. Solution phase melt temperatures were determined to be lowered for the PNA chiralbox/DNA hybrids, indicating reduced thermodynamic stability. However, no evidence of hybridization of DNA sequences containing a single base mismatch with the PNA chiralbox could be observed within 10 °C of the melt temperature. This indicated that the PNA chiral sequences could selectively bind fully complementary targets in competition with target sequences containing a single base mismatch. Furthermore, hybridization energetics were shown to be superior in comparison to both the DNA and achiral PNA sequences such that the chiral PNA/DNA hybrids were less stable when hybridization occurred at an interface rather than in solution [27].

A more sensitive method for detecting biomolecules at the SPR metal interface involves the use of nanoparticle-labelled target oligonucleotides. Metal nanoparticles provide for substantial alterations in refractive index (i.e. optical density) at the metal surface with respect to those provided by oligonucleotides alone [28]. Nicewarner-Peña et al. [29] have exploited this fact to demonstrate that nanoparticle tags associated with oligonucleotide targets can provide increases in assay sensitivity by up to three orders of magnitude with respect to similar assays done without the use of nanoparticle tags.

In general, for practical application of the SPR method, the interface presented to the sample solution must be made intrinsically resistant to non-specific adsorption as adsorbed material cannot be differentiated from actual target molecule binding in these systems. Response is based solely on alterations in the refractive index and thickness of the sensing layer and not on the chemical identity of the species on the substrate surface. To this end, oligonucleotide films should therefore be deposited/assembled at high density so that non-specific adsorption is minimized [13], or stringency conditions must be carefully controlled to effectively remove non-specifically bound or otherwise adsorbed materials from the surface while maintaining the integrity of the properly hybridized complexes at the surface for quantitative determinations.

5 Particles as Labels for Transduction of Interfacial Target Binding

Baeumner et al. [30] have developed a nucleic acid biosensor based on fluorescence emission from a fluorogenic dye that is encapsulated in a liposomal nano-vesicle. A sandwich "dip-stick" assay was developed that was sensitive to nanomolar concentrations (1 fmol total sample) of complementary target material with a dynamic range covering 1–750 nM (1–750 fmol) [30]. The detection scheme involved synthesis of a liposome vesicle in which sulforhodamine B (SRB) dye molecules were trapped. The liposomes were filtered to select for structures of roughly equal size and to separate the liposomes from free dye. A 5′-maleimide functionalized dC-rich 20-mer oligonucleotide sequence was immobilized on the outer surface of the liposome by coupling to sulfhydryl groups that were present on the liposome. Unreacted SH groups on the liposome were capped and the liposomes purified by gel filtration to eliminate unreacted probe sequences. The sandwich assay involved mixing the liposome-functionalized oligonucleotides with a reporter strand. The reporter strand contained a region that was complementary to the strand that was attached to the liposome, and also another region that was complementary to the target strand. The synthetic target sequences (STS) used in this application were the following: *E. coli* (*clp*B STS), *B. anthracis* (*Atx*A STS), and *C. parvum* (*hsp*70 STS) [30]. A capture probe was also employed to permit immobilization of the sandwich complex on a test strip. The capture probe was comprised of a sequence that was also complementary to a region on the target nucleic acid and was functionalized with a biotin moiety at the 5′-terminus. The probe, target and capture strand were incubated together at 41 °C for 20 min in a test tube [30] prior to "development" of the test strip. The test strip was created by deposition of streptavidin at one end of a polyethersulfone membrane to create a capture zone. The non-functionalized end of the test strip was inserted into the test tube and the mixture in the tube was allowed to adsorb to the surface of the test strip. Running buffer was then added to the tube allowing the analyte mixture to travel up the membrane. Once the running buffer had run the entire length of the strip, a signal was measured at the capture zone using a reflectometer that was set at a wavelength that was suitable for determination of sulforhodamine dye (560 nm) encapsulated in the liposomes.

This method provided a simple and fast procedure to identify the presence of a nucleic acid sequence with good sensitivity. On average, the liposomes entrapped several thousand dye molecules, which provided intrinsic amplification of individual binding events. A qualitative measure of nucleic acid hybridization by visualization of the colour on the dip stick was intrinsically provided and quantitative measurements were also possible by use of a reflectometer. This approach resulted in a limit of detection of $\sim$ 1 fmol and a dynamic range of over two orders of magnitude. Maximum signal-to-noise was obtained with 0.2 mol% of probe sequence immobilized on the liposomes.

Metal nanoparticles have been used in optical absorbance assays as they have absorptivity coefficients that are many orders of magnitude greater than most organic fluorescent dyes due to the effect of plasmon excitation within the conduction band of the metal particles [31, 32]. The spectrum of a nanoparticle can be designed by selection of particle size, shape and composition. The density of nanoparticles bound at a surface can be determined by use of absorption measurements when a transparent sensor surface is used [33, 34]. Methods similar to the development of photographic film have also been used by Taton et al. [35] to detect nanoparticle-labelled DNA hybridization events. An aqueous solution of silver salt and hydroquinone was used as a development solution, and hybridized material on the surface catalysed a reaction of silver deposition that increased the size of the particles, which were observed using optical microscopy [36, 37]. More recently, Nam et al. [38] have combined this approach with magnetic particle sample collection and a bio-barcode functionalized nanoparticle detection scheme to permit for scanometric determination of the bio-barcode DNA (each in 100-fold abundance and indicative of the presence of a specified target nucleic acid) with PCR-like sensitivity (i.e. 5–10 target copy sensitivity) within the $\sim$ 5 h duration of the assay.

Methods of immobilizing probe oligonucleotides to silica nanoparticles have also been developed. Hilliard et al. [39] immobilized 20-mer probe oligonucleotides to silica nanoparticles by means of disulfide chemistry. The hybridization of 3′-fluorescein labelled nucleic acid targets to the immobilized capture probes was done by allowing the nanoparticles and targets to incubate for 30 min in hybridization buffer [39]. A spectrofluorimeter was used to detect the fluorescence emission from the fluorescein label on the target. Multiple washings of the DNA-nanoparticle solution with buffer ensured that signal from non-specifically adsorbed target material was minimized. These nanoparticles were able to detect target sequences in the nanomolar concentration range. However, stability of the nanoparticles was poor, as 25% of the probes had detached from the particle surface after 24 h, and there was very little signal difference between fully complementary and single base-pair mismatch sequences. The hybridization reaction was also lengthy, taking approximately 30 min to reach saturation [39].

6 Fluorescent Dyes for Detection of DNA Hybridization

6.1 Fluorescent Labels

Optical nucleic acid biosensors and microarrays based on fluorescence detection make use of fluorescent dyes to provide a measurable signal for target/probe hybridization. Ideally, fluorophores used for detecting nucleic acid hybridization should exhibit large molar absorptivity, resistance to photobleaching, quantum yields that approach unity and the ability to produce a resolvable signal at low concentrations both quickly and reproducibly [40].

Fluorescein and its derivatives are one set of fluorescent labels that are commonly used to monitor DNA hybridization, and are often used as labels but do not selectively associate with the structure of double-stranded DNA. The most popular fluorescein derivative for nucleic acid labelling is FAM (carboxyfluorescein). The structure of FAM is shown in Fig. 3.

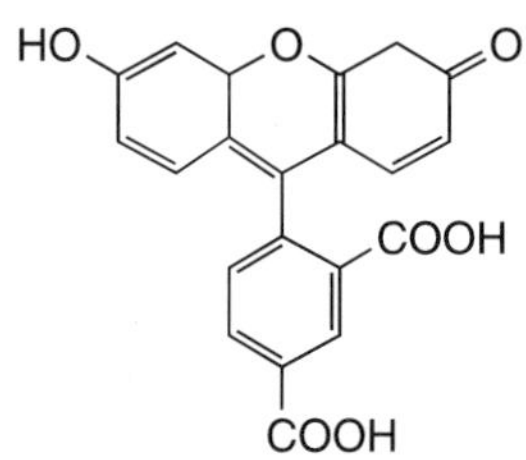

Fig. 3 Structure of Carboxyfluorescein (FAM)

Fluorescein is useful in that its excitation maximum aligns well with the 488 nm line provided by argon-ion lasers and can be synthesized in high purity. However, fluorescein does provide for some practical limitations, including a high rate of photo-bleaching, pH-sensitive fluorescence and quenching of fluorescence upon conjugation to biopolymers [41]. Some fluorescein derivatives can be used for genetic analysis and these include: JOE (6-carboxy-4′,5′-dichloro-2′,7′-dimethoxyfluorescein), TET (6-carboxy-2′,4,7,7′-tetrachlorofluorescein) and HEX (6-carboxy-2′,4,4,′,5,7,7′ hexachlorofluorescein) [41]. These dyes are commonly used in conjunction with rhodamine dyes or with fluorescence quenching molecules (DABCYL or black hole quenchers) as FRET pairs that may be used in molecular beacon structures, duplex probes and on hybridization probes as used in real-time PCR assays based on 5′-endonuclease activity of the polymerase for decomposition of the hybridization probe and release of the quenched FRET pair. Rhodamine dyes such as ROX (carboxy-X-rhodamine) or TAMRA (carboxytetramethylrhodamine) have excitation maxima that overlap with the fluorescence emission of fluorescein derivatives such that non-radiative energy transfer can occur. DABCYL (4-((4′-dimethylamino)phenyl)azo)benzoic acid exhibits a broad absorption spectrum in the range from 400 to 560 nm, and it is a useful acceptor in FRET applications for monitoring conformational changes and other dynamic spatial movements [41].

Much attention has been focussed lately on the family of asymmetric cyanine dyes for use in fluorescence detection of nucleic acids. These dyes show a significant enhancement in fluorescence intensity (100- to 1000-fold) upon binding to double-stranded DNA as compared to that from the fluorophore in solution. Use of cyanine fluorophores may be advantageous for use in assay design and sensor applications with respect to some of the more commonplace dyes, such as ethidium bromide and Hoechst 33342, as these latter dyes exhibit significant fluorescence intensity as background when in solution and have significantly lower enhancement in emission intensity [42].

Asymmetric cyanines are composed of two aromatic ring systems that are bridged by either a mono or polymethine bond which is part of the conjugated system [43]. Several dyes of this type have been synthesized and studied in recent years, and their interactions with single-stranded DNA (ssDNA) and dsDNA have been extensively investigated [44–47]. The cyanine dyes have been observed to exhibit three main modes of binding that include intercalation, groove-binding, and via electrostatic interactions with the phosphate backbone [48, 49]. Intercalating cyanines exhibit binding constants (K_b) that do not usually exceed 10^7 M^{-1} [48]. Groove-binding dye complexes are stabilized by hydrogen bonding with the phosphate backbone and water, and consequently exhibit K_b values that can sometimes be as high as 10^8–10^9 M^{-1} [48]. Two of the more common cyanine dyes used for nucleic acid studies are Cy3 and Cy5, which are trimethine and pentamethine cyanine dyes, respectively. These fluorophores serve only as high efficiency labels to indicate the presence of a particular sequence as the fluorescence properties of these dyes are not significantly perturbed by association with nucleic acids. The structures of Cy3 and Cy5 are shown in Fig. 4.

NMR and FRET data have shown that Cy3 attaches onto the end of a double helix in a manner that mimics the addition of another base pair [50]. Although Cy3 is commonly used, the trimethine bridge allows for many degrees of freedom that result in non-radiative energy transfer losses and decreased quantum yield [40]. The difference in the structure of the methine bridge between Cy3 and Cy5 shifts the wavelengths of excitation and emission for Cy5. This limitation may affect the use of Cy3 at very low concentrations as fluorescence signals may be difficult to discern above background noise. One way to ameliorate this limitation is to make the polymethine bridging chain more rigid. This has been achieved by the introduction of alkyl substituents into the chain.

Yarmoluk et al. [51] have synthesized Cyan 2, a trimethine dye derivative with a methyl substituent at the β-carbon of the polymethine chain. The structure of this dye is shown in Fig. 5.

The introduction of the alkyl substituent decreased the fluorescence intensity of the free dye in solution without sacrificing fluorescence intensity upon binding to dsDNA. The K_b value was determined to be 3.6×10^4 M^{-1}, which

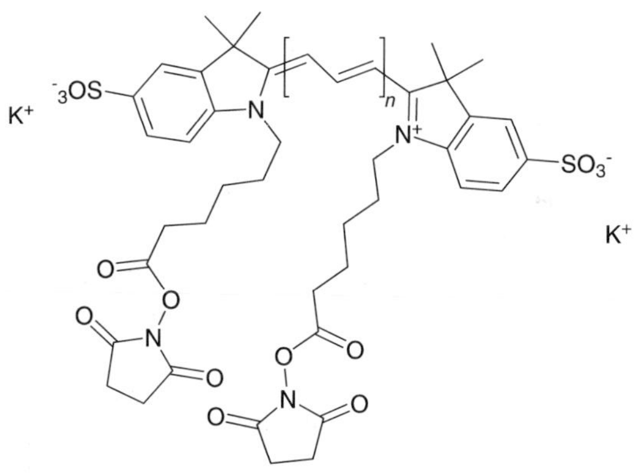

Fig. 4 Structure of Cy3 ($n = 1$) and Cy5 ($n = 2$)

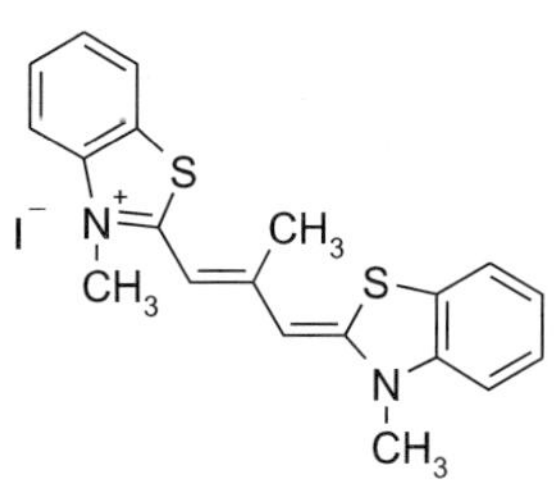

Fig. 5 Structure of Cyan 2

indicates a stable complex is formed with dsDNA and a mode of binding consistent with intercalation [48].

Cyan 2 is quenched by its iodide anion in solution, and quenching is also dependent on ionic strength. Fluorescence quenching of free dye in solution was much higher than dye associated with dsDNA. In particular, quenching was minimized when the dye was bound to poly(dGdC/dGdC), indicating that the anion of the dye in solution may also aid in the low fluorescence of the free dye in solution and that a high ionic strength buffer leads to an increased stability of the complex formed with dsDNA.

6.2
Fluorescence from Dyes that Associate with DNA

Perhaps the most well-recognized fluorescent dye for detection of DNA hybridization is ethidium bromide (EtBr). EtBr is a cationic phenanthridinium compound that can bind to DNA by intercalation. This dye has an excitation maxima at 518 nm when bound to double-stranded DNA (dsDNA). Excitation of EtBr is often done by use of an argon ion laser, making this fluorophore a viable choice for applications in optical sensors as well as confocal scanning laser microscopy and fluorometry [41]. The structure of ethidium bromide is shown in Fig. 6.

Intercalators associate with dsDNA by insertion between the stacked base pairs of DNA [52]. EtBr binds to dsDNA with little to no sequence specificity, with one dye molecule inserting for every 4–5 base pairs [53]. It also binds weakly via a non-intercalative binding mechanism only after the intercalative sites have been saturated [54]. Propidium iodide (PRO) is structurally similar to ethidium bromide, and both dyes show a fluorescence enhancement of approximately 20–30 fold upon binding to dsDNA [41]. As well, their excitation maxima shift ~30–40 nm upon binding due to the environment change associated with intercalation into the more rigid and hydrophobic interior of the double-stranded nucleic acid structure relative to aqueous solution [41].

Other commonly used dyes include Hoechst 33258 and DAPI (4′,6-diamidino-2-phenylindole), both of which show selectivity towards poly A-T sequences. This selectivity limits the utility of these dyes as generic transducers in sensor applications. These dyes associate with double-stranded nucleic acids in a groove-binding motif, more specifically, via binding in the minor groove of B-DNA [41].

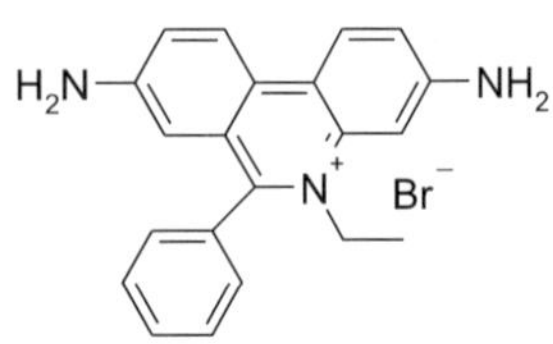

Fig. 6 Structure of ethidium bromide (EtBr)

Intercalating dyes are usually cationic and exist as salts. As a result, they can bear differing amounts of charge depending on their structure. The number of positive charges on the dye bears an impact on the quantum yield of fluorescence. Deligeoriev et al. [55] have synthesized variously charged intercalating asymmetric monomethine cyanine dyes bearing one, two or three positive charges. It was found that increasing the number of positive charges led to an increase in quantum yield upon association with dsDNA. It was also shown that the sensitivity of the dyes to dsDNA became greater as a function of increasing the number of positive charges associated with the dye. The major limitation to this set of dyes was that the dye could not distinguish between ssDNA and dsDNA despite the increase in quantum yield of the dye bearing three positive charges. The dye was observed to electrostatically bind to the phosphate groups on a ssDNA backbone, however, did not show a shift in the wavelength of the fluorescence intensity maximum between ssDNA and dsDNA.

Thiazole orange (TO) and its derivative oxazole yellow (YO) have been used as intercalative transduction agents in nucleic acid hybridization assays. TO is a non-planar chromophore composed of a benzothiazole derivative and a quinolinium ring linked via a monomethine bridge. TO has been reported to provide ~3000-fold fluorescence intensity enhancement upon DNA binding [56]. This is a significant improvement over ethidium bromide (20- to 25-fold) [57]. However, TO and YO have also been reported to undergo an increase in quantum yield of 18 900 and 700 times, respectively, upon binding to DNA [58]. Clearly there is inconsistency in the literature concerning the degree of fluorescence enhancement upon binding for TO. However, it is worth noting that differing experimental conditions can significantly influence the degree of fluorescence enhancement that is observed. Despite the seemingly broad range of fluorescence enhancement values reported in the literature, the fact that TO has a significantly higher intensity enhancement upon binding to DNA in comparison to ethidium bromide is evident. The structures of TO and YO are shown in Fig. 7.

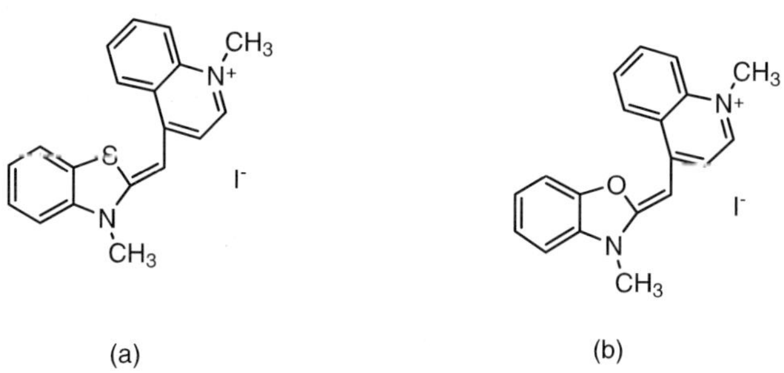

Fig. 7 Structure of **a** thiazole orange (S) and **b** oxazole yellow (O)

The increase in fluorescence intensity of TO upon binding to dsDNA is due to the restriction of rotation around the monomethine bridge upon intercalation of the dye into the double helical structure as the benzothiazole and quinolinium rings adapt to the propeller twist of the base pairs [49]. The monomethine bridge has a low energy barrier to rotation and hence is free to rotate in solution, allowing for the electronically excited dye to relax by non-radiative decay [49]. The quantum yield of free TO in solution has been reported to be 2×10^{-4} at 25 °C [43]. The binding constant for TO is $10^6\ M^{-1}$ while that of ethidium bromide is $1.5 \times 10^5\ M^{-1}$ [59].

Nygren et al. [43] have reported that although TO binds with highest affinity to dsDNA, it also weakly binds to single-stranded polypurines (5–10 times more weakly) and single-stranded polypyrimidines (10–1000 times more weakly). It was also reported that TO binds as a monomer to dsDNA and to poly(dA) sequences. However, it binds as both a monomer and a dimer to poly(dG), and binds as a dimer to poly(dC) and poly(dT) sequences. The quantum yield of TO was determined to be 0.1 when bound to mixed composition dsDNA, 0.1 when bound to single-stranded poly(dA), 0.4 when bound to single-stranded poly(dG), 0.06 for single-stranded poly(dC) and 0.01 for single-stranded poly(dT). The seemingly high quantum yield observed for the single-stranded poly(dG) sequences could be a result of the formation of higher order "guanine quartet" structures in association with the fluorophore [60, 61].

The TO and YO dyes are also available in homodimeric forms, commonly named TOTO and YOYO, respectively, and as shown in Fig. 8.

The dimeric form of TO shows only ~ 100- to 1000-fold fluorescent enhancement upon binding to DNA. However, it shows a nucleic acid binding affinity that is 100 times greater than its monomeric form [41]. One significant limitation in using TOTO is that it binds irreversibly to dsDNA and also binds to ssDNA and RNA. Monomeric TO binds reversibly to DNA, and this is an attractive feature when considering TO as a dye candidate for a self-contained biosensor. The mode of binding of TOTO has been shown to follow a bis-intercalation motif. Each TO unit intercalates between the bases, with the benzothiazolium rings inserted between the pyrimidines, and the quinolinium rings between the purine bases, and the linker interacting in the minor groove area [62].

Viscometric investigations reported by Bordelon et al. [49] indicated that the binding was intercalative. Intercalators provide for an overall lengthening of the nucleic acid strands upon insertion, which increases the viscosity of the solution in contrast to groove binders, which do not cause strand lengthening [63, 64]. The length of the linker between the two TO monomers also affects the spacing of the intercalation and hence the quantum yield. Staerk et al. [65] have shown that a longer alkyl chain linker between TO monomers in TOTO derivatives increased the quantum yield upon intercalation of the dye into an 8-mer oligonucleotide sequence. Weak non-specific interaction

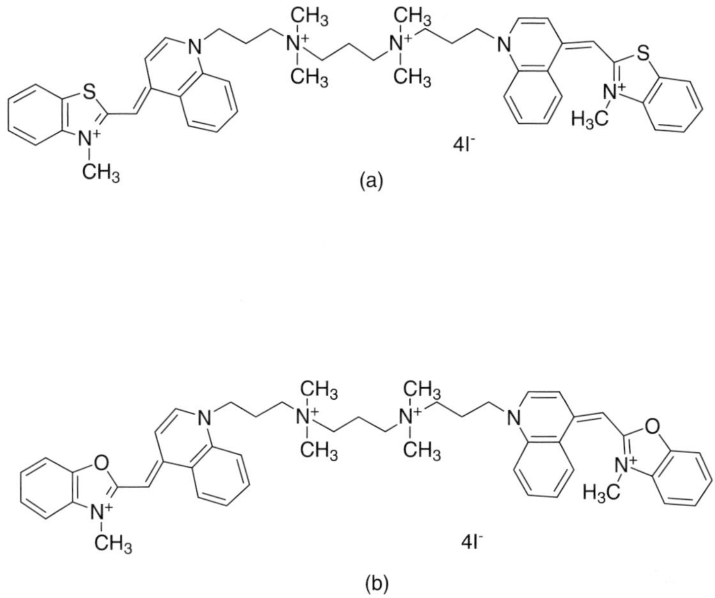

Fig. 8 Structure of **a** TOTO and **b** YOYO

of charged planar aromatic cations with the polyanionic phosphate backbone can occur at high concentrations [66] and can be ameliorated by working at low salt (≤ 0.1 M) and low dye concentrations ($< 100\ \mu$M) [67].

7 Fluorescence-Based Nucleic Acid Hybridization Assays

Solution-based nucleic acid hybridization assays represent a class of analytical methodologies that provide for detection of target·probe hybridization events. Solution-based assays offer several advantages in comparison to configurations that use surface-bound probes, primarily from the standpoint of negating the requirement for immobilization of a probe sequence to a solid surface. Therefore, thermodynamic, kinetic and adsorptive effects that are relevant in terms of consideration of hybridization at interfaces are not encountered. Solution-based hybridization assays provide a simple way of detecting hybridization events in real-time.

7.1 Hybridization Assays

Thiazole orange derivatives have also been covalently attached to oligonucleotides to generate a fluorescently labelled probe oligonucleotide capable of transducing binding interactions with complementary sequences. Privat et al. [68] have synthesized fluorescent oligonucleotide probes using a TO derivative that was covalently linked to the oligonucleotide sequence by both an alkyl and polyethylene glycol tether [68, 69]. The linker was attached to the TO through the quaternary nitrogen on the quinolinium ring. The TO-linker moieties were attached to the eighth phosphate group of a 15-mer poly d(A) sequence by a short chain thiol-terminated amino alkyl group that was bonded to the phosphorus on the backbone of the oligonucleotide via a phosphoramidate linkage.

The Rp and Sp stereoisomers (chirality at the phosphorus) of the dye-tethered oligonucleotides were examined independently to determine whether either isomer had an effect on the intercalation of the dye into the double helix upon hybridization of a d-CACT$_{15}$CAC sequence where the T$_{15}$ portion of the sequence was complementary to the probe sequence. The dye-linker complex was also tethered to the tenth phosphate of 20-mer icosathymidylates, where the thymine bases were either in the α-anomeric or β-anomeric carbon position. The bond to the phosphate was again a phosporamidate linkage, and the Rp and Sp isomers were once again isolated. Only the polymethylene linker was used in these experiments. The complementary sequence in this case was a poly r(A) sequence. Using this experimental design, the nature of the linker (hydrophobic alkyl chain versus hydrophilic polyethylene glycol chain) and the stereochemistry of the dye-linker attachment to the sequence could be studied to address whether the linkage of an intercalator to an oligonucleotide would have a stabilizing effect on the duplex formed, depending on the site of linkage (internucleotidic positions versus 5′-end labelling) and the chemistry of the linking chain [70].

The results showed that an increase in T_m was observed for hybridization events where TO was linked to the oligonucleotide sequence by the polyethylene glycol and the polymethylene linker. The increases were greater for the polyethylene glycol linker (+8.5 °C and +7.5 °C for the Rp and Sp isomers, respectively) as compared to the polymethylene linker (+8 °C and +5.5 °C for the Rp and Sp isomers, respectively). It was suggested that the stabilization was associated with the orientation of the isomers with the minor groove. This argument was consistent with the fact that the TOTO dimer intercalates with the orientation of the quinolinium nitrogen towards the minor groove [69]. The polyethylene (tetraethylene) glycol linker was longer than the polymethylene (decyl) linker, leading to full intercalation for both isomers and hence a similar change in melt curve temperature for the Rp and Sp isomers as for the polyethylene glycol linker. The hydrophilicity of the polyethylene glycol chain could lead to a greater range of motion as well, so that the dye could readily intercalate into the double helix.

There was no increase in stability observed for the icosathymidine-linked derivatives upon binding of the complementary polyr(A) sequence, suggesting that no intercalation of the dye took place. The thymine residue is a pyrimidine and the TOTO dimer has been reported to intercalate with the benzothiazole ring oriented towards the pyrimidine bases. Given that the dye was linked to the phosphate backbone of the polypyrimidine strand through the quinolinium ring, this orientation may not have facilitated an intercalative interaction.

Further work by Privat et al. [71] involving the poly(dA) sequence functionalized with a TO tethered at the central internucleotidic position by the same polyethylene and polymethylene linkers showed that there was a significant increase in quantum yield for the dsDNA hybrid over the ssDNA oligonuclotide in solution using both the hydrophilic and hydrophobic linker conjugates. The quantum yield enhancement was shown to be higher for the alkyl chain linker species than the polyethylene glycol linker. It is interesting to note that a study of dsDNA of the same base sequence was done using TO-PRO-1 in bulk solution (i.e. dye was not tethered to the sequence). The quantum yield enhancements were significantly larger than had been observed for ssDNA-free dye experiments. The quantum yield was 7×10^{-3} for the ssDNA sample and 0.17 for dsDNA with free dye in a 1 : 1 ratio of DNA : dye. The quantum yield was 0.25 for dsDNA with free dye in a 10 : 1 ratio of DNA : dye, and was 0.35 and 0.26 for the alkyl chain tethered dsDNA conjugates (*R* and *S* isomers), and 0.10 and 0.06 for the polyethylene glycol chain dsDNA conjugates (*R* and *S* isomers) [71]. These results are significant in that the quantum yields for the alkyl chain tethered dsDNA conjugates exhibited higher values than free dye in solution. Furthermore, the TO-PRO-1 fluorophore bears two positive charges, which are proposed to enhance the quantum yield by imparting increased binding stability [72], whereas the TO-linker species bears only one positive charge. It was also proposed that the Sp stereoisomeric form of dye linkage lead to a lower quantum yield because the linker had to cross over the sugar-phosphate backbone. In the Rp stereoisomeric form, the dye was oriented towards the minor groove and this orientation facilitated intercalation as shown through molecular modelling simulations [71].

Several complexities must be addressed in order to fully understand the mode(s) of binding and the effects of the linker and dye on the stability of the duplex. The fact that the greatest stability was observed with an alkyl chain linker may be somewhat problematic in a "dirty" sample, as might be encountered in the analysis of crudely prepared "real-world" samples. In such cases, the alkyl chain could provide sites at which proteins and lipids might adsorb. Quantum yields are sufficient such that polyethylene glycol tethers could be used in real-world sensor configurations.

7.2 Strategies for Fluorescence Transduction of Hybridization Based on Use of Labelled Binding Material

7.2.1 Fluorescent Cationic Polymers

Recently, polymeric polyelectrolyte materials have been synthesized that can detect the presence of a hybridization event. Doré et al. [73] have designed fluorescent polymeric transducers, and have reported detection of zeptomole concentrations of nucleic acids. The mechanism of function of the cationic polythiophene transducer is based on electrostatic interactions between the polyanionic backbones of nucleic acid strands and the positively charged polymer material. A single-stranded oligonucleotide probe is essentially wound around the positively-charged polymer material and fluorescence of the polymer material is quenched ($\Phi = 0.03$) since these structures tend to form aggregates in solution [74, 75]. Upon introduction of target oligonucleotide material, hybridization takes place. This increases the solubility of the complex and the aggregates no longer form so that the polymeric material can emit fluorescence. A "triplex" polyelectrolyte is formed in solution as a result of the interaction between the two polyanionic oligonucleotides and the polycationic polymer. There is a specific emission wavelength associated with this triplex that is monitored and this is the premise for the detection of double-stranded DNA complexes. Kinetic differences observed between perfect match and base pair mismatched target sequences allows the discrimination of single base pair mismatches. This method can also detect different sequence lengths of target material and leads to higher fluorescence signals for longer oligonucleotide sequences, likely owing to the requirement for more polymeric material, as would be required to intertwine with the longer probe sequence. The detection limit for a 20-mer system was estimated to be 3.6×10^{-18} M and 2.4×10^{-18} M for a 50-mer sequence, for a reaction time of 60 min [73]. This label-free detection system does not require the amplification of target sequence by PCR, and the polymeric material also has the potential for use as a "molecular wire" in electrochemical nucleic acid biosensor applications [73].

7.2.2 Molecular Beacons

One of the limitations associated with the use of fluorogenic dyes as labels for nucleic acids is the relatively high background fluorescence intensity in solution. This problem can be ameliorated by the use of asymmetric cyanine dyes. Another approach that can achieve low fluorescence background is based on the use of FRET pairs. If the acceptor and donor are in close proximity to one another, the acceptor can absorb the emission energy of the donor in a non-radiative mechanism as long as the pair are within the Förster distance for that particular set of donor and acceptor species [76]. The donor and acceptor can both be fluorescent dyes where the excitation wavelength for the acceptor spectrally overlaps with the fluorescence emission wavelength of the donor. Alternatively, the acceptor can be a "dark absorber" which does not fluorescence. In this case the fluorescence that is observed is from the donor, and fluorescence is observed only when the donor is sufficiently distanced from the acceptor species. Methods that use FRET pairs include molecular beacons [77], duplex probes [78], scorpion primers [79] and fluorescence polarization assays [80].

Molecular beacons (MB) are single-stranded probe oligonucleotides that possess a donor and an acceptor bound to the terminal ends of the sequence [77]. A specified number of bases at the 3′- and 5′-ends of the sequence (usually 4–7 bases) are complementary to one another allowing the MB to adopt a hairpin or "stem-loop" conformation. This conformation allows the donor and acceptor to be in close proximity to one another and efficient quenching of the donor species by the acceptor species results. The MB maintains this conformation until a target sequence of sufficient complementarity to the MB is introduced, which allows the system to overcome the energetics associated with maintenance of the stem-loop structure via the terminal nucleotide pairs. The hairpin structure is then dissociated as the MB binds to its complementary sequence. This binding event elongates the MB sequence and this increases the distance between the donor and acceptor species beyond the Förster distance so that the fluorescence of the donor species is restored, as illustrated in Fig. 9.

Several combinations of donor/acceptor pairs have been utilized including two fluorophores [77], a fluorophore with a quencher which does not emit fluorescence radiation [81, 82], a fluorophore with an intercalator [83], and a fluorophore with a gold particle [84]. There are several mechanisms available for the acceptor to quench the fluorescence of the donor species including static quenching, dynamic quenching and contact-mediated quenching [85]. Length and the chemical composition of the linker species involved in tethering the donor and acceptor moieties to the 3′- and 5′-ends of the oligonucleotide are also factors that must be considered when optimizing an MB system. Zhang et al. [77] designed MB species and reported limits of detection of 1.7×10^{-10} M using only a simple spectrophotometer detection scheme. The use of automated solid phase oligonucleotide synthesis methods in conjunction with special phosphoramidites containing the donor and quencher species of interest permits facile assembly of MBs [86]. One limitation to using automated synthesis is that the MB, especially the donor/acceptor pair comprised of two fluorophores, is synthesized in low yield as a result of difficulties associated with synthon coupling and purification of these complex molecules [86].

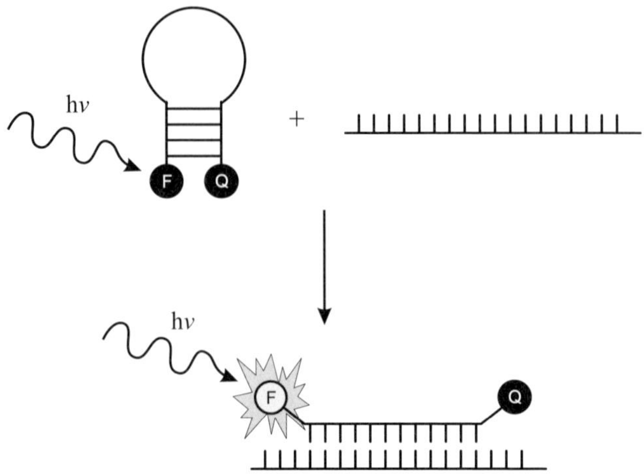

Fig. 9 Binding of a MB to a complementary target nucleic acid. In the closed form, the fluorescence of F is quenched by Q. Upon hybridization with the target, the F–Q pair are no longer in close proximity and the fluorescence of F is restored

Limitations exist with the use of MB technology in the analysis of single-nucleotide polymorphisms (SNPs). This is owing to the fact that the destabilization of the duplex formed on binding of MP to the SNP-containing target is insufficient to prevent denaturation of the stem, particularly since the stem is only held together by a small number of base pairs. This will result in restoration of the fluorescence of the donor species, which gives rise to an analytical signal that may not be indicative of the presence of a SNP within the target of interest. Also, the MB donor and acceptor are both on the same oligonucleotide sequence, which can lead to partial quenching of the donor species by non-collisional mechanisms as they are still in a somewhat close proximity [78, 87].

7.2.3 Duplex Probes

Duplex probes are two complementary oligonucleotides of different length, where the longer sequence is labelled at the 5′-end with a fluorescence donor species and the 3′-end of the shorter sequence is labelled with a quencher species [78]. The longer strand containing the fluorophore acts as the probe sequence that is complementary to the target material. In the absence of target material, the long and short labelled strands hybridize, and this places the fluorophore and the quencher in close proximity so fluorescence is quenched. In the duplex probe scheme, the fluorophore should be chosen so that it exhibits minimal fluorescence in solution when the duplex probe is denatured and the fluorescent probe is not hybridized to its target. If the duplex probe is used in a PCR experiment, the duplex probe sequence is designed so that the probe T_m is higher than the anneal-extend temperature but lower than the probe/amplicon duplex melt temperature [78]. This ensures that the duplex probe conformation is maintained by excess probe so that its fluorescence is quenched [78]. When primer extension reaches the fluorescent probe, the probe and the fluorophore are hydrolysed, allowing fluorescence to be cumulative over all of the cycles of amplification so that real-time monitoring of PCR can be achieved [78].

The duplex probe configuration provides for a close proximity of the donor/acceptor pair and may lead to two mechanisms of quenching: FRET and direct transfer (contact-mediated) quenching [78]. Placement of the fluorophore and quencher toward the centre of the duplex probe sequence may further add to the quenching efficiency as the ends of duplexes are known to "breathe" and are not as tightly bound as internal base pairs. Duplex probes are relatively easy to synthesize as the fluorophore and quencher moieties do not have to be incorporated into the same strand.

If the duplex probe was to be used in a sensor configuration, the system could be custom-designed with differing number and position of base mismatches and differing lengths of donor- and acceptor-labelled strand so that the energetics could be optimized for development of fluorescence signal upon probe/target recognition.

8
Detection of DNA Hybridization at Surfaces by Fluorescence

The use of fluorescently labelled oligonucleotides for detection of nucleic acid hybridization events at surfaces is well-established and continues to evolve as new detection schemes and improvements in surface chemistry are developed. The emission of fluorogenic dyes provides for high sensitivity detection of hybridization events at interfaces. Methods involving fluorescent detection at surfaces commonly use glass slides or optical fibres as the substrate platform. Slides offer a geometric simplicity that is conducive for microarrays, while optical fibres provide excellent sensitivity and a compact optical design owing to the mode of coupling of the fluorescence radiation back through the fibre for delivery to the filtering optics and detectors.

Molecular beacon technology was first developed in solution and has recently been applied to applications involving surface immobilized chemistry. Ramachandran et al. [18] covalently immobilized molecular beacons on both aldehyde glass and hydrogel slides to provide probe oligonucleotides for interfacial hybridization experiments. The stem-loop structure of the molecular beacon helped to minimize background fluorescence. An amino-terminated C6-alkyl chain linker attached to the 5-methyl position on the aromatic ring of the 3′ terminal T nucleotide was covalently linked to a glass slide that was functionalized with aldehyde moieties. A Blackhole quencher was linked to a terminal 3′-phosphate group attached to the sugar of the same T base so that the quencher was situated at the 3′-end of the sequence. The Cy3 fluorophore was positioned at the 5′-end by a linkage to the 5′-terminal phosphate group. The last four bases at both the 3′-end and the 5′-end were complementary so that a hairpin conformation was adopted. There were 18 bases, which comprised the loop of the hairpin, for a total of 26 bases that were available for hybridization on the probe oligonucleotide. The target sequences were designed such that they were complementary to 18 bases of the loop. Two different molecular beacons of this design were synthesized, each being complementary to different areas of part of the genome of the bacterium *Francisella tularensis* [18]. Target oligonucleotides were synthesized and were 20 base pairs in length, corresponding to the 18 bases in the probe loop and two on the stem portion. Targets of full complementarity as well as one, two, three and four base-pair mismatches were tested. Experiments done using the surface immobilized molecular beacons were also duplicated in solution phase experiments in order to study the differences between hybridization in solution and hybridization at the interface. Solutions containing the molecular beacon probes were spotted onto the slides using a robotic printer, and then the surface was blocked using bovine serum albumin (BSA). Target solutions were then spotted onto a cover slip and the cover slip was inverted and placed on top of the slide so that the target spots were aligned to interact with the probe spots. Hybridization was done at room temperature for 5 min after which the slides were washed with buffer and then allowed to dry for 10 min [18].

In general, background fluorescence intensities were observed to be of consequence, with the aldehyde-functionalized slides providing the most intense background signal [18]. In some cases, the signals observed from single base-pair mismatched samples were greater than signals observed from the completely complementary sequence. It was demonstrated that the magnitude of the fluorescence signal intensity from the mismatched sequences was dependent on the position of the mismatch within the sequence. The signal was greatest for the mismatch located at the terminal base on the 5′-end of the target. This is consistent with the notion that the bases at the termini of the hybrids are more weakly associated than those in the central portion of the double-stranded nucleic acid structure. However, it is interesting to note that the signal from mismatched hybrid was greater than that of the signal from the binding of the fully complementary sequence.

Thermal denaturation profiles were done using one of the molecular beacon structures in solution, without the alkyl chain linker attached to the 3′-thymidine residue. The fully complementary and single base pair mismatched sequence targets were investigated. The differences of T_m from the melt profiles showed that single base pair mismatches could not be discriminated from the complementary target material. Melt curves were also obtained for the same molecular beacon probe immobilized at a surface when exposed to the target oligonucleotide sequences containing two, three and four base-pair mismatches; however, reliable differences in melt temperature differences were not observed. The sensors were also tested in terms of their reusability after a 0.4 M NaOH solution was used to wash the slide surface. The background between the first and the second use was decreased. However, only 75% of the signal was regenerated upon the second use.

Cordek et al. [20] have immobilized molecular beacons by use of an avidin-biotin motif between a linker species immobilized on the surface of optical fibres and a probe molecular beacon. The beacon used the donor/acceptor pair comprised of tetramethylrhodamine (TMR) and DABCYL, respectively. The target species was a 204 nucleotide RNA sequence and the two probe sequences were 30-mer and 28-mer oligonucleotides. The fibres were cleaned, silanized and then functionalized with a biotinylated linker chain to impart mobility to the nucleic acid probes to be immobilized via coupling to the linker terminus. The fibres were then immersed in a streptavidin solution and the biotinylated molecular beacons were permitted to bind to the immobilized streptavidin. No fluorescence was observed for non-complementary RNA sequences and a fluorescence signal was observed within 10 min following introduction of complementary target RNA. In this configuration, a discrimination between complementary and single base pair mismatch (placed in the middle of the target sequence) was observed at the early stages of hybridization, prior to the establishment of equilibrium binding ($\sim$ 3 min). It was also shown that the molecular beacon could be regenerated by exposure to 90% formamide in TE buffer. However, the fluorescence signal upon target/probe hybridization decreased with increasing cycles of application, indicating that the molecular beacon was likely removed in-part from the surface during each regeneration procedure.

Ionic strength influences are well known with respect to the rate and energetics of nucleic acid hybridization [17]. Charge and ionic radius are both important in terms of stabilizing the structure of the duplex as well as stabilizing the stem portion of the molecular beacon [17]. The stem structure stability was increased when a divalent cation was incorporated into the hybridization buffer solution [17]. It was reported that cations were best at stabilizing the duplex formed upon hybridization in the order $Ca^{2+} > Mg^{2+} \gg K^{+} > Na^{+}$. The ultimate detection limit of the sensor configuration was calculated to be 1.1 nM [17].

Linear oligonucleotide probes have been immobilized along the length of fibres as well as on the distal end of fibres. Healey et al. [88] described the immobilization of three different oligonucleotide probes in acrylamide polymer matrices at the distal ends of optical fibres. Base pair mismatches could be distinguished on the same sensor after a 20-min hybridization with a detection limit of 0.2 nM [88]. More recently, zeptomole ($\sim$600 molecule) detection limits have been obtained by using etched distal ends of optical fibres, which provide microwells for fluorescent microspheres functionalized with nucleic acid probe molecules [89]. A 500 μm fibre bundle contained 6000 optical fibres of 3 μm diameter, each capable of holding one microsphere. A 5′-amino-C6 linker modified 23-mer probe was covalently bonded to the microspheres and unreacted amine groups were capped with succinic anhydride. Hybridization experiments with fluorescein-tagged target nucleic acids were done over a 12 h time period. The DNA sequences were chosen based on sections of the interleukin 2, interleukin 6 and F508C mutation in the gene that has been linked to cystic fibrosis [89]. Different concentrations of target could be detected as a function of hybridization time. After 1 h, femtomolar concentrations could be distinguished [89]. The number of microsensors incorporated into the array was decreased to ten per probe type and used to determine that target concentrations of 100 aM (10^{-21} mol or 600 molecules) could be detected from a 10 μL sample [89]. Quantitative analyses at the femtomolar concentration level showed good reproducibility, while determinations done in the attomolar concentration regime demonstrated a precision of about 20% [89]. Introduction of short chain (10–125 base pair) salmon sperm DNA reduced the fluorescence signal observed from the complementary target material [89].

A sensitive optical biosensor device was reported by Kleinjung et al. [90]. An evanescent wave-based fibre optic sensor provided a detection limit of 3.2 attomoles (1.9×10^6 molecules) of non-labelled fully complementary nucleic acid target sequence in the presence of a double-strand selective fluorochrome with full regenerability for over 60 cycles of application without discernible loss of sensitivity. Capture oligonucleotides were immobilized onto the optical fibre substrate by either an avidin-biotinylated oligomer approach or by covalent attachment of the capture oligomer by carbonyldiimidazole activation. Hybridization experiments were done using a protocol involving exposure of the sensor situated in a continuous flow apparatus (100 $\mu L\,min^{-1}$ flow rate, 100 μL volume exposed to the fibre-optic sensor to the target nucleic acid in hybridization buffer for 60–180 s, followed by a 120–180 s wash with buffer, 45–60 s treatment with a solution of fluorochrome (either Pico-Green or YOYO-1), 30 s wash with buffer (at 2.0 $mL\,min^{-1}$), and finally, immersion in buffer for 210–255 s (steady state for data collection). Experiments that used biotin-avidin immobilized oligonucleotides provided a hybridization efficiency of 55%; however, the sensors could not be regenerated owing to the high stability of the double-stranded oligonucleotide fluorochrome complex. Thermal regeneration of the sensors at high temperature (90 °C) could denature the nucleic acid fluorochrome complex ($T_m = 88$ °C for Pico-Green/dsDNA, $T_m > 90$ °C for YOYO-1/dsDNA, $T_m = 45$ °C for dsDNA alone) and regenerate the sensors; however, the avidin-biotin linkage of the oligonucleotide to the waveguide substrate would also be compromised. This led the researchers to abandon the avidin-biotin immobilization protocol for covalent attachment methods, where complete regeneration was possible using thermal denaturation by a stepwise heating to 90 °C following hybridization and staining. Single-base mismatch discrimination was also investigated and it was demonstrated by Kleinjung et al. that under appropriate conditions of temperature, pH and ionic strength, a high degree of selectivity for detection of a central single-base (T-T) mismatch could be achieved for a tridecanucleotide.

Piunno et al. [91, 92] have reported a method of detecting nucleic acid hybridization using probe oligonucleotides that were immobilized onto optical fibres. Very recently, rapid detection of SNPs have been established based on a total internal reflection scheme [13]. Probe oligonucleotides were designed using sequences associated with mutations of the SMN1 gene, which are diagnostic of the childhood neurodegenerative disease known as spinal muscular atrophy [13]. Probe sequences that were 19 nucleotides in length were covalently immobilized to the surfaces of fused silica optical fibres by solid phase oligonucleotide synthesis. The fibres were first functionalized with a hexaethylene glycol linker in high density, and then oligonucleotides were grown in-situ using carefully controlled conditions to achieve specific average densities of probe immobilization. Target sequences of 202 base pair length obtained as asymmetric PCR amplicons were labelled at the 5′-end with Cy5 fluorogenic dye. By choosing an operating temperature at which single base mismatched target material did not bind to the probe sequence, rapid and quantitative SNP analysis could be achieved in less than 2 min based on the time derivative of fluorescence intensity changes. High probe immobilization density resulted in more organized nucleic acid films. The use of increased probe density was instrumental in increasing the selectivity of the probe sequences as well as decreasing the amount of non-selective adsorption. This allowed experiments to be done in high ionic strength buffer (1 M) in order to maximize hybrid yields while still maintaining the selectivity for discrimination between the fully matched target sequence. It was also shown that T_m values could be greater for interfacial hybrids than for T_m values of similar oligonucleotide systems in solution [13]. Regeneration of sensor surfaces was accomplished by washing the sensors with formamide and deionized water, and 80 cycles of use were reported.

One limitation that plagues every detection strategy that requires fluorogenic labels is that the labels must be linked to the target strand prior to analysis. A significant advancement for such biosensor devices would be to associate the fluorescent label to the probe strand so that target strands do not require labelling prior to analysis. It would also be advantageous to use an intercalating fluorescent dye as these dyes are dependent on formation of stable hybrids.

Krull and Wang [93] have reported preliminary work on the immobilization of oligonucleotide probes on the surface of an optical fibre where the probe molecules were functionalized by covalent attachment of intercalating dye. Probe sequences that were 20 nucleotides in length were covalently immobilized to the surfaces of fused silica optical fibres functionalized with a flexible hexaethylene glycol linker by solid phase oligonucleotide synthesis. TO was linked to the 5′-phosphate of the terminal nucleotide by multiple units of diethylene glycol linkers, which in turn were linked to the quinolinium ring of the TO. The length of the polyethylene glycol linkers was chosen such that they would allow intercalation to occur below the 3–4 base pairs at the end of the hybridized sequence which "breathe" and are not necessarily as securely bound as the other, more centrally located base pairs [94]. Fully matched targets provided a fluorescence signal that was observed to be associated with hybridization, and clear discrimination between base pair mismatches was reported.

In further work, PCR products (286 and 344 bp), and plasmids of about 3k-bp were prepared. The two plasmids were designed to contain a gene target that was selected from *Erwinia herbicola* (EH). The PCR products were made by amplifying the specific regions of the plasmid that contained the EH insert. TO-tethered to 15-mer probes were used for hybridization, and these probes were immobilized onto optical fibres. The targets provided a fluorescence signal that was observed to be associated with hybridization, although the response was slower (minutes) than previously observed for hybridization in the absence of tethered dye. The assay could also be done in the presence of significant quantities of non-selective background DNA without interference.

9 Summary

The emerging field of optical biosensor and biochips for nucleic acid analysis has developed significantly over the last decade. The key limitations faced by the technologies developed to date rest with issues of sensitivity and achievement of selectivity without compromise of the speed of analysis and the reusability of the devices. Significant advances have been made with respect to the enhancement of device selectivity. Several groups have determined that the chemistry used for the immobilization of films of nucleic acid probes greatly influences the selectivity of response from the standpoint of the physical chemistry within the nucleic acid film (which is substantially different from that of bulk solution). Thermodynamics of hybridization may therefore be tuned via appropriate selection of the immobilization chemistry and the resultant organization of the immobilized nucleic acid probes. Ultimately a balance must be struck between the benefits of high immobilization density to enhance thermodynamic selectivity and for reduction of non-selective binding interactions versus the onset of nearest-neighbour interactions that can reduce the availability and stability of hybrids, which would limit the sensitivity of a given hybridization assay.

Many device designs have also been investigated with the goal of creating simple, practical optical biosensor technologies. SPR detection systems have been developed and provide the intrinsic design advantage that assays may be conducted without the use of labels, with the caveat that the interfacial chemistry must be made intrinsically resistant to non-specific adsorption, as adsorbed material cannot be differentiated from target binding.

Several approaches based on fluorescent signal evolution on optical substrates have been described. Some assays are based on the use of fluorophores strictly as labels to indicate the presence of a biomolecule on an interface. Though sensitive detection may result, such an approach often necessitates the requirement to pre-label the target sequence, which imparts considerable preparative work to the assay protocol and takes away from the concept of a true biosensor. Fluorophores or systems of fluorophores have also been developed that provide indications of structural change associated with hybrid formation. A "molecular beacon" approach has been explored via the immobilization of stem-loop FRET probes on optical substrates. The disruption of the stem-loop structure of the native beacon on association with target molecules occurs to form a linear duplex, with the terminal fluorophore and quencher moieties of the beacon well distanced so that quenching by FRET is eliminated. Though the MB avoids the requirement of pre-labelling the target species prior to detection, limitations exist with the use of MB technology in the analysis of single-nucleotide polymorphisms (SNPs). This is owing to the fact that the destabilization of the duplex formed on binding of the MP to the SNP-containing target is insufficient to prevent denaturation of the stem, particularly since the stem is only held together by a small number of base pairs.

The use of intercalant dyes tethered to nucleic acids has also been explored to provide for immobilized probes that, like molecular beacons, allow reagent-less determinations of selective interfacial binding events without the requirement to pre-label the target species. Issues with respect to high background intensities from the fluorophores in the presence of single-stranded oligonucleotides exist owing to limitations in quantum yield alteration between the bound and unbound state of the intercalant. This limits the ultimate sensitivity offered by the approach using tethered intercalant. Additional complexities exist with respect to acquiring a more detailed understanding of the mode(s) of tethered dye binding, the effects of the linker and dye on the stability of the duplex, and how the properties of the fluorophore associated with immobilized nucleic acid are affected as a function of immobilization chemistry, density and physical chemistry of the interface.

Advancements in the area of optical biosensors for nucleic acid determinations have been substantial and many design concepts and surface chemistries have been explored. A practical solution that provides for all of the desired biosensor attributes – reusability, robustness, reliability, selectivity, sensitivity and speed of response – in a single device has not yet been identified. The key limitations are centred on the issues of sensitivity and selectivity while not compromising the speed of analysis and the reusability of the devices. It is anticipated that work in these areas of device design and selectivity will be a development priority.

References

1. Walker NJ (2002) Science 296:5567
2. Shchepinov MS, Case-Green SC, Southern EM (1997) Nucleic Acids Res 25:1155
3. Petersen JL, Small GD (2001) Nucleic Acids Res 29:4472
4. Oh SJ, Cho SJ, Kim CO, Park JW (2002) Langmuir 18:1764
5. Watterson JH, Piunno PAE, Wust CC, Krull UJ (2000) Langmuir 16:4984
6. Watterson JH, Piunno PAE, Wust CC, Krull UJ (2001) Sens Actuators B 74:76
7. Herne TM, Tarlov MJ (1997) J Am Chem Soc 119:8916
8. Fotin AV, Drobyshev AL, Proudnikov DY, Perov AN, Mirzabekov AD (1998) Nucleic Acids Res 26:1515
9. Peterson AW, Heaton RJ, Georgiadis RM (2001) Nuceic Acids Res 29:5163
10. Piunno PAE, Watterson JH, Wust CC, Krull UJ (1999) Anal Chim Acta 400:73
11. Vainrub A, Pettitt BM (2003) J Am Chem Soc 125:7798
12. Watterson JH, Piunno PAE, Wust CC, Raha S, Krull UJ (2001) Fresenius J Anal Chem 369:601
13. Watterson JH, Raha S, Kotoris CC, Wust CC, Gharabaghi F, Jantzi SC, Haynes NK, Gendron NH, Krull UJ, Mackenzie AE, Piunno PAE (2004) Nucleic Acids Res 32:18
14. Piunno PAE, Watterson JH, Kotoris CC, Krull UJ (2005) Anal Chim Acta 534:53
15. Eggins B (1996) Biosensors. Wiley, New York
16. Love WF, Button LJ, Slovacek RE (1991) Optical characteristics of fiberoptic evanescent wave biosensors: theory and experiment. In: Wise LB, Wingard DL (eds) Biosensors with fiberoptics. Humana, New Jersey, pp 139–180
17. Liu X, Tan W (1999) Anal Chem 71:5054
18. Ramachandran A, Flinchbaugh J, Ayoubi P, Olah GA, Malayer JR (2004) Biosens Bioelectron 19:727
19. Anerson GP, Golden JP, Ligler FS (1994) IEEE Trans Biomed Eng 41:578
20. Cordek J, Wang X, Tan W (1999) Anal Chem 71:1529
21. Henke L, Piunno PAE, McClure AC, Krull UJ (1997) Anal Chim Acta 344:201
22. Ebersole RC, Miller JR, Moran JR, Ward MD (1990) J Am Chem Soc 112:3239
23. Liu X, Tan W (1999) Mikrochim Acta 131:129
24. Homola J (2003) Anal Bioanal Chem 377:528
25. Corradini R, Feriotto G, Sforza S, Marchelli R, Gambari R (2004) J Mol Recognit 17:76
26. Sforza S, Galaverna G, Dossena A, Corradini R, Marchelli R (2002) Chirality 14:591
27. Neilson PE, Haaima G (1997) Chem Soc Rev 26:73
28. Lyon LA, Musick MD, Natan MJ (1998) Anal Chem 70:5177
29. Nicewarner-Peña SR, Raina S, Goodrich GP, Fedoroff NV, Keating CD (2002) J Am Chem Soc 124:7314
30. Baeumner AJ, Pretz J, Fang S (2004) Anal Chem 76:888
31. Yguerabide J, Yguerabide EE (1998) Anal Biochem 262:157
32. Kreibig U, Vollmer M (1995) Optical properties of metal clusters. Springer, Berlin Heidelberg New York
33. Kelly KL, Coronado E, Zhao LL, Schatz GC (2003) J Phys Chem B 107:668

34. Grabar KC, Freeman RG, Hommer MB, Natan MJ (1995) Anal Chem 67:735
35. Taton TA, Mirkin CA, Letsinger RL (2000) Science 289:1757
36. Fritzsche W, Taton TA (2003) Nanotechnology 14:R63
37. Csáki A, Kaplanek P, Möller R, Fritzsche W (2003) Nanotechnology 14:1262
38. Nam J, Stoeva SI, Mirkin CA (2004) J Am Chem Soc 126:5932
39. Hilliard LR, Zhao X, Tan W (2002) Anal Chim Acta 470:51
40. Cooper M, Ebner A, Briggs M, Burrows M, Gardner N, Richardson R, West R (2004) J Fluoresc 14:145
41. Haugland R (2002) Handbook of molecular probes and research products, 9th edn. Molecular Probes, Eugene
42. Timtcheva I, Maximova V, Deligeorgiev T, Zaneva D, Ivanov I (2000) J Photochem Photobiol A 130:7
43. Nygren J, Svanik N, Kubista M (1998) Biopolymers 46:39
44. Timcheva II, Maximova VA, Deligeorgiev TG, Gadjev NI, Sabnis RW, Ivanov IG (1997) FEBS Lett 405:141
45. Yue S, Johnson I, Huang Z, Haugland RP (1994) US Patent 5321130
46. Roth B, Millard P, Yue S, Wells S, Haugland RP (1995) PCT Int Appl WO9424913
47. Deligeorgiev TG, Gadjev NI, Drexhage KH, Sabnis RW (1995) Dyes Pigm 29:315
48. Yarmoluk SM, Lukashov SS, Losytskyy MY, Akerman B, Kornyushyna OS (2002) Spectrochim Acta Part A 58:3223
49. Bordelon JA, Feierabend KJ, Siddiqui SA, Wright LL, Petty JT (2002) J Phys Chem B 106:4838
50. Norman DG, Grainger RJ, Uhrin D, Lilley DMJ (2000) Biochemistry 39:6317
51. Yarmoluk SM, Kovalska VB, Lukashov SS, Slominskii YL (1999) Bioorg Med Chem Lett 9:1677
52. Garrett RH, Grisham CM (1995) Biochemistry. Saunders College and Harcourt Brace College, Toronto
53. Waring MJ (1965) J Mol Biol 13:269
54. Bloomfield VA, Crothers DMC, Tinoco I (1974) Physical chemistry of nucleic acids. Harper and Row, New York
55. Deligeorgiev T, Timtcheva I, Maximova V, Gadjev N, Drexhage K (2000) J Fluoresc 12:225
56. Lee LG, Chen CH, Chiu LA (1986) Cytometry 7:508
57. Kricka LJ (2002) Ann Clin Biochem 39:114
58. Rye HS, Yue S, Wemmer DE, Quesada MA, Haugland RP, Mathies RA, Glazer AN (1992) Nucleic Acids Res 11:2803
59. Rye HS, Quesada MA, Peck K, Mathies RA, Glazer AN (1991) Nucleic Acids Res 19:327
60. Sen D, Gilbert W (1992) Methods Enzymol 211:191
61. Williamson JR, Raghuraman MK, Cech TR (1989) Cell 59:871
62. Spielmann HP, Wemmer DE, Jacobsen JP (1995) Biochemistry 34:8542
63. Cohen G, Eisenberg H (1969) Biopolymers 8:45
64. Wakelin LPG, Waring MJ (1976) Biochem J 157:721
65. Staerk D, Hamed AA, Pedersen EB, Jacobsen JP (1997) Bioconjugate Chem 8:869
66. Laib S, Seeger S (2004) J Fluoresc 14:187
67. Wilson WD (1999) In: Kool ET (ed) DNA and aspects of molecular biology. Comprehensive natural products chemistry. Pergamon, Amsterdam, p 427
68. Privat D, Asseline U (2001) Bioconjugate Chem 12:757
69. Jensen LK, Gotfredsen CH, Bondensgaard K, Jacobsen JP (1998) Acta Chem Scand 52:641
70. Asseline U, Thuong NT, Helene C (1997) New J Chem 21:5

71. Privat E, Melvin T, Merola F, Schweizer G, Prodhomme S, Asseline U, Vigny P (2002) Photochem Photobiol 75:201
72. Petty JT, Bordelon JA, Robertson ME (2000) J Phys Chem B 10:7221
73. Doré K, Dubus S, Ho H-A, Lévesque I, Brunette M, Corbeil F, Boissinot M, Boivin M, Bergeron MG, Boudreau D, Leclerc M (2004) J Am Chem Soc 126:4240
74. Ho H-A, Boissinot M, Bergeron MG, Corbeil G, Doré K, Boudreau D, Leclerc M (2002) Angew Chem 41:1548
75. Nilsson KPR, Inganas O (2003) Nat Mater 2:419
76. Atkins P, dePaula J (2002) Physical chemistry, 7th edn. Freeman, New York
77. Zhang P, Beck T, Tan W (2001) Angew Chem 113:416
78. Kong D-M, Huang Y-P, Zhang X-B, Yang W-H, Shen H-X, Mi H-F (2003) Anal Chim Acta 491:135
79. McKeen CM, Brown LJ, Nicol JTG, Mellor JM, Brown T (2003) Org Biomol Chem 1:2267
80. Kwok P-Y (2002) Hum Mutat 19:315
81. Li JJ, Fang X, Schuster SM, Tan W (2000) Angew Chem Int Ed 39:1049
82. Tyagi S, Kramer FR (1996) Nat Biotechnol 14:303
83. Yamene A (2002) Nucleic Acids Res 30:97
84. Dubertret B, Calame M, Libchaber AJ (2001) Nat Biotechnol 19:365
85. Marras SAE, Kramer FR, Tyagi S (2002) Nucleic Acids Res 30:122
86. Mullah B, Livak K (1999) Nucleotides Nucleosides 18:1311
87. Solinas A, Brown LJ, McKeen C, Mellor JM, Nicol JTG, Thelwell N, Brown T (2001) Nucleic Acids Res 29:26
88. Healey BG, Matson RS, Walt DR (1997) Anal Biochem 251:270
89. Epstein JR, Lee M, Walt DR (2002) Anal Chem 74:1836
90. Kleinjung F, Bier FF, Warsinke A, Scheller FW (1997) Anal Chim Acta 350:51
91. Piunno PAE, Krull UJ, Hudson RHE, Damha MJ, Cohen H (1994) Anal Chim Acta 288:205
92. Piunno PAE, Krull UJ, Hudson RHE, Damha MJ, Cohen H (1995) Anal Chem 67:2635
93. Wang X, Krull UJ (2002) Anal Chim Acta 470:57
94. Jakeway SC, Krull UJ (1999) Can J Chem 77:2083

Springer Ser Chem Sens Biosens (2005) 3: 261–277
DOI 10.1007/5346_008

Published online: 5 October 2005

Gold Nanoparticles in Bioanalytical Assays and Sensors

Nguyen Thi Kim Thanh[1,2] · Aude Vernhet[3] · Zeev Rosenzweig[1] (✉)

[1]Department of Chemistry, University of New Orleans, New Orleans, LA 70148, USA
ntkthanj@liv.ac.uk, zrosenzw@uno.edu

[2]Centre for Nanoscale Science, Department of Chemistry & School of Biological Sciences, University of Liverpool, Crown Street, Liverpool L69 7ZD, UK
ntkthanj@liv.ac.uk

[3]UMR Sciences pour l”nologie, Campus ENSAM/INRA, bât. 28, Maître de Conférences ENSA-Montpellier, 2 place Viala, 34060 Montpellier, France

Abstract In this review we report two major applications of gold nanoparticles in the field of bioassay and sensing. The first application is a unique, sensitive, and highly specific immunoassay system for antibodies using gold nanoparticles. The assay is based on the aggregation of gold nanoparticles that are coated with protein antigens in the presence of their corresponding antibodies. Aggregation of the gold nanoparticles results in an absorption change at 620 nm that is used to calibrate the amount of antibodies. The effects of pH, temperature, and the concentration of protein A-coated gold nanoparticles on the sensitivity of the assay were investigated. A dynamic range of two orders of magnitude and a limit of detection of 1 μg/mL of anti-protein A were observed.

The second application of nanoparticles is luminescence nanosensors, which have their potential use as site-specific probes in samples of limited dimensions. Novel methods of nanosensor fabrication to obtain nanosensors with improved analytical properties are reported. A new approach for controlled synthesis of fluorescence nanosensors for pH measurements is also presented. Gold nanoparticles were used as a supportive matrix for the sensing component. Polymer layers that include the active sensing element were deposited on the gold nanoparticles surface using an electrostatic-based layer by layer deposition method. Polymer layers of alternating charges were deposited on the particle surface through attractive electrostatic interactions. Such method enabled a more precise control of the size, size distribution and density of fluorophores on each particle. The study shows that this is an effective way to fabricate particle-based fluorescent nanosensors that are stable and effective in measuring the pH in aqueous media.

1
Introduction

The ability to sense and detect the state of biological systems and living organisms has been significantly improved by the development of nanotechnology in the last decade. The emerging ability to modulate matters on the nanometer length scale will result in the near future in new types of sensors for biological studies. These new systems will be capable of sensing at the single-molecule level in living cells and be capable of parallel integration for detecting multiple signals from a large number of cells simultaneously [1]. Molecular events in biology or intricate chemical cascades which affect the environment will be sensed and detected using three main formats: optical detection, magnetic detection, and electrical detection. This review focuses on the synthesis and application of gold nanoparticles in optical-based bioanalytical assays and sensing applications.

Gold nanoparticles have been used for immunolabeling and imaging of cells, biomolecules, and other biological components since the 1970s. These works were extensively reviewed in the book edited by Hayat in 1989 [2]. Colloidal gold particles are still commonly used as tracers in electron microscopic studies of cellular biological samples primarily owing to their high electron density and biocompatibility [3–7]. The spectroscopic properties of gold nanoparticles have enabled their use in colorimetric measurements [8–14]. Gold nanoparticles are red in color because of the Mie absorption by their surface-plasmon oscillation that peaks at 520 nm [15]. The aggregation of gold nanoparticles leads to broadening of their absorption peak owing to electric dipole–dipole interactions and coupling between the plasmons of neighboring particles in the aggregates formed. Nanoparticle aggregates with interparticle distances substantially greater than the average particle diameter appear red, but as the interparticle distance in these aggregates decreases to less than approximately the average particle diameter, the color of the aggregates turns purple [16]. Mirkin et al. recently developed a colorimetric DNA hybridization assay using the spectral properties of gold nanoparticles [10–13]. In their experiments, gold nanoparticles averaging 13 nm in diameter were coated with single-stranded DNA oligonucleotides. The gold nanoparticles agglutinated in the presence of target oligonucleotides with a matching sequence as indicated by a color change of the sample from red to purple. They were able to detect femtomolar levels of matching oligonucleotides using this technique. Recently, Otsuka et al. [9] used gold nanoparticles modified with α-lactosyl-ω-mercaptopoly(ethylene glycol) to induce selective aggregation of the particles in the presence of lectin. The intensity of the plasmon resonance absorption band at 520 nm was used to monitor the aggregation of the particles. This led to a long assay time of about 8 h, since the intensity of the plasmon band decreased only when large aggregates were formed. In a similar study, Kim et al. [8] used functionalized gold nanoparticles to sense spectroscopically silent heavy metal ions via an ion-chelation-induced aggregation process. In our laboratory we showed for the first time that Mirkin's approach to DNA hybridization assays could be used to quantify the level of antibodies in aqueous and serum samples on the basis of the aggregation of gold nanoparticles [14]. Gold nanoparticles coated with protein A were used to determine the level of antiprotein A in aqueous and serum solutions. The rate of aggregation of the protein A coated gold nanoparticles in the presence of antiprotein A was monitored by measuring the absorption of the gold colloid suspension at 620 nm. The aggregation process was also monitored by transmission electron microscopy (TEM). The analytical capabilities of the new assay are discussed later in this chapter.

Nanoparticle-based sensors offer several important advantages when applied in volume-limited samples like single cells [17–20]. Chemical information is obtained from a large number of sensors and cells simultaneously. The technique is minimally invasive, allowing intracellular measurements while maintaining cell viability. The confinement of the sensing dyes to the particles prevents dye compartmentalization and enables the differentiation of the nanosensor location from autofluorescence centers in the observed cells; however, a few problems still remained to be solved. Unless hydrophilic indicators are covalently bound to the particles, a high leaking rate is seen, which decreases the stability and sensitivity of these sensors. The size distribution of the particle-based nanosensors is rather large, often exceeding 100%. This causes a large fluorescence intensity distribution and limits intensity-based sensors to applications in which only relative signal changes are of interest. Finally, the sensors exhibit varying degrees of photostability depending on the fluorophore used for analysis. A low-power excitation must be used to minimize the photobleaching rate of the fluorophores. This in turn limits the analytical properties of these sensors owing to inherent signal-to-noise problems. Recent advances in detection technology, more specifically the introduction of new and more sensitive CCD detectors has enabled the measurement of luminescence signals from individual molecules and luminescent nanoparticles [21, 22]. This new detection capability enabled us to develop new fluorescence nanosensors, which are based on gold nanoparticle technology. These new nanosensors are described in this chapter.

2 Experimental

2.1 Materials and Reagents

Colloidal gold nanoparticles (10 nm) labeled with protein A, monoclonal and polyclonal antibodies to protein A, antihuman albumin, bovine serum, phosphate-buffered saline at pH 7.4, citric acid (trisodium salt, dihydrate), hydroxylamine, mercaptosuccinic acid, 3-mercaptopropionic acid (MPA), sodium 2-mercaptoethane sulfonate, hydrogen tetrachloroaurate(III) trihydrate (99.9%), 16-mercaptohexadecanoic acid (MHA), 11-mercaptoundecanoic acid (MUDA), poly(allylamine hydrochloride) (PAH), M_W 70 000, and poly(diallyldimethylammonium chloride) (PDDA), M_W 20 000 and 200 000, were purchased from Sigma-Aldrich (St. Louis, MO, USA). The protein A coated gold nanoparticle suspensions were kept under light-tight conditions at – 20 °C to protect the gold nanoparticles from oxidation [23]. Dextran fluorescein anionic (3000, 10 000, 40 000, and 70 000 Da), dextran fluorescein polyanionic (70 000 Da), and amino dextran (70 000 Da) were purchased from Molecular Probes (Eugene, OR, USA). All reagents were used as received without further purification. Lab-Tek chambered coverglass was purchased from Nalge Nunc International (IL, USA). The deionized water used in all experiments was prepared in a NANOpure UV purification system (Barnstead, USA) (resistance higher than 18.2 Ω cm).

2.2 Instrumentation

UV/vis absorption spectra were obtained using a Cary 500 UV/vis/near – IR spectrophotometer. Fluorescence spectra of fluorescent nanoparticles were obtained using a PTI model QM1 spectrofluorometer (Photon Technology International, ON, Canada). Digital fluorescence images and spectra of the nanoparticles were obtained using a fluorescence imaging microscopy and spectroscopy system that consisted of an inverted fluorescence microscope (Olympus IX-70) equipped with high-resolution CCD cameras for imaging and spectroscopy measurements. TEM images of gold nanoparticles were obtained using a high-resolution transmission electron microscope (JEOL EM 2010, Tokyo, Japan). ζ-potentials and electrophoretic mobilities of bare and coated gold nanoparticles were measured with a Zetasizer HS 3000 (Malvern Instruments, UK).

2.3 Synthesis of Gold Nanoparticles

Since our studies focused on the use of gold nanoparticles in aqueous solutions we only used synthetic methods that yielded dispersible particles in aqueous media. Citrate reduction is the most commonly used method to produce gold nanoparticles [24]. In an early report, Fens [25] described the synthesis of gold nanoparticles ranging between 10 and 150 nm in diameter by varying the reaction conditions, particularly the citrate concentration. In our studies we used a modified method to form 20-nm gold nanoparticles by using sodium citrate to reduce Au^{3+} following the method of Grabar et al [26]. The gold nanoparticles were stabilized using mercaptocarboxylic acids as capping ligands. These groups provided the surface of gold nanoparticles with negative charges, which were needed for subsequent deposition of polyelectrolytes when polymer-coated gold nanoparticles were formed. The pH nanosensors that will be described later in this chapter were formed using a layer-by-layer (LbL) deposition approach. A layer of the positively charged polymers PAH (70 000 Da) and PDDA (20 000 and 200 000 Da) was deposited on the negatively charged gold nanoparticles. The particles were then coated with negatively charged dextran molecules (3000, 10 000, 40 000, and 70 000 Da) that were labeled with fluorescein, a pH-sensitive fluorophore. A final layer of positively charged amino dextran (70 000 Da) was deposited on the particles to complete the synthesis of the gold nanoparticle based pH sensors.

3
Results and Discussion

As previously mentioned, this section describes two examples of analytical applications involving the use of the unique spectroscopic properties of gold nanoparticles. The use of changes in the plasmon resonance absorption of gold nanoparticles as a basis for the development of improved aggregation-based immunoassays is then discussed.

3.1
Gold Nanoparticle Aggregation Immunoassays

In the newly developed gold nanoparticle aggregation immunoassays the level of antibodies in serum samples was determined based on a color change that resulted from aggregation of antigen-coated gold nanoparticles in the presence of antibody molecules in the analyte sample. Antiprotein A was used in these experiments as a model target analyte. The aggregation of commercially available 10-nm-sized protein A coated gold nanoparticles in the presence of antiprotein A took several hours to complete. However, a color change from red to purple was observed within 30 min. To maximize the absorption change at 620 nm, we used an assay time of 1 h. At longer times, precipitation of large aggregates and a clear supernatant were observed. As a result, the absorption intensity was no longer concentration-dependent. The UV/vis absorption spectra of solutions containing protein A coated gold nanoparticles and antiprotein A at different times during the assay are shown in Fig. 1. Spectrum a, measured at $t = 0$, is a typical spectrum of a gold nanoparticle solution showing a plasmon resonance peak at 520 nm. The aggregation of gold nanoparticles led to an increase in absorbance at longer wavelength (600–750 nm) as seen in spectra b ($t = 30$ min) and c ($t = 60$ min). The larg-

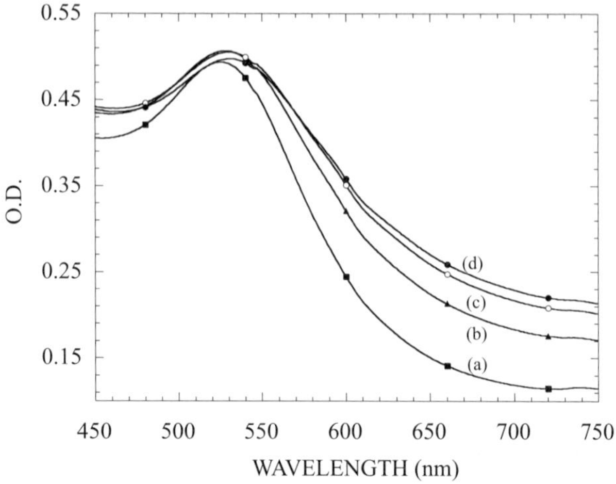

Fig. 1 UV/vis absorption spectra following the aggregation of protein A coated gold nanoparticles in the presence of antiprotein A for *a* the control—prior to adding antiprotein A, *b* at 30 min, and *c* at 60 min. The solution contained 7.8×10^{11} gold nanoparticles/mL and 0.32 μg/mL antiprotein A in phosphate-buffered solution at pH 7.0

est change in absorption during the aggregation of the protein A coated gold nanoparticles was at 620 nm.

Figure 2 shows TEM images (×100000) of the 10-nm protein A coated nanoparticles prior to and following adding antiprotein A to the sample.

The TEM images were in agreement with the results obtained using UV/vis absorption spectroscopy. The images indicated increasing particle density in the field of view at 30 min, clear formation of aggregates at 60 min, and the appearance of even larger aggregates at 120 min (not shown).

The immunoaggregation process involved binding of the antigen-coated particles to antibody molecules followed by reaction between particles to form aggregates. Since both reactions involve charged species, it was expected that the pH in the analyte solution would affect the aggregation rate. Indeed, the aggregation rate peaked at approximately pH 7, while lower rates were observed at lower or higher pH possibly owing to repulsive forces between the charged particles. It should be noted, however, that the optimum pH would be system-dependent and should, therefore, be determined for each aggregation assay. The aggregation rate was found to be temperature-dependent and in fact increased with decreasing temperature. The rate of aggregation at 20 °C was about 2.5 times higher than the rate of aggregation at 37 °C. Although the rate of chemical reactions usually increases with temperature, the aggregation reaction involves a negative change in entropy ($\Delta S < 0$), therefore, increasing temperatures result in an increase of the free-energy change, ΔG, of the aggregation process, and in a decreased stability of the aggregate formed. On the other hand, decreasing the temperature fur-

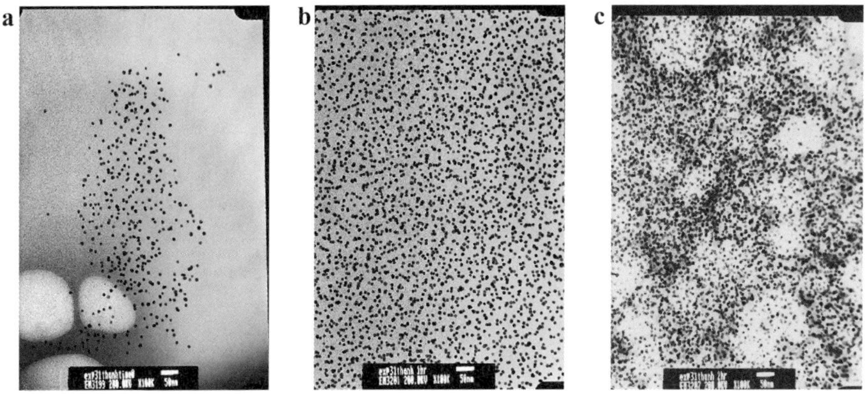

Fig. 2 Transmission electron microscopy (*TEM)* images (×100000) during the immunoassay of antiprotein A: **a** control—prior to adding antiprotein A, **b** 30 min, and **c** 60 min following the addition of antiprotein A to the sample

ther to 4 °C significantly increased the rate of nonspecific aggregation. This temperature also presented instrumental complications, because temperature control was needed throughout the assay. While it is relatively simple to carry out the experiments at 4 °C under laboratory conditions, this requirement would complicate and possibly preclude the use of the technique in field applications. We, therefore, decided to characterize the assay performance at ambient temperature. While variations in temperature clearly affected the assay, it could be conducted in an unregulated temperature environment with variation in aggregation rates of about 5%. Several control experiments were carried out to validate the selectivity of the aggregation assay. The results of these experiments are shown in Fig. 3.

Curve a shows the absorption intensity at 620 nm of a phosphate buffer solution at pH 7.4. Curve b shows the absorption intensity at 620 nm of a solution containing polyclonal antiprotein A in the absence of protein A coated gold nanoparticles. Curve c shows the absorption intensity at 620 nm of a solution containing protein A coated nanoparticles in the absence of antiprotein A. The absorption intensity at 620 nm remained constant for 2 h in these control experiments. Curve d shows the absorption intensity at 620 nm of a solution containing protein A coated nanoparticles and a monoclonal antiprotein A. No absorption change was observed for 2 h, indicating that aggregation did not take place. This was attributed to the lack of multiple binding sites and the inability of monoclonal antibodies to form particle aggregates. Curve e shows another control experiment in which protein A coated nanoparticles were mixed with anti-albumin. No absorption change was observed, indicating that no nonspecific aggregation and no cross-reactivity of the antigen occurred. Curve f shows a positive response of the assay when protein A coated gold nanoparticles and an-

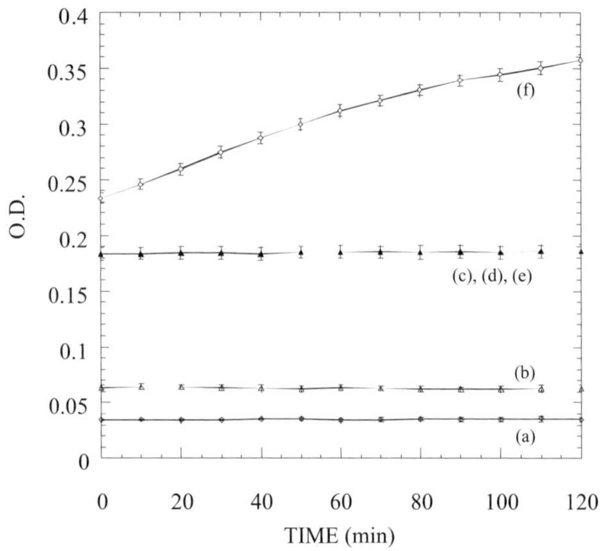

Fig. 3 Control experiments to validate the selectivity of the aggregation process. Changes in the absorption intensity at 620 nm are used to monitor the rate of aggregation of the gold nanoparticles. *a* Phosphate buffer solution at pH 7.0, *b* 0.32 mg/mL polyclonal antiprotein A, *c* 7.8×10^{11} protein A coated gold nanoparticles/mL, *d* 7.8×10^{11} protein A coated gold nanoparticles/mL and 0.6 mg/mL monoclonal antiprotein A, *e* 7.8×10^{11} protein A coated gold nanoparticles/mL and 0.72 mg/mL anti-albumin, and *f* 7.8×10^{11} protein A coated gold nanoparticles/mL and 0.32 mg/mL polyclonal antiprotein A. All experiments were carried out in phosphate-buffered solution at pH 7.0 for 2 h

tiprotein A were present in the analyte sample. As expected, a noticeable absorption change at 620 nm occurred, indicating aggregation of the protein A coated gold nanoparticles. A calibration curve describing the absorption change at 620 nm vs. the concentration of antiprotein A in 10% bovine serum is shown in Fig. 4.

The antiprotein A solution concentrations ranged from 0 to 50 μg/mL. A dynamic range of 2 orders of magnitude in concentration was observed. At higher concentration, the aggregation process was inhibited, which may be due to the blocking of active sites. Each data point represented the average of triplicate measurements. The relative standard deviation was 0.3–2.6%. The limit of detection of the assay in serum samples was 1 μg/mL of antiprotein A. The sensitivity and limit of detection of the aggregation assay were comparable with those of other immunological methods such as ELISA. However, the assay was simpler to perform and was completed in only 1 h.

The application of gold nanoparticles for quantitative immunoassays was a departure from common practice. Previously, gold nanoparticles were only used as additives in aggregation assays that involve larger particles to provide color to the aggregates formed. In our immunoassays the aggregation rate

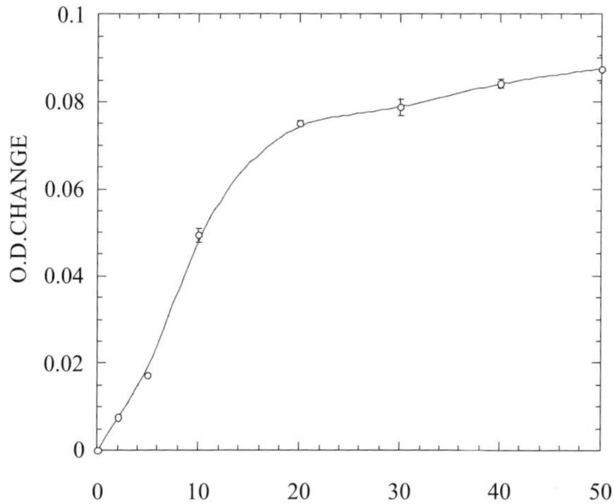

Fig. 4 A calibration curve describing the absorption change at 620 nm against antiprotein A concentration in 10% bovine serum solution. A dynamic range of 2 orders of magnitude and a limit of detection of 1 μg/mL antiprotein A were observed

and antibody concentration were measured based on a color change due to a shift in the plasmon resonance of gold nanoparticles rather than on increasing absorption intensity due to aggregation. This enabled the development of highly quantitative immunoassays. The next section describes the use of gold nanoparticles as a supportive matrix for the fabrication of highly controlled fluorescence pH sensors. The use of gold nanoparticles as a supportive matrix for the sensor fabrication addressed two major problems in the fabrication of nanometric sensors. The first is the poor control of the size and the size distribution of fluorescent nanosensors. Since the number of fluorescent indicators in individual nanosensors is proportional to the sensor size, controlling the size and the size distribution in nanosensor formulations is imperative to obtain quantitative data, particularly if intensity-based sensors are used for sample analysis. The second problem is the tendency of nanoparticles to aggregate owing to hydrophobic effects. As described in the previous section the plasmon resonance absorption of gold nanoparticles is extremely sensitive to aggregation. This enabled us to carefully monitor and optimize the reaction conditions during the fabrication of fluorescent nanosensors to minimize aggregation of the nanoparticles.

3.2
Gold Nanoparticle Based pH Nanosensors

The fluorescence-based nanosensors were prepared by coating gold nanoparticles with polymeric layers that contain fluorescence-sensing components. Coating of the gold nanopaticles was based on alternate deposition of positively and negatively charged polymer layers on the surface of the nanoparticles. Decher et al. [27,28] first introduced this electrostatic LbL (ELbL) coating technique in the early 1990s. Since then, there have been a number of reports on the use of the ELbL deposition method to fabricate layer-coated gold films or gold electrodes [29–35]. These films were formed via electrostatic attraction between adjacent polymeric layers of opposite charge. For example, Caruso et al. [35] deposited thin organic films of PAH and poly(styrenesulfonate) successively on gold surfaces. Schmitt et al. [36] used the ELbL technique to deposit 15-nm gold particles as periodic strata on a silicon wafer using various combinations of polyelectrolytes. Recently, Grittins and Caruso [37,38] used the ELbL deposition method to form nanometric capsules through the alternate deposition of negatively and positively charged polymer layers on the surface of gold nanoparticles. Then, the gold core was dissolved to form self-supporting hollow nanocapsules. In our study, particle-based nanosensors were prepared by the alternate deposition of the positively charged polymer, PAH, M_W 70 000, and the negatively charged polymer dextran fluorescein, which is a pH-sensitive indicator on the surface of gold nanoparticles.

Gold nanoparticles of various diameters were synthesized following well-established synthetic protocols. The particles exhibited excellent monodispersity and relatively narrow size distribution with a relative standard deviation of less than 10%. The gold nanoparticle solution was stable for up to 6 weeks at room temperature. The concentration of the gold nanoparticles synthesized was calculated on the basis of their crystal structure. Gold has a cubic close-packed, face-centered cubic structure with each unit cell containing four atoms with the cubic size of 4.078 Å [39]. The number of gold atoms per particle was estimated as 4 times the division product between the average diameter of the particles as determined by TEM and the size of a unit cell. The concentration of the gold nanoparticle solution was calculated by dividing the total number of gold atoms in the solution by the number of gold atoms per particle and by the volume of the solution. Typically, the concentration of gold nanoparticles in our samples was on the order of 10^{13} particles/mL.

Self-assembly of monolayers of alkanethiols and their derivatives on gold planar surfaces has been studied extensively [40–42]. We optimized the conditions to coat the surface of our 15–100-nm gold nanoparticles with thiol groups. We found that it was imperative to conduct the thiol-coating reaction under mild alkaline conditions (pH 9) to prevent aggregation. It has been reported that long-chain thiols (longer than C8) are less likely to desorb from gold surfaces than short chain thiols (shorter than C8) [43]. However, long-chain thiols like MHA and MUDA are insoluble in aqueous solution even at a low concentration of 0.1 mM, owing to their relatively high hydrophobicity. To overcome this problem, Gittins and Caruso [38] previously used sodium 3-mercaptopropane sulfonate to coat gold nanoparticles. We used MPA as a thiol source. MPA has a pK_a of 3.6 and it is therefore negatively charged even under mild acidic conditions. We found MPA to be an effective source for thiol coating of the gold nanoparticles when its concentration was limited to 1 mM. Higher concentration of MPA resulted in aggregation of the particles.

Multiple parameters affect the efficiency of ELbL polymer deposition on gold nanoparticles. These include the polymer type, its molecular weight and concentration, the solvent used in the reaction, the surface charge of the particles and their functionalities, the ionic strength of the solution, the charge density of the polymer, the deposition time, and temperature. PAH and PDDA have been commonly used as cationic polymers in ELbL deposition on planar surfaces and particles [43, 44]. PAH formed a thicker and stabler polymeric layer than PDDA on the surface of the particles in our experiments. A high-resolution TEM image of a single PAH coated gold nanoparticle is shown in Fig. 5. The spherical particle appears to be evenly coated by the polymeric layer.

The PAH polymeric layer played an important role in our fluorescence sensor design. First, its positive charges enabled the deposition of anionic dextran that was labeled with the pH indicator fluorescein on the surface of the nanoparticles. More importantly, the PAH polymeric layer separated between the fluorescein molecules and the metal particle. In fact, the thickness of the polymeric layer was over 10 nm, which is larger than the Förster distance required for efficient energy transfer between the fluorophore and the metallic gold particles.

Fluorescein-labeled high molecular weight dextrans were used as anionic polymers in the ELbL method to coat the gold nanoparticles and provide them with pH-sensing capabilities. Fluorescein (pK_a = 6.4) has been used in

Fig. 5 A high-resolution (×400000) TEM image of a single 30-nm gold nanoparticle that was deposited with a layer of poly(allylamine hydrochloride) polymer using the electrostatic layer-by-layer method

pH fluorescence-sensing applications owing to its high emission quantum yield and pH sensitivity in the physiological pH range. Dextrans are water soluble and relatively inert. Owing to their biologically uncommon α-1,6-polyglucose linkages dextrans are resistant to cleavage by most endogenous glycosidases and therefore could be used as a coating matrix of nanosensors in biological samples and living cells. Fluorescein-labeled anionic dextrans with molecular masses ranging from 3000 to 70 000 Da were investigated for their affinity to the positively charged PAH coated gold nanoparticles. We found that dextrans of higher molecular weight formed thicker and stabler coating layers on the surface of the PAH-coated gold nanoparticles. The fluorescence intensity of particles coated with high molecular weight fluorescein labeled dextran (e.g. 70 000) was higher than the fluorescence intensity of particles coated with lower molecular weight fluorescein-labeled dextrans. This was attributed to the larger Stoke's radius of higher molecular weight dextrans [45]. For example, the Stoke's radius of 70 000-Da dextran is 55 Å, while the Stoke's radius of 3000-Da dextran is only 22 Å. The radius (R) of the particle was calculated from the diffusion coefficient (D) via the Stokes–Einstein equation:

$$D = \frac{kT}{f} = \frac{kT}{6\pi\eta R},$$

where k is the Boltzmann constant, T is the temperature, k is the solvent viscosity, and $f = 6\pi\eta R$ is the frictional coefficient for a compact sphere in a viscous medium. The larger spacing between the gold core and the fluorescein molecules decreased the efficiency of nonradiative energy transfer between fluorescein and the gold cores, which led to increasing brightness of the particles. We also found that dextrans with multiple anionic charges formed a stabler coating on the particles, which was attributed to increasing attraction between the dextran coating and the PAH-coated particles. A typical digital fluorescence image of the fluorescein-labeled dextran-coated gold nanoparticles is shown in Fig. 6. The fluorescence signal-to-background ratio of the particles in randomly chosen fields of view ranged from 30 to 40, which is a significant improvement in the fluorescence intensity distribution of particle-based fluorescence nanosensors fabricated using less controlled coating techniques.

ζ-potential measurements were used to monitor the coating of the gold nanoparticles with alternate charged polymers. Uncoated gold nanoparticles showed a negative ζ-potential of – 17.4 mV owing to citrate residues on their surface. MPA-coated gold nanoparticles were characterized with a negative ζ-potential of – 28.6 mV since the pK_a of MPA is 3.6. The ζ-potential of PAH-coated particles was 38.1 mV, which indicated the abundance of positive charges on the surface, owing to the positive charge of the cationic polymer PAH. The ζ-potential decreased to – 27.3 mV when a fluorescein-labeled anionic dextran coating was formed on the surface of the PAH-coated

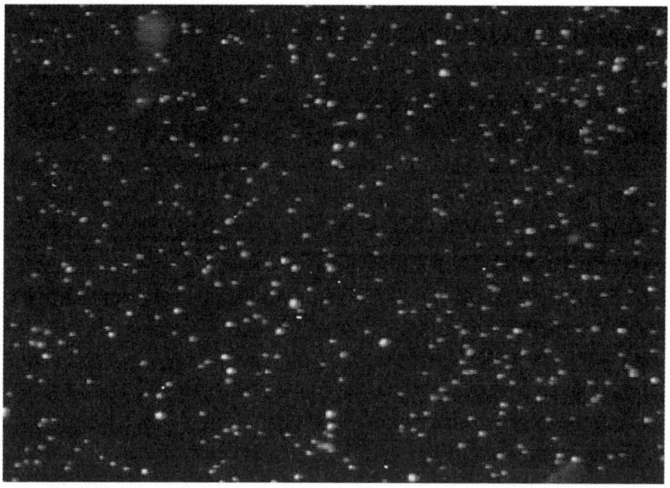

Fig. 6 ×100 fluorescence image of dextran fluorescein coated gold nanoparticles. The sensing particles were suspended in aqueous solution. A 470–490-nm excitation filter, a 505-nm dichroic mirror, and a 515-nm cut-on emission filter were used to collect the image. The exposure time is 0.5 s. The nanosensors are uniform in size and evenly coated. No aggregation was observed

nanoparticles. When amino dextran was coated as an outer layer of the nanoparticles, the ζ-potential was again positive, owing to the positive charge of the amino groups; however, it was lower than the ζ-potential of the PAH layer. The ζ-potential measurements confirmed that the ELbL method is an effective technique in the fabrication of gold particle based polymer coated nanosensors.

To test the photostability of the particles, samples of particles were placed on the microscope stage and illuminated continuously at 480 nm. The fluorescence intensity of fluorescein decreased by approximately 30% during the first 5 min of continuous illumination. To overcome the inherent instability of fluorescein, we limited the exposure time and the number of exposures of the particles to the excitation light during our kinetics measurements. The fluorescent gold particles were exposed to the excitation light for 0.1 s in each measurement. The number of exposures of the nanosensor was limited to 30. Under these illumination conditions, the fluorescein-containing nanosensors remained photostable throughout the experiment. The fluorecein-labeled gold nanoparticles were stored at room temperature in the dark in water. The solutions were stable for at least 1 week under these storage conditions. No precipitation or significant changes in UV/vis absorption were observed, which indicated that the particles did not aggregate under these conditions. No significant changes in the fluorescence intensity of the solutions and in the signal-to-background ratio in the fluorescence images were seen, which indicated that fluorophores did not leak from the particles and that no degrading

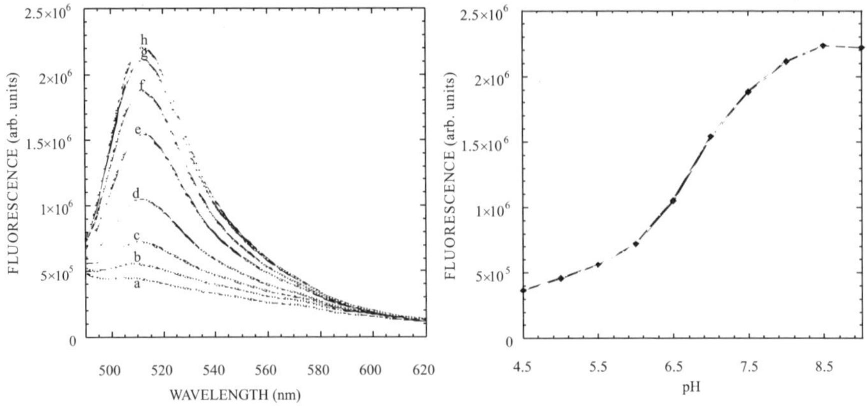

Fig. 7 **a** Fluorescence spectra of individual fluorescein-containing nanoparticles in solutions of increasing pH: *a* pH 5.0, *b* pH 5.5, *c* pH 6.0, *d* pH 6.5, *e* pH 7.0, *f* pH 7.5, *g* pH 8.0, *h* pH 8.5. **b** A pH calibration curve of the fluorescent nanoparticles. The fluorescence intensity at 525 nm ($\lambda_{ex} = 470$ nm) increases with the pH

chemical reactions, like oxidation, occurred. The pH sensitivity of the gold nanoparticles is demonstrated in Fig. 7.

Figure 7a shows the fluorescence spectra of solutions containing fluorescein-labeled gold nanoparticles in solution of increasing pH values. Figure 7b shows a pH calibration curve constructed by plotting the maximum fluorescence intensity of the solutions at 525 nm against the pH of standard phosphate buffer solutions. The pH calibration curve was similar to the one obtained for free fluorescein in aqueous media. This indicated that the alternating growth of polymers on the gold nanoparticles did not affect the pH-sensing properties of the fluorescent indicator, in this case fluorescein. The dynamic range of the particles was found to be between pH 5.5 and 8.0 with a pH sensitivity of 0.1 pH units. The response time of the pH-sensing nanoparticles was less than 1 s.

4 Summary and Conclusions

The review described two new applications of gold nanoparticle technology in the field of bioanalysis. The first describes the development of a new aggregation immunoassay technique based on the use of gold nanoparticles. The rate of aggregation of antigen-coated gold nanoparticles in the presence of antibodies toward the antigen was measured by monitoring the absorption change of the gold nanoparticles upon aggregation. The assays were conducted at ambient temperature where a significant absorption signal change occurred in 1 h. The new gold nanoparticle based assay was sensitive to concentrations of antiprotein A as low as 1 μg/mL, which was comparable with the sensitivity of ELISA. However, the new assay is much simpler to perform, because it involves only one step, compared with ELISA assays, which require multiple washing steps and separation between bound and unbound labeled macromolecules. Both ELISA and our new gold nanoparticle based aggregation assay are capable of analyzing a large number of samples simultaneously using a microplate reader format; however, the gold nanoparticle based assay would be easier to automate because of the small number of steps involved in the aggregation assay. Although gold is an expensive metal, the cost of the assay is kept low as a result of the use of small numbers of gold nanoparticles. To further reduce cost and increase simplicity, we have already developed a simple diode laser based absorption spectrometer that could eventually replace the full-frame absorption spectrometer used in these studies. The new system could be suitable for field applications for aggregation-based immunoassays.

The second application of gold nanotechnology involved the development of new gold nanoparticle based sensors. We developed a novel technique for the fabrication of fluorescence-based nanosensors with improved properties. The study addressed successfully two problems that were encountered in past attempts to fabricate particle-based fluorescent nanosensors. The use of gold nanoparticles as the core material of the particles enabled tight monitoring of the aggregation of the particles during the fabrication of the sensors and during their storage. It also enabled the optimization of the synthetic and storage conditions required to minimize aggregation of the particles, which would obviously lead to a substantial degradation in their sensing properties and temporal and spatial resolution. The second problem that was addressed in this study was the often poor consistency in size, size distribution, and intensity distribution of fluorescence-based nanosensors fabricated using less controlled conditions. In this study we took advantage of the tight size control and narrow size distribution of gold nanoparticles that were synthesized following well-established literature protocols. Coating of the nanoparticles with thin layers of alternately charged polymers did not increase significantly their size or size distribution. The variation in the fluorescence intensity of the particles as observed in digital images was attributed more to technical difficulties associated with particles moving in and out of the focal plane and to heterogeneity in the excitation field rather than to a significant variation in the number of fluorophores per particle. Our study showed that the ELbL technique is an effective way to fabricate particle-based fluorescent nanosensors that are stable and effective in measuring the pH in aqueous media. Currently we are exploring ways to attach cell-penetrating peptides to the surface of the particles to enable their delivery into nonphagocytotic cells and their application as intracellular pH nanosensors.

In conclusion, gold nanoparticles provide a very promising platform for bioanalytical application owing to their unique optical properties, monodispersity in aqueous media, and biocompatibility. While gold nanoparticles have been widely applied as labels in electronic microscopy studies owing to their high electron density, their application in bioanalytical assays is a new trend in bioanalysis that is sure to grow in coming years.

Acknowledgements The work described in this chapter was supported by DARPA grant no. HR0011-04-C-0068 and by NSF grant CHE-0314027.

References

1. Alivisatos P (2004) Nat Biotechnol 22:47
2. Hayat MA (ed) (1989) Colloidal gold, principles, methods and applications. Academic, New York
3. Bendayan M (2000) Biotech Histochem 75:203
4. Hainfeld JF, Powell RD (2000) J Histochem Cytochem 48:471
5. Jurgens L, Nichtl A, Werner U (1999) Cytometry 37:87
6. Neagu C, Vanderwerf KO, Putman CAJ, Kraan YM, Degrooth BG, Vanhulst NF, Greve J (1994) J Struct Biol 112:32
7. Horisberger M, Clerc MF (1985) Histochemistry 82:219
8. Kim YJ, Johnson RC, Hupp JT (2001) Nano Lett 1:165
9. Otsuka H, Akiyama Y, Nagasaki Y, Kataoka K (2001) J Am Chem Soc 123:8226
10. Mirkin CA (2000) Inorg Chem 39:2258
11. Reynolds RA, Mirkin CA, Letsinger RL (2000) Pure Appl Chem 72:229
12. Storhoff JJ, Elghanian R, Mucic RC, Mirkin CA, Letsinger RL (1998) J Am Chem Soc 120:1959
13. Elghanian R, Storhoff JJ, Mucic RC, Letsinger RL, Mirkin CA (1997) Science 277:1078
14. Thanh NTK, Rosenzweig Z (2002) Anal Chem 74:1624
15. Takeuchi Y, Ida T, Kimura K (1996) Surf Rev Lett 3:1205
16. Kreibig U, Genzel L (1985) Surf Sci 156:678
17. Ji J, Rosenzweig N, Griffin C, Rosenzweig Z (2000) Anal Chem 72:3497
18. Park EJ, Brasuel M, Behrend C, Philbert MA, Kopelman R (2003) Anal Chem 75:3784
19. Xu H, Aylott JW, Kopelman R (2002) Analyst 127:1471
20. Clark HA, Barker SLR, Brasuel M, Miller MT, Monson E, Parus S, Shi ZY, Song A, Thorsrud B, Kopelman R, Ade A, Meixner W, Athey B, Hoyer M, Hill D, Lightle R, Philbert MA (1998) Sens Actuator B 51:12
21. Chen YF, Rosenzweig Z (2002) Anal Chem 74:5132
22. Pierce DW, Vale RD (1999) Methods in cell biology, vol 58. Academic, London
23. Takeuchi Y, Ida T, Kimura K (1997) J Phys Chem B 101:1322
24. Turkevich J, Stevenson PC, Hillier J (1951) Discuss Faraday Soc 11:55
25. Fens G (1973) Nat Phys Sci 241:20
26. Grabar KC, Freeman RG, Hommer MB, Natan MJ (1995) Anal Chem 67:735
27. Decher G (1997) Science 277:1232–1237
28. Decher G, Hong JD (1991) Makromol Chem Macromol Symp 46:321–327
29. Forzani ES, Otero M, Perez MA, Teijelo ML, Calvo EJ (2002) Langmuir 18:4020–4029
30. Baba A, Kaneko F, Advincula RC (2000) Colloids Surf A 173:39–49
31. Li WJ, Wang Z, Sun CQ, Xian M, Zhao MY (2000) Anal Chim Acta 418:225–232
32. Cheng YF, Murtomaki L, Corn RM (2000) J Electroanal Chem 483:88–94
33. Lu ZQ, Lvov Y, Jansson I, Schenkman JB, Rusling JF (2000) J Colloid Interface Sci 224:162–168

34. Lvov YM, Kamau GN, Zhou DL, Rusling JF (1999) J Colloid Interface Sci 212:570–575
35. Caruso F, Niikura K, Furlong DN, Okahata Y (1997) Langmuir 13:3422–3426
36. Schmitt J, Decher G, Dressick WJ, Brandow SL, Geer RE, Shashidhar R, Calvert JM (1997) Adv Mater 9:61
37. Gittins DI, Caruso F (2001) J Phys Chem B 105:6846–6852
38. Gittins DI, Caruso F (2000) Adv Mater 12:1947–1955
39. Maeland A, Flanagan TB (1964) Can J Phys 42:2364
40. Palegrosdemange C, Simon ES, Prime KL, Whitesides GM (1991) J Am Chem Soc 113:12–20
41. Abbott NL, Whitesides GM (1994) Langmuir 10:1493–1497
42. Hasan M, Bethell D, Brust M (2002) J Am Chem Soc 124:1132–1133
43. Caruso F, Mohwald H (1999) J Am Chem Soc 121:6039–6046
44. Sukhorukov GB, Donath E, Lichtenfeld H, Knippel E, Knippel M, Budde A, Mohwald H (1998) Colloids Surf A 137:253–266
45. Lang I, Scholz M, Peters R (1986) J Cell Biol 102:1183–1190

Springer Ser Chem Sens Biosens (2005) 3: 279–301
DOI 10.1007/5346_009

Published online: 27 September 2005

Reverse Symmetry Waveguide for Optical Biosensing

Róbert Horváth (✉) · Nina Skivesen · Niels B. Larsen · Henrik C. Pedersen

Optics and Plasma Research Department, Risø National Laboratory, Building 128,
Frederiksborgvej 399, P.O. Box 49, 4000 Roskilde, Denmark
robert.horvath@risoe.dk

Abstract The present chapter deals with a novel design of planar optical waveguide biosensors. The principle of reverse symmetry is based on making the refractive index (RI) of the waveguide substrate less than the RI of the medium covering the waveguiding film, which is usually an aqueous solution (RI ~ 1.33). This is opposed to the conventional sensor geometry, where the substrate is glass or polymers with RIs of approximately 1.5. The reverse configuration can be used to tune the penetration depth of the evanescent electromagnetic field into the cover medium up to infinity; thus the waveguide can be tailor-made so that biological objects with any size can be probed by the evanescent field. This is an important improvement compared with, for example, surface plasmon resonance sensors, where the penetration depth is fixed by the choice of metal.

1 Introduction

For almost 20 years, the optical waveguide sensor has been used for the label-free, evanescent-field detection of chemical or biological reactions taking place in the close vicinity of the waveguide surface. This includes numerous applications within the fields of chemical and biological sensing, where in most cases the aim has been to monitor interactions on a molecular or macromolecular level [1–8]. For these purposes the waveguide sensor has proven to be a highly sensitive tool, which relies on the fact that the evanescent field is restricted to probe the nearest 100–150 nm into the sample volume in which the chemical and biological interactions take place.

When dealing with the detection of interactions involving larger substances, like the surface attachment of bacterial cells [9] or the attachment and proliferation of eukaryotic (e.g., human) cells [10–15], the conventional waveguide technology reaches a limitation in the sense that the upper limit of the probing depth is typically 100–150 nm [16–18]. This implies that, say, an adhering bacterial cell, being typically (0.5–5 μm) in size, is only partially sensed by the evanescent field, resulting in a limited sensitivity [19, 20]. This problem gets even more serious when detecting eukaryotic cells, partly because they are even bigger (5–50 μm) and partly because typically they leave an aqueous gap between the cell membrane and the surface, a gap that does not contribute to the sensor signal [21]. The key problems are illustrated in Fig. 1a.

The physical limit of the probing depths for conventional waveguide sensors originates from the geometry of the three-layer waveguide consisting of substrate, film, and cover medium. In conventional waveguides the substrate is usually glass or polymer, which have refractive indices (RIs) larger

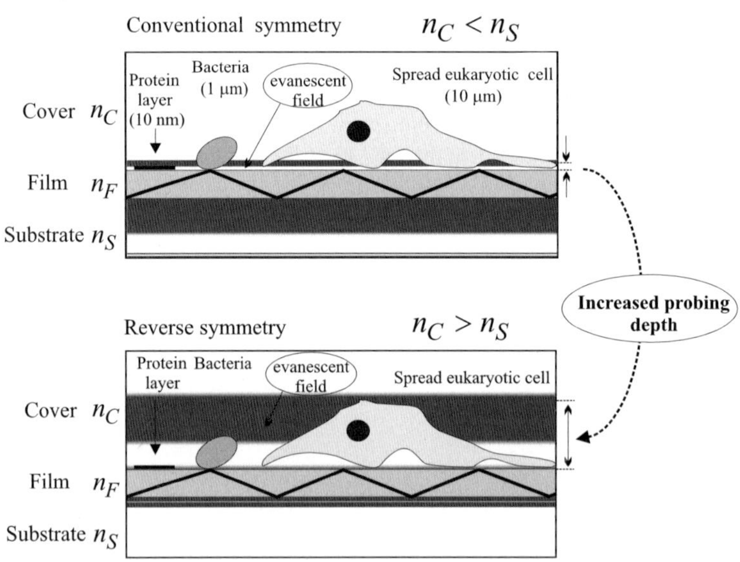

Fig. 1 Waveguide sensor geometries comprising substrate (refractive index, *RI*, n_S), film (RI n_F), and cover medium (RI n_C) using **a** conventional symmetry ($n_C < n_S$) and **b** reverse symmetry ($n_C > n_S$)

than that of the (aqueous) cover sample being investigated. This causes the evanescent field to extend deeply into the substrate and slightly into the cover. However, by choosing a substrate with a RI *less* than the RI of the cover the probing depths are swapped, resulting in a large cover penetration depth as illustrated in Fig. 1b. As a result, large, micron-sized objects can also be enclosed by the probing evanescent field, which improves the sensitivity [22]. In the following the reverse symmetry waveguide sensor will be described theoretically and experimentally.

2 Waveguide Sensor Basics in Brief

2.1 Basic Physics

A three-layer, planar optical waveguide consists basically of a high-RI film with RI n_F that is sandwiched between two lower-refractive-index materials referred to as the substrate and the cover medium with RIs n_S and n_C, respectively; Fig. 2. If Maxwell's equations are solved for such a layered structure, the following solutions for the electromagnetic field ψ can be found [23]:

$$\psi = \begin{cases} a_\mathrm{C} \exp\left[\mathrm{i}\left(k_x x + k_{\mathrm{C}z} z\right)\right] & \text{(in the cover)}\,, \\ a_\mathrm{F}^{+} \exp\left[\mathrm{i}\left(k_x x + k_{\mathrm{F}z} z\right)\right] + a_\mathrm{F}^{-} \exp\left[\mathrm{i}\left(k_x x - k_{\mathrm{F}z} z\right)\right] & \text{(in the film)}\,, \\ a_\mathrm{S} \exp\left[\mathrm{i}\left(k_x x - k_{\mathrm{S}z} z\right)\right] & \text{(in the substrate)}\,, \end{cases} \tag{1}$$

with $k_x^2 + k_{jz}^2 = k^2 n_j^2$, where $j = \mathrm{S, F, C}$. a_j are the individual wave amplitudes, k_x and k_{jz} are the wave vector components along x and z (Fig. 2), and k is the vacuum wavenumber of the light.

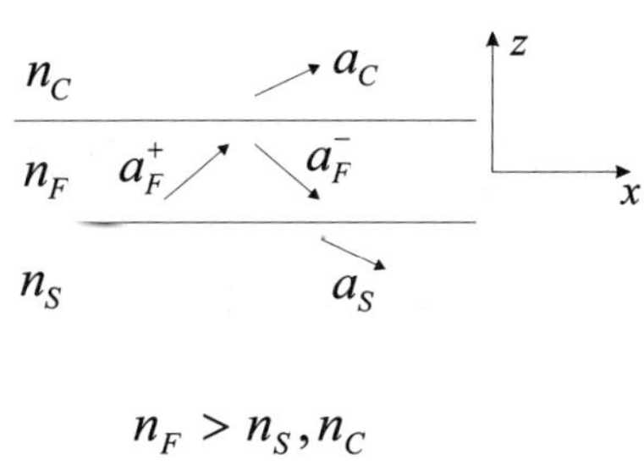

Fig. 2 Waveguide consisting of three layers with RIs n_S, n_F, and n_C. a_j ($j = \mathrm{S, F, C}$) are the electromagnetic wave amplitudes

For the solution Eq. 1 to be non-trivial, the following mode equation must be fulfilled [16, 23]:

$$m\pi = d_F k \sqrt{n_F^2 - N^2} - \arctan\left[\left(\frac{n_F}{n_S}\right)^{2\rho}\left(\frac{N^2 - n_S^2}{n_F^2 - N^2}\right)^{0.5}\right] - \arctan\left[\left(\frac{n_F}{n_C}\right)^{2\rho}\left(\frac{N^2 - n_C^2}{n_F^2 - N^2}\right)^{0.5}\right], \quad (2)$$

where $m = 0, 1, 2, \ldots$ is the mode number, d_F is the film thickness, $N = k_x/k$, and ρ is a polarization index that equals 0 or 1 in the case of TE or TM polarized light, respectively.

Using Eq. 2 the solutions N_m are plotted versus film thickness d_F in Fig. 3 for the three lowest-order modes $m = 0, 1, 2$. It is seen that N_m varies between the values $n_{max} = \max\{n_C, n_S\}$ at $d_F = d_{cutoff}$ and n_F at $d_F \to \infty$. d_{cutoff} is defined as the film thickness at which the field's propagation angle in the film gets below the critical angle at either the film–cover or the film–substrate interface, in which case the field escapes from the waveguide and Eq. 1 is no longer a solution. Formally, d_{Cutoff} can be expressed as [22]

$$d_{cutoff} = k^{-1}\left(n_F^2 - n_{max}^2\right)^{-1/2} \times \left\{\arctan\left[\left(\frac{n_F}{n_{min}}\right)^{2\rho}\left(\frac{n_{max}^2 - n_{min}^2}{n_F^2 - n_{max}^2}\right)^{1/2}\right] + m\pi\right\}, \quad (3)$$

where $n_{min} = \min\{n_C, n_S\}$.

The fact that $n_{max} < N_m < n_F$ implies that $k_{Cz} = k\left(n_C^2 - N_m^2\right)^{1/2}$ and $k_{Sz} = k\left(n_C^2 - N_m^2\right)^{1/2}$ are always purely imaginary and the electromagnetic fields in the substrate and the cover medium are therefore evanescent. In the film, however, we have a real value of $k_{Fz} = k\left(n_F^2 - N_m^2\right)^{1/2}$, meaning that the elec-

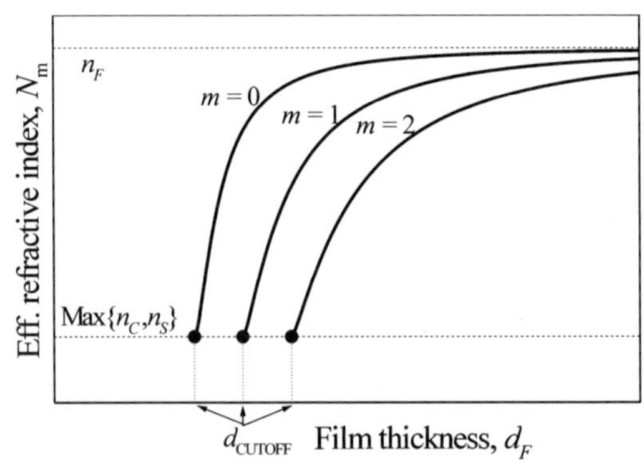

Fig. 3 Effective RI N_m versus film thickness for the lowest-order modes

tromagnetic field here consists of two propagating plane waves. Hence, all in all, the solutions assume the characteristic mode power profiles shown in Fig. 4.

From Eq. 1 it is seen that the mode's phase velocity along x is given by $v_m = c/N_m$, which is why N_m is referred to as the effective RI of the mode.

The reason why N_m varies between n_{max} and n_F is further elucidated in Fig. 5, which displays the penetration depths $d_{p,j} = 1/|k_{jz}|$ (j = S, C), into the substrate and the cover versus film thickness. Here, it is seen that when $d_F \to d_{cutoff}$, the penetration depth into that medium (substrate or cover) with highest RI goes to infinity, in which case practically all the mode power flows in this medium. This is why the mode's phase velocity $v_m \to c/n_{max}$ and $N_m \to n_{max}$. The penetration depth at cutoff into the lowest-RI medium has a finite value of $d_{p,a} = k^{-1}\left(n_{max}^2 - n_{min}^2\right)^{-1/2}$.

For infinite film thickness the two waves in the film will just act as freely propagating waves with minimal influence from the surrounding medium; hence, practically all the mode power will be flowing in the film, resulting in $v_m \to c/n_F$ and $N_m \to n_F$. However, the total internal reflection at the boundaries still results in evanescent fields with penetration depths of $d_{p,b} = k^{-1}\left(n_F^2 - n_{max}^2\right)^{-1/2}$ into the highest-RI medium and $d_{p,c} = k^{-1}\left(n_F^2 - n_{min}^2\right)^{-1/2}$ into the lowest-RI medium.

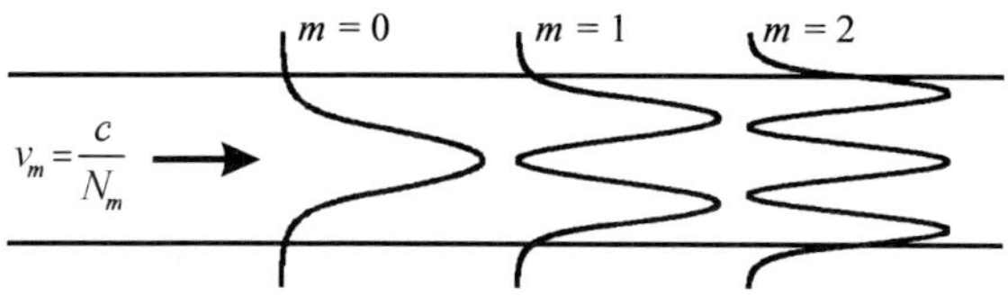

Fig. 4 Three lowest-order mode-power profiles in a three-layer waveguide

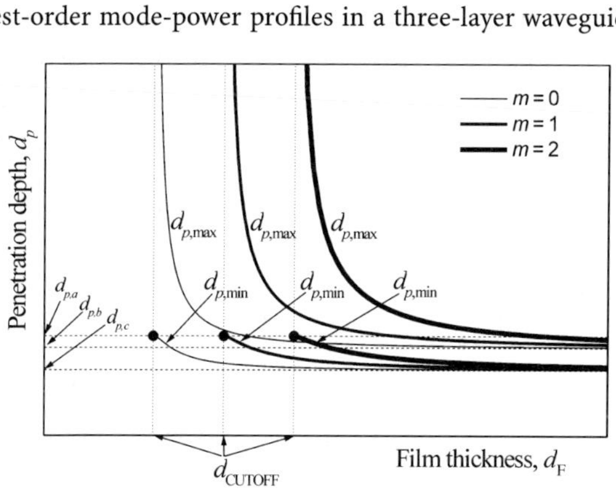

Fig. 5 Penetration depths d_p into substrate and the cover versus film thickness d_F for the three-lowest order modes. $d_{p,max}$ is the penetration depth into that medium (substrate or cover) with the largest RI and vice versa. The limiting values $d_{p,a}$, $d_{p,b}$, and $d_{p,c}$ are given in the text

2.2
Sensor Issues

When used as a biosensor, light is typically coupled into the waveguide via a diffraction grating as shown in Fig. 6a. Via an external fluid handling system, the cover medium, which is usually aqueous, is brought in contact with the film surface. By scanning the angle of incidence while monitoring the level of in-coupled light by a detector, a resonance peak is identified at a resonance angle α_m (Fig. 6b) which is given by

$$n_S \sin\left(\alpha_m\right) = N_m - K/k\,, \tag{4}$$

where K is the grating number and N_m is the effective RI of the mode being coupled into the waveguide. Hence, from Eqs. 2 and 4, n_C can be derived from the measured value of α_m.

The sensitivity of the sensor is usually defined by the change in N_m per change in n_C, i.e., $\partial N_m/\partial n_C$. An expression for this for, say, TE polarized light can be derived by using perturbational analysis of the mode equation (Eq. 2) [16]:

$$\frac{\partial N_m}{\partial n_C} = \frac{n_C}{N_m}\left(\frac{n_F^2 - N_m^2}{n_F^2 - n_C^2}\right)\frac{d_{p,C}}{d_F + d_{p,C} + d_{p,S}}, \quad d_{p,j} = k^{-1}\left(N_m^2 - n_j^2\right)^{-1/2}. \tag{5}$$

The sensitivity is plotted in Fig. 7 versus film thickness for the two cases $n_S > n_C$ and $n_S < n_C$. The first-mentioned case corresponds to a conventional symmetry in which the substrate is a solid material like glass or polymer ($n_S = 1.5$), the film is a typical metal oxide ($n_F = 1.75$), and the cover is water ($n_C = 1.33$). In this case it is seen that the sensitivity is rather poor, at maximum 0.1. For the reverse symmetry case, in which $n_S = 1.2$, the sensitivity goes to 1 at cutoff. Hence, with the second geometry having the reverse symmetry as compared with the conventional waveguide geometry, the sensitivity is improved by more than an order of magnitude. The reason for this is that if $n_S < n_C$, the penetration depth into the cover $d_{p,C}$ goes to infinity at cutoff (Fig. 5), meaning that practically all

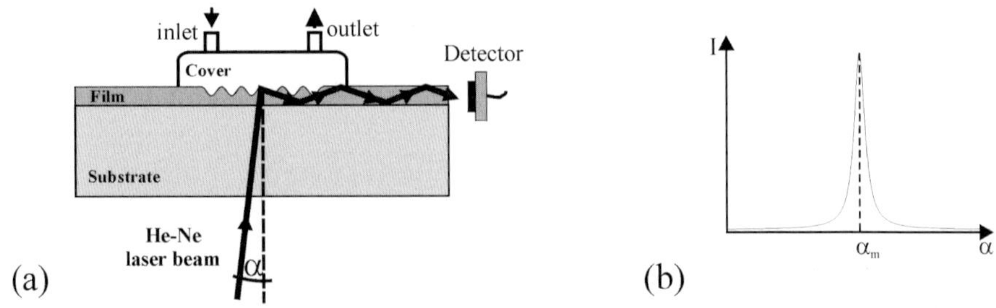

Fig. 6 **a** General waveguide sensor configuration; **b** typical sensorgram

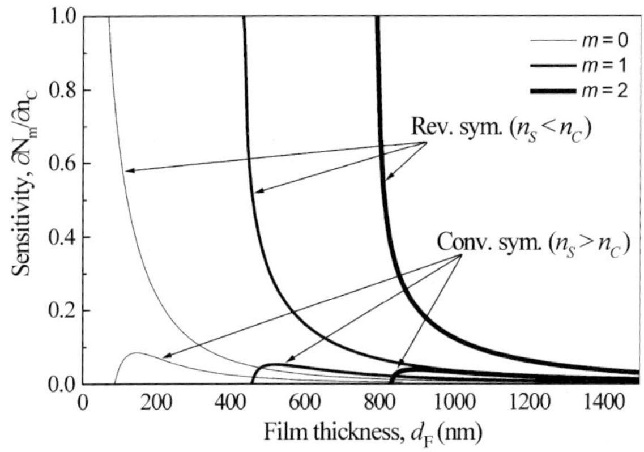

Fig. 7 Sensitivity $\partial N_m/\partial n_C$ versus d_F for the three lowest-order TE modes, i.e., $\rho = 0$, $m = 0, 1, 2$. For the conventional symmetry ($n_S > n_C$) the parameters $n_S = 1.5$, $n_F = 1.75$, and $n_C = 1.33$ were used, whereas for the reverse symmetry $n_S = 1.2$, $n_F = 1.59$, and $n_C = 1.33$. In all cases a light wavelength of 633 nm was used

the mode power flows in the cover and therefore achieves optimum sensitivity to n_C changes. Similar relations apply to the TM modes.

In many biosensor applications the biological events to be monitored occur as binding events between an affinity layer immobilized at the waveguide surface and certain biological agents being looked for in the aqueous sample. The agents range in size from lipids and antibodies (a few nanometers) to bacterial cells (approximately 1 μm) to whole eukaryotic cells (approximately 10 μm). In each of these cases it is beneficial to have as high a portion of the guided mode power flowing in a sensing layer close to the waveguide surface with a layer thickness that matches the size of the objects. This can be analyzed by using the four-layer mode equation developed by Tiefenthaler and Lukosz [16]:

$$2\pi m = 2k_{z,\mathrm{F}}d_\mathrm{F} + 2\arctan\left(i\frac{1-r_{\mathrm{F,S}}}{1+r_{\mathrm{F,S}}}\right) + 2\arctan\left[i\frac{\left(1-r_{\mathrm{F,A}}\right)}{\left(1+r_{\mathrm{F}},\mathrm{A}\right)}\frac{\left[1-r_{\mathrm{A,C}}\exp\left(i2k_{z,\mathrm{A}}d_\mathrm{A}\right)\right]}{\left[1+r_{\mathrm{A,C}}\exp\left(i2k_{z,\mathrm{A}}d_\mathrm{A}\right)\right]}\right], \quad (6)$$

$$r_{j,j'} = \frac{\dfrac{k_{z,j}}{n_j^{2\rho}} - \dfrac{k_{z,j'}}{n_{j'}^{2\rho}}}{\dfrac{k_{z,j}}{n_j^{2\rho}} + \dfrac{k_{z,j'}}{n_{j'}^{2\rho}}},$$

where $r_{j,j'}$ is the amplitude reflection coefficient between layer j and layer j'. On the basis of this equation, the change in effective RI per change in sensing layer RI n_{SL}, $\partial N_m/\partial n_{SL}$ can be calculated and plotted versus sensing layer thickness d_{SL}, as shown in Fig. 8. Here it is seen that the reverse symmetry waveguide is superior to the conventional one for any sensing layer thickness; however, the sensitivity improvement from conventional to reverse symmetry sensing increases and saturates at a factor of 4 for increasing sensing layer depths. It should be noted, however, that quite often large biological bodies like human cells only attach to the waveguide surface at a few small attachment points (as illustrated in Fig. 1), which leaves an aqueous gap of typically 50–100 nm between most parts of the cell membrane and the waveguide surface [24]. In these cases the advantages of the reverse symmetry waveguide are expected to be even more pronounced.

In recent years, the measurement of thick adlayers of up to 400 nm has received increasing interest in the field of so-called polyelectrolyte multilayer films [25, 26]. Such large thicknesses are obviously not easily monitored using a conventional symmetry waveguide because of the limited penetration depth

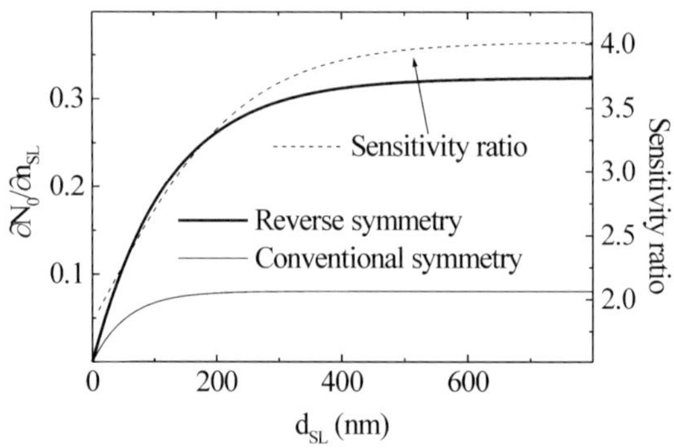

Fig. 8 Sensing layer sensitivities against sensing layer thickness for the conventional and reverse symmetry waveguides considered in Fig. 7. The *dashed curve* represents the ratio of the reverse symmetry sensitivity to the conventional symmetry sensitivity

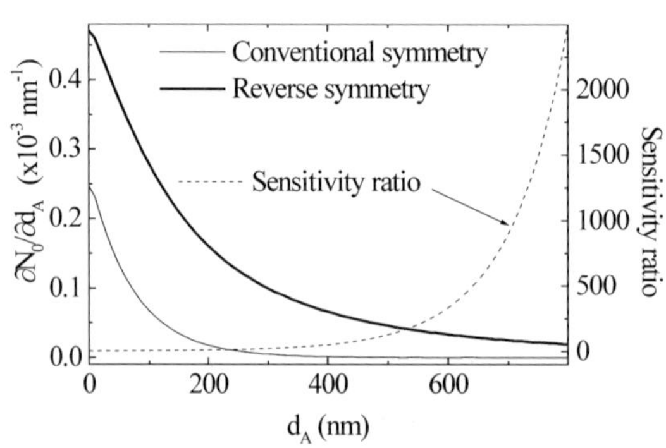

Fig. 9 Adlayer thickness sensitivity against adlayer thickness for the conventional and reverse symmetry waveguides considered in Fig. 7. The adlayer RI equals 1.5

into the adlayer. To analyze this we have calculated the change in effective RI per change in adlayer thickness versus adlayer thickness for the conventional and reverse symmetry waveguides, as shown in Fig. 9. Again, it is seen that the reverse symmetry waveguide exhibits an improved sensitivity for all adlayer thicknesses; however, the sensitivity improvement is seen to increase exponentially with adlayer thickness. Hence, at 400 nm the sensitivities are 1.5×10^{-6} and 6.6×10^{-5} nm^{-1} for the conventional and the reverse symmetry waveguides, respectively, meaning that the reverse symmetry waveguide is 50 times more sensitive to adlayer thickness changes.

2.3 Multimode Operation

As can be seen from Figs. 5 and 7 the improved cover penetration depths and sensitivities are not restricted to single-mode waveguides. Also, higher-order-mode waveguides in reverse symmetry achieve infinite penetration depth close to cutoff. This may be exploited using very simple designs where thin glass or polymer plates of, say, 50–100-μm thickness may be used as free-standing waveguides employing air as the "substrate". In this case $n_S = 1$ is indeed less than n_C, resulting in a reverse symmetry design.

In Fig. 10 the multimode solutions to Eq. 2 for a 50-μm-thick glass plate are shown in the case where n_C is changed in three steps. Hence, if a grating with grating number K is applied to the waveguide (either etched into the glass surface or applied via a polymer coating), the corresponding resonance peaks in the sensorgram will appear at angles α_m given by

$$\sin\left(\alpha_m\right) = N_m - K/k\,. \tag{7}$$

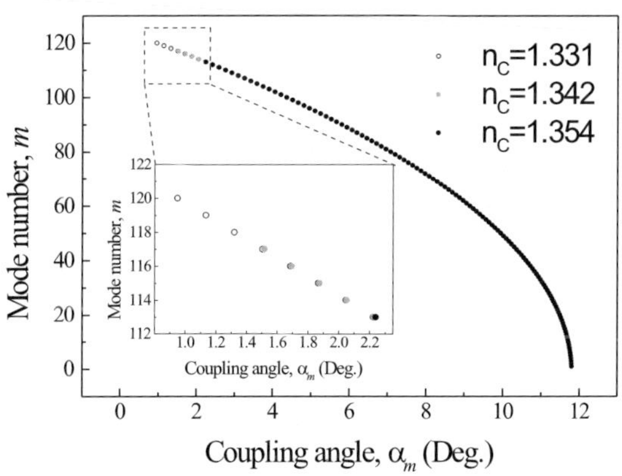

Fig. 10 Multimode solutions for a 50-μm glass plate with cover RIs of 1.331, 1.342, and 1.354 [35]

Again, we have for the highest-order mode that $N_m \rightarrow n_C$; hence, simply by measuring the highest-order peak angle $\alpha_{m,\max}$, n_C can easily be found:

$$n_C = \sin\left(\alpha_{m,\max}\right) + K/k\,. \tag{8}$$

In this manner it is possible to construct a very simple refractometer. It should be mentioned that very sensitive devices using the multimode free-standing waveguide can be achieved by monitoring only the intensity of the highest-order mode. Here detection limits down to 3×10^{-5} hve been demonstrated [27].

3
Reverse Symmetry Designs and Applications

In practice, the reverse symmetry biosensor is not very obvious to realize, because of the difficulty in finding a solid substrate with RI less than 1.33. We managed, however, to fabricate three different versions shown schematically in Fig. 11, the nanoporous waveguide, the thin-plate waveguide, and the microstructured waveguide. In all three cases the introduction of air into the substrate is the key element in reducing the RI. The detailed designs, manufacturing procedures, and sensing features are discussed in the following.

3.1
Nanoporous Waveguide

The design of the nanoporous waveguide is shown both schematically and as a photograph in Fig. 12. It consists of an 8 mm × 12 mm × 1.6 mm soda lime glass substrate, spin-coated with a 1-μm-thick layer of nanoporous silica (Nanopore, NM, USA) and then coated with a 160-nm-thick layer of polystyr-

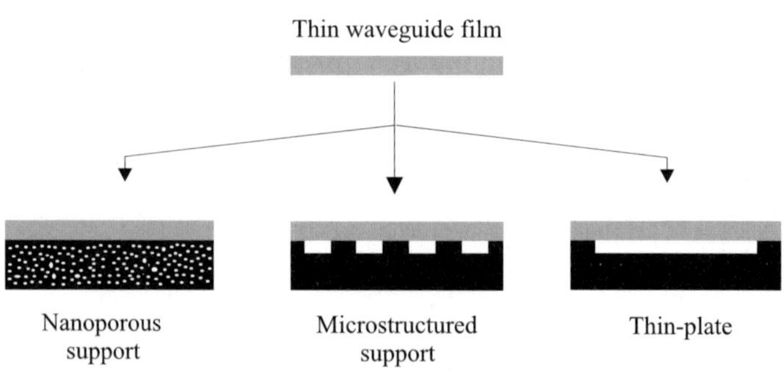

Fig. 11 Three different kinds of reverse symmetry waveguides: *a* nanoporous waveguide, *b* microstructured waveguide, and *c* thin-plate waveguide

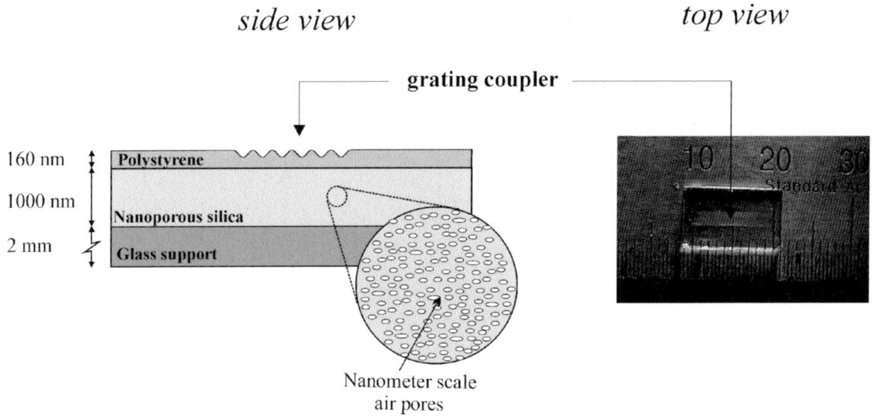

Fig. 12 Schematic design and photograph of the nanoporous waveguide chip

ene. In the polystyrene film a surface grating with a period of approximately 480 nm and an amplitude of 10–15 nm is imprinted by using soft lithography replication from a poly(dimethylsiloxane) (PDMS) grating stamp [28]. The nanoporous silica and the polystyrene layers have RIs of 1.193 and 1.587, respectively. It should be noted that the nanoporous silica layer is so thick that the evanescent field practically vanishes in the glass support; hence, the mode only experiences the nanoporous silica, the film, and the cover medium, resulting in a pure monomode operation in reverse symmetry. We did try to construct waveguides based on bulk nanoporous substrates, but it was not possible to obtain samples with a satisfactory surface quality.

3.1.1 Refractometry

In the first test, the nanoporous waveguide chip was inserted in the setup shown in Fig. 13. Here the waveguide chip is rotated by a high-precision rotation stage while a fixed He-Ne laser beam with wavelength $\lambda = 632.8$ nm illuminates the coupling grating positioned at the center of rotation. During rotation, the in-coupled light intensity is measured by fiber-coupled photoreceivers. In this way the in-coupled light intensity is monitored as a function of illumination angle α [5, 28]. In order to shift between TE and TM illumination, the polarization of the incoming laser light is controlled by a polarizer and a half-wave plate. A cuvette with tubes connected to a peristaltic pump is positioned on top of the waveguide, which makes it possible to flow various liquids across the surface of the waveguide.

In Fig. 14 four sensorgrams are shown for air and water cover media and TE/TM polarized light. The peaks are correctly centered at the angles given by the grating equation $\sin(\alpha_0) = N_0 - \lambda/\Lambda$ and the half-widths are approximately 0.05 deg. It is seen that the shift in peak position from air to water is

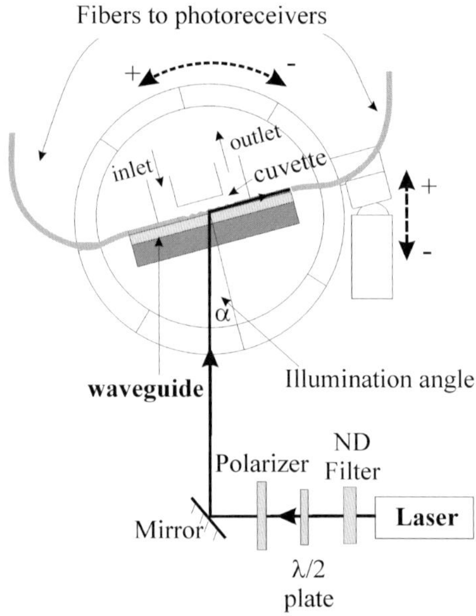

Fig. 13 Experimental waveguide sensor setup using a rotation stage for angular interrogation [28]

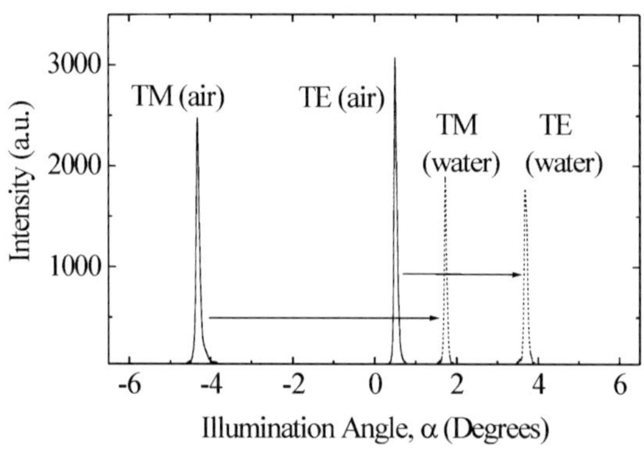

Fig. 14 TE and TM sensorgrams for the nanoporous waveguide using air and water as cover media [28]

largest for the TM mode, which is due to the fact that the TM mode is closest to cutoff and therefore exhibits the largest sensitivity, according to Fig. 7. The cutoff thicknesses are 63 nm for the TE mode and 93 nm for the TM mode.

By altering the cover medium between pure water and a water–glycerol solution (corresponding to RIs of 1.331 and 1.355, respectively), the TE and TM peaks shift according to Fig. 15. It is seen that compared with the data obtained with a commercial waveguide chip the nanoporous waveguide has

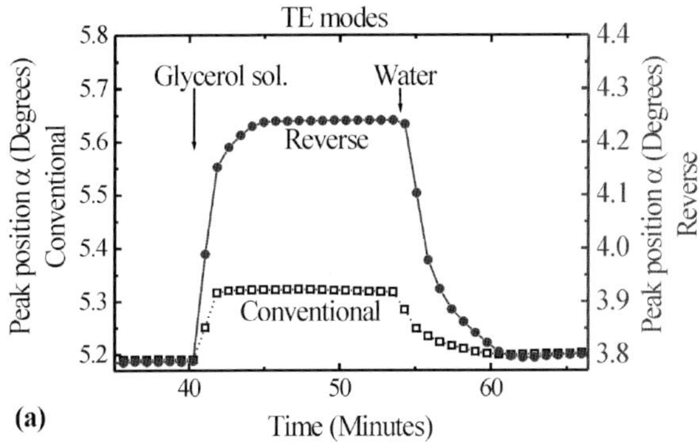

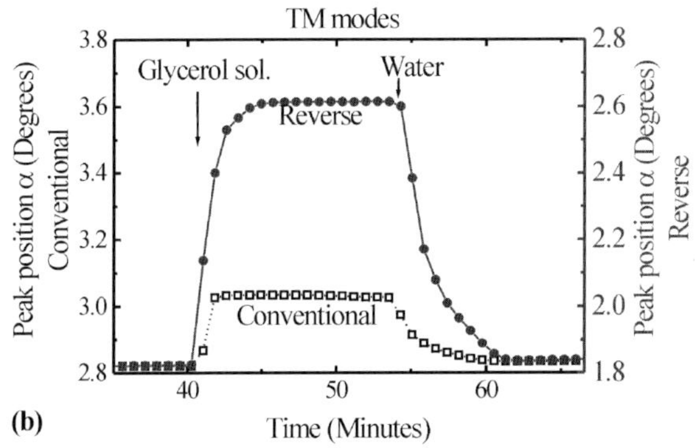

Fig. 15 TE (**a**) and TM (**b**) peak positions versus time after exposing the surfaces of reverse and conventional symmetry waveguides to pure water and a water–glycerol solution. The conventional waveguide was an ASI-2400 μV chip from MicroVacuum, Hungary, with the data $n_S = 1.5246$, $n_F = 1.79$, and $d_F = 184.6$ nm [28]

a sensitivity (i.e., shift in peak angle per change in cover RI) approximately 3.5 times larger, which agrees well with the theoretical predictions from Fig. 7.

3.1.2 Bacterial Detection

In order to attach bacterial cells to the waveguide surface, the nanoporous chip was first plasma-treated and coated with a thin layer of poly-L-lysine, which is a thin protein layer that binds unspecifically all types of bacterial cells [29]. After this, a bacterial solution consisting of phosphate-buffered saline (PBS) with a concentration of 3×10^7 *Escherichia coli* K12 cells per milliliter was pumped through the cuvette. The resulting changes in TE and TM peak positions are shown in Fig. 16. It is seen that after exposing the waveguide surface to the bacterial solution, the peak positions shift grad-

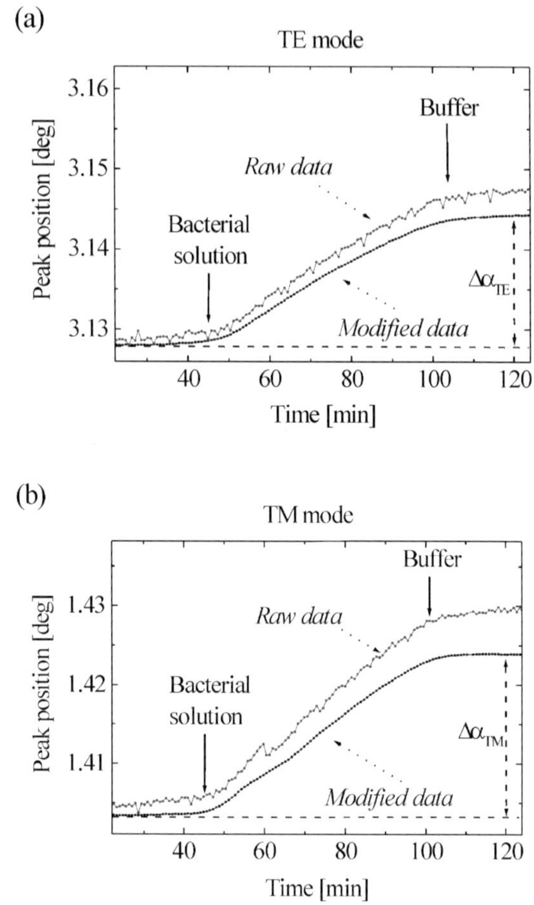

Fig. 16 Measured peak shift versus time due to bacterial adhesion to the waveguide surface. The processed data have been modified to compensate for the constant drift and to minimize the noise [31]

ually to higher values, as bacterial cells attach to the surface. The shift to higher coupling angles reflects the fact that the RI of a bacterial cell is slightly larger than the water RI [30]. After rinsing the cuvette system with PBS again, the peak position levels off, indicating that the bacterial cells already attached to the surface stay immobilized. The waveguide chip was afterwards investigated under a microscope, where clear images of the bacterial cells, 1 μm × 3 μm in size, were observed; Fig. 17. The surface density was estimated to be 12 400 cells mm^{-2}, leading to surface cell sensitivities of 1.3×10^{-6} and 1.65×10^{-6} deg mm^2 $cell^{-1}$ for the TE and TM mode operations, respectively. In view of the noise level of 10^{-4} deg these sensitivities correspond to detection limits of 78 and 60 cells /mm^{-2} for TE and TM mode detection [31].

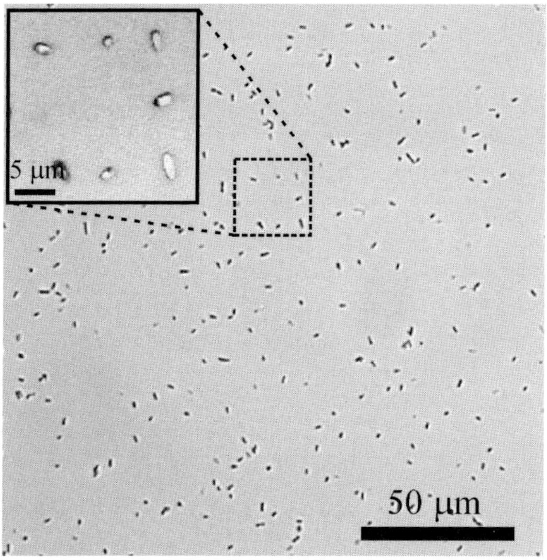

Fig. 17 Microscope image of the waveguide surface after the exposure to *Escherichia coli* cells [31]

3.1.3
Whole-Cell Detection

The nanoporous waveguide was also tested for monitoring the attachment of human dermal fibroblast cells to the surface. As before, the waveguide surface was coated with a thin layer of poly-L-lysine to improve cell attachment and spreading.

The resulting sensorgrams are shown in Fig. 18a. It is seen that the pure buffer solution gives the expected sharp in-coupling peaks, but as the fibroblast cells start to attach to the surface, the peaks change considerably. It is seen that the peaks are both broadened and also tend to split into two peaks. We believe that the main reason for not observing a simple peak shift, as is the case for protein attachment and bacterial attachment, is due to the size of the cells, which are 10–100 times larger than the wavelength of the light. This implies that rather than averaging the RI of the cover layer exposed to the evanescent field, the electromagnetic field considers the locations with cells as local waveguide areas with their own distinct cover RI. This type of peak splitting due to inhomogeneous coverage was predicted theoretically by Horvath et al. [32].

In Fig. 18b the center of mass of the peaks are monitored versus time. It is seen that the rates of change in peak position are not identical for the TE and TM modes. This may be due to the higher cover index sensitivity of the TM mode (approximately a factor of 2 larger than for the TE mode), but also the fact that the probing depths are different for the two modes (approxi-

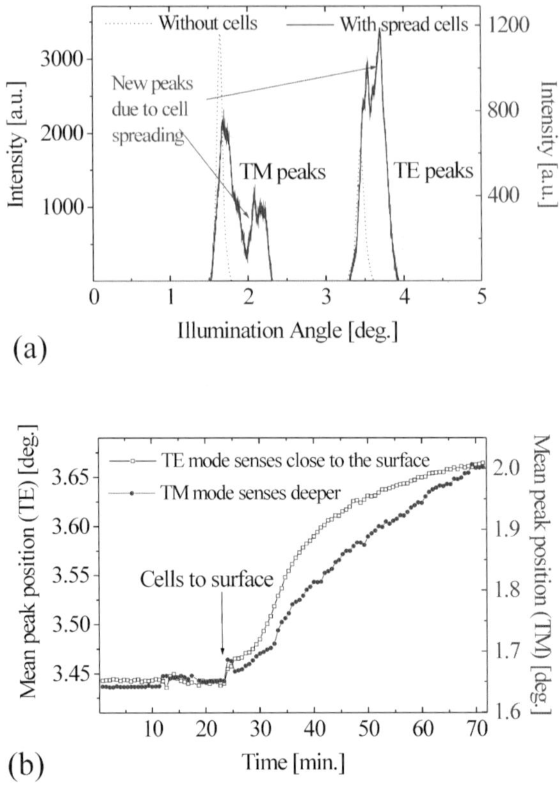

Fig. 18 **a** TE and TM sensorgrams obtained before and after cell attachment to the waveguide surface, **b** Mean TE and TM peak positions versus time after exposing the waveguide surface to the cell solution

mately 300 nm for the TE mode, approximately 600 nm for the TM mode); hence, the TE mode is mostly affected by changes occurring close to the waveguide surface, whereas the TM mode is capable of sensing changes deeper in the cells [33]. In Fig. 19 a microscope image shows the fibroblast cells on the waveguide surface after spreading.

3.1.4
Amplitude Detection

The main advantage of using the reverse symmetry for extending the penetration depth into the cover medium is based on operating the waveguide close to the cover cutoff, i.e., close to the point at which the evanescent wave in the cover transforms into a propagating wave. Obviously, in that transition another phenomenon takes place, namely, that the guided mode vanishes. This is illustrated in Fig. 20, where the sensorgrams for seven different cover solu-

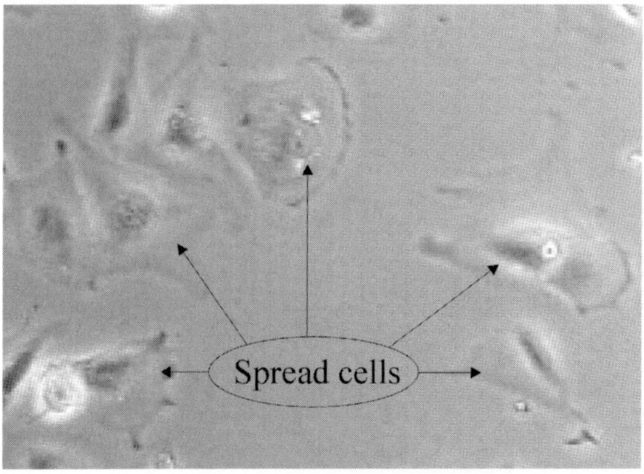

Fig. 19 Phase-contrast microscope image of human dermal fibroblast cells attached to and spread on the surface of the waveguide chip

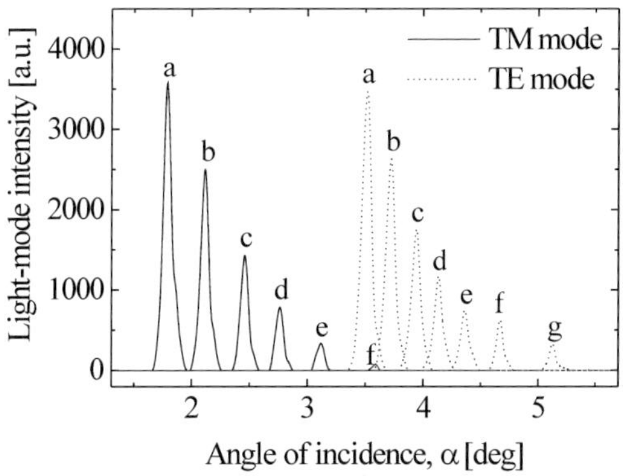

Fig. 20 TE and TM sensorgrams for the nanoporous waveguide obtained for the cover RIs: *a* 1.331, *b* 1.342, *c* 1.354, *d* 1.363, *e* 1.373, *f* 1.383, and *g* 1.403. The cover consists of water with various concentrations of NaSCN [34]

tions are shown. Apart from the expected shifts in peak positions, it is also seen that the peaks decrease significantly in height, as the modes get closer to cutoff (the theoretical cutoff cover RIs for the TE and TM modes are 1.452 and 1.415, respectively). This height change may be used as a transduction parameter alternative to the well-known peak position. In Fig. 21 the two transduction mechanisms are compared by alternately changing the RI of the cover medium. The main difference in performance between the two mechanisms is seen to be the noise levels. Hence, TM detection limits of 5×10^{-6} and 6×10^{-5} RI units are obtained for the peak-angle and peak-height measurements [34]. However, one has to emphasize that measuring the intensity of a signal is generally much easier than measuring an angle, so to keep the instrumentation price down, the peak-height detection method may be a nice alternative.

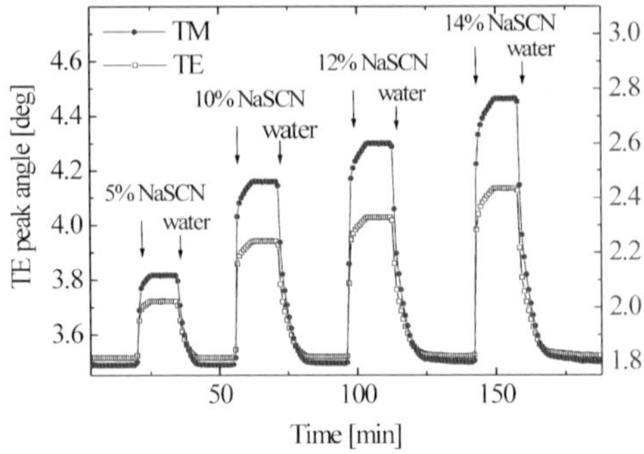

(a)

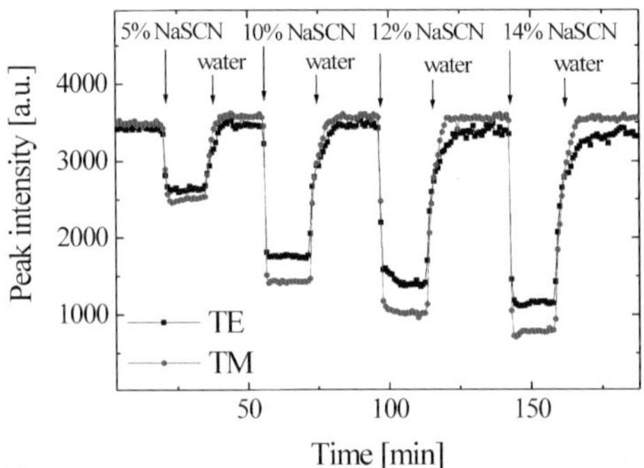

(b)

Fig. 21 **a** TE and TM peak angles measured for alternating the cover medium between pure water and water containing increasing concentrations of NaSCN. **b** The corresponding peak heights [34]

3.2 Multimode Thin-Plate Waveguide

As discussed in Sect. 2.3, a simple, freestanding thin glass plate may be used as a reverse symmetry waveguide sensor. In our case, we purchased a couple of thin plates of BK7 with a thickness of 52 μm and a RI of 1.52. In order to couple light into the plate, a thin polystyrene layer of 40 nm was spin-coated on top of the plate after which a grating with a period of 481 nm and a depth of 25 nm was stamped into it from a PDMS grating stamp using soft lithography [35]. The plate was then positioned in a rotational stage setup identical to the setup described earlier.

The sensorgrams obtained after exposing the surface of the waveguide to pure water and to two concentrations of NaSCN are shown in Fig. 22. The multimode operation is clearly manifested by the multiple peaks in the sensorgrams. The effect of increasing the cover RI is that the highest-order-mode propagation angles (corresponding to the smallest angles of incidence in Fig. 22) get

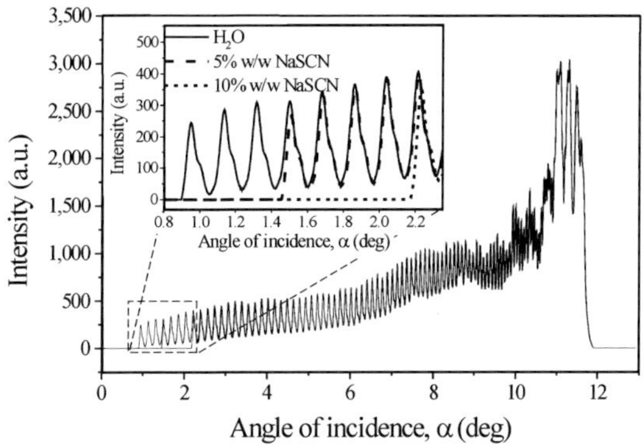

Fig. 22 Thin-plate waveguide sensorgrams obtained with a grating-coupled, 52-μm-thick glass plate for three different aqueous cover media, $n_C = 1.331$ (pure water), 1.342 (water with 5% NaSCN), and 1.354 (water with 10% NaSCN) [35]

below the critical angle between the glass plate and the cover medium, causing these modes to disappear. Hence, as is seen, three and four modes disappear as a result of the changes from pure water to the NaSCN solutions.

By using the grating Eq. 7 the cover RI can be deduced from the angular position of the highest-order modes. Using this, cover RI values of 1.332, 1.342, and 1.355 were obtained, which match very well with the table values. The main advantages of the thin-plate waveguide sensor chip in the configuration used here are that it is very simple to fabricate and it has a large detection range. The detection limit is, however, quite limited ($\Delta n_C \approx 0.001$) for the plate used here, but by carefully monitoring the height of the vanishing peak, detection limits down to 3×10^{-5} have been demonstrated [27].

3.3 Microstructured Waveguide

The basic idea of using a microstructured substrate is to enclose sealed air cavities under parts of the film, for example, in the shape of grooves. In this case the electromagnetic field will regard the air cavities as substrate areas with RI equal to 1, leading to the reverse symmetry. In Fig. 23 the fabrication steps of such a groove-based waveguide are illustrated [36]. The resulting waveguide is all in one material (UV resin). In Fig. 24a the microstructured waveguide has been photographed, where the grooves are clearly seen owing to light diffraction off the gratings. Confocal fluorescence laser scanning microscopy, taking advantage of the autofluorescence of the UV-cured resin can visualize the buried channels of intact waveguide units. Figure 24b shows a typical cross section of a single material (NOA71) unit having 50-μm-wide and 8-μm-deep grooves below the continuous waveguide film. Discontinuities in the emitted fluorescence were not observed in the sealing plane at places other than the groove

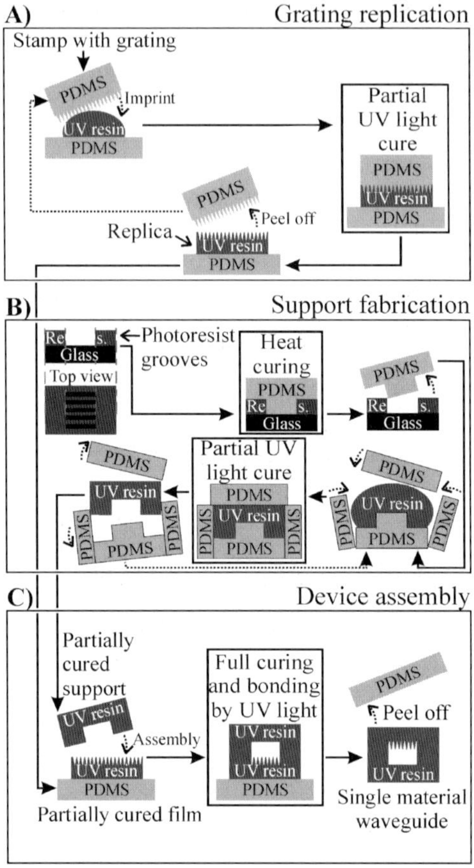

Fig. 23 Fabrication procedure for microstructured waveguide: **a** fabrication of waveguiding film including imprinted surface grating, **b** fabrication of support, and **c** assembly of film and support into final waveguide. *PDMS* poly(dimethylsiloxane) [36]

locations. The light waveguiding ability of the assembled waveguide module is demonstrated in Fig. 24c. Here, a 1-mm-wide He–Ne laser beam is coupled into the freestanding waveguides and the incoupled light is observed at the edge of the module using a CCD camera. Considering the ease of processing the polymeric materials into operational modules, the suggested waveguide configuration could be a basis for cheap and disposable sensor modules.

4
Summary

On the basis of the enhanced penetration depth into the analyte medium, the reverse symmetry waveguide sensor shows great potential in terms of high-

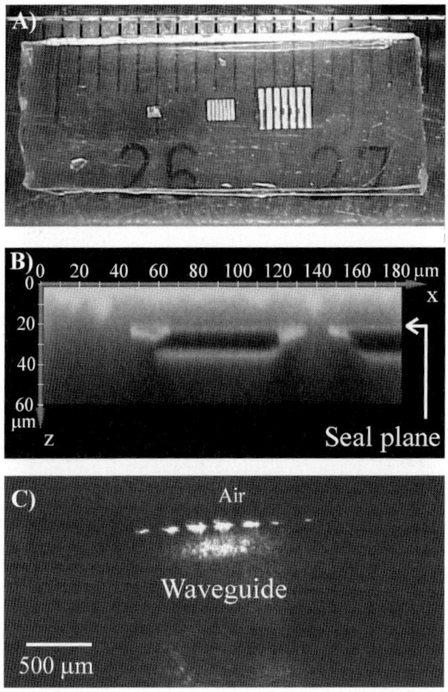

Fig. 24 **a** Photograph of the single material polymer freestanding waveguide [22], **b** Confocal micrograph of the microstructured waveguide in a cross section, and **c** butt-end outcoupling of light from the six channels [36]

sensitivity detection. In particular, the detection of larger biological objects such as bacterial cells (approximately 1 μm) and eukaryotic cells (approximately 10 μm) is seen to benefit significantly from the reverse symmetry design. This is so because unlike the conventional waveguides the reverse symmetry waveguide is capable of embedding biological bodies of any size in the probing evanescent field. Hence, simply by choosing the right film thickness, the reverse symmetry waveguide can be tailor-made with respect to detecting agents of a certain size.

We have demonstrated three ways of realizing the reverse symmetry waveguide, one relying on the application of a nanoporous silica layer between the support and the film, another relying on the introduction of air grooves under the film, and a third based on a simple, freestanding glass plate. The sensor performances were tested to show the expected sensitivity improvements for refractometry and bacterial detection. Also promising sensor qualities regarding the detection of eukaryotic cell attachment and spreading were demonstrated. We believe that the reverse symmetry waveguide sensor may be particularly suited for the monitoring of cell behavior, which is probably one of the fastest growing biosensing activities at the moment (toxicological screening, drug testing, etc.).

Acknowledgements The authors wish to express their special thanks to the following coworkers: L.R. Lindvold, J. Stubager, L. Knudsen, E. Papp, G. Meszena, B. Sass, D. Selmeczi, L. Hubert, and S. Wallace. Financial support from the Danish Technical Research Council, grant nos. 26-01-0211 and 26-03-0272, and from the Graduate School of Biophysics (NBI, Copenhagen University) is also gratefully acknowledged.

References

1. Tiefenthaler K, Lukosz W (1984) Proc Soc Photo-Opt Instrum Eng 514:215–218
2. Tiefenthaler K, Lukosz W (1985) Thin Solid Films 126:205–211
3. Lukosz W (1995) Sens Actuators 29:37–50
4. Voros J, Ramsden JJ, Csucs G, Szendro I, Paul SM, Textor M, Spencer ND (2002) Biomaterials 23:3699–3710
5. Horvath R, Fricsovszky G, Papp E (2003) Biosens Bioelectron 18:415–428
6. Kuhlmeier D, Rodda E, Kolarik LO, Furlong DN, Bilitewski U (2003) Biosens Bioelectron 18:925–936
7. Brusatori MA, Van Tassel PR (2003) Biosens Bioelectron 18:1269–1277
8. Ramsden JJ (1999) Chimia 53:67–71
9. Hirmo S (1999) J Microbiol Methods 37:177–182
10. Ramsden JJ, Li SY, Prenosil JE, Heinzle E (1994) Biotechnol Bioeng 43:939–945
11. Ramsden JJ, Li SY, Heinzle E, Prenosil JE (1995) Cytometry 19:97–102
12. Hutchinson AM (1995) Mol Biotechnol 3:47–54
13. Ruiz L (1999) J Biomater Sci Polym Ed 10:931–955
14. Hirmo S, Artursson E, Puu G, Wadstrom T, Nilsson B (1998) Anal Biochem 257:63–66
15. Voros J, Graf R, Kenausis GL, Bruinink A, Mayer J, Textor M, Wintermantel E, Spencer ND (2000) Biosens Bioelectron 15:423–429
16. Tiefenthaler K, Lukosz W (1989) J Opt Soc Am B 6:209–220
17. Lukosz W (1995) Sens Actuators B 29:37–50
18. Kunz RE (1997) Sens Actuators B 38:13–28
19. Fratamico PM, Strobaugh TP, Medina MB, Gehring AG (1998) Biotechnol Tech 12:571
20. Watts HJ, Lowe CR, Pollard-Knight DV (1994) Anal Chem 66:2465
21. Kurrat R (1998) PhD dissertation no 12891, ETH, Zurich
22. Horvath R, Lindvold LR, Larsen NB (2002) Appl Phys B 74:383–393
23. Tien PK (1977) Rev Mod Phys 49:61–420
24. Giebel KF, Bechinger C, Herminghaus S, Riedel M, Leiderer P, Weiland U, Bastmeyer M (1999) Biophys J 76:509–516
25. Picart C, Lavalle P, Hubert P, Cuisinier FJG, Decher G, Schaaf P, Voegel JC (2001) Langmuir 17:7414–7424
26. Picart C, Gergely C, Arntz Y, Voegel JC, Schaaf P, Cuisiner FJG, Senger B (2004) Biosens Bioelectron 20:553–561
27. Qi ZM, Matsuda N, Santos JH, Takatsu A, Kato K (2002) Opt Lett 27:689–691
28. Horvath R, Pedersen HC, Larsen NB (2002) Appl Phys Lett 81:2166–2168
29. An YH, Friedman RJ (1998) J Biomed Mater Res 43:338
30. Walkenburg JAC, Woldringh CL (1984) J Bacteriol 160:1151–1157
31. Horvath R, Pedersen HC, Skivesen N, Selmeczi D, Larsen NB (2003) Opt Lett 28:1233–1235
32. Horvath R, Voros J, Graf R, Fricsovszky G, Textor M, Lindvold LR, Spencer ND, Papp E (2001) Appl Phys B 7:441–447

33. Horvath R, Pedersen HC, Skivesen N, Selmeczi D, Larsen NB (2005) Appl Phys Lett 86:071101
34. Horvath R, Skivesen N, Larsen NB (2004) Appl Phys Lett 84:4044–4046
35. Skivesen N, Horvath R, Pedersen HC (2003) Opt Lett 28:2473–2475
36. Horvath R, Lindvold LR, Larsen NB (2003) J Micromech Microeng 13:419

Springer Ser Chem Sens Biosens (2005) 000: 303–322
DOI 10.1007/5346_010

Published online: 27 September 2005

Materials for Luminescent Pressure-Sensitive Paint

Yumi Takeuchi · Yutaka Amao (✉)

Department of Applied Chemistry, Oita University, Dannoharu 700, 870-1192 Oita, Japan
amao@cc.oita-u.ac.jp

Abstract Pressure-sensitive paint (PSP) is applied to the areodynamics measurement. PSP is optical sensor based on the luminescence of dye probe molecules quenching by oxygen gas. Many PSPs are composed of probe dye molecules, such as polycyclic aromatic hydrocarbons (pyrene, pyrene derivative etc.), transition metal complexes (ruthenium(II), osumium(II), iridium(III) etc.), and metalloporphyrins (platinum (II), palladium(II), etc.) immobilized in oxygen permeable polymer (silicone, polystyrene, fluorinated polymer, cellulose derivative, etc.) film. Dye probe molecules adsorbed layer based PSPs such as pyrene derivative and porphyrins directly adsorbed onto anodic oxidised aluminium plat substrate also developed. In this section the properties of various oxygen permeable polymer for matrix and various dye probes for PSP are described.

1 Pressure-Sensitive Paint

Surface pressure distribution measurement is of fundamental importance in the experimental study of aerodynamic problems in the fields of avionics, car, rocket, aerospace, and aircraft design [1]. The conventional methods based on pressure taps or transducers have a number of limitations. The most serious problem is that their very nature limits them to providing information only at discrete points on the surface of a substrate. A new approach to surface pressure distribution measurement, the use of pressure-sensitive paint (PSP), has recently developed that offers the potential of revolutionizing the nature of such measurements in the field of aerodynamics. This method employs the oxygen sensitivity of fluorescent materials in the form of a paint, in conjunction with image processing techniques, to map the pressure field over aerodynamic surfaces. PSP has many advantages over the conventional point measurement using pressure taps or transducers. As a pressure sensor it is unique in that it is a field measurement, pressure being measured over the entire surface of a model simultaneously at a very high spatial resolution. With pressure data available over the entire model surface, flow anomalies at any point on the surface are immediately obvious, rather than having to be deduced from discrete tap data. Without a large number of pressure taps, wind tunnel models could be constructed faster. Only recently has the potential of the oxygen sensitivity of photoluminescent materials for surface pressure field measurement been recognized and pursued by a continually growing number of groups worldwide.

2
Principles of PSP

As mentioned above, PSP is an optical surface-pressure measurement method based on the luminescence quenching by oxygen gas [2]. The PSP with an appropriate imaging system is attractive to aerodynamic researchers because one can easily obtain the pressure distribution on the whole surface of a wind-tunnel model with an incredible spatial resolution, as shown in Fig. 1. The principle of an optical oxygen sensor is based on oxygen-quenched luminescence (fluorescence and phosphorescence). As shown in the Jablonski diagram (Fig. 2), luminescence (spontaneous emission of light) is produced in certain molecules following excitation by light from their ground state to higher energy states. When the excited molecule returns to the ground state, it emits light (fluorescence or phosphorescence) at a longer wavelength than the original simulated light (Stokes' shift). Fluorescence is a short-lived emission from the singlet state with the electrons spin-paired. As the transition from the singlet state (S_1) to the ground singlet state (S_0) is spin-allowed (i.e., with no change in multiplicity), it has a high probability of occurrence and the decay time of fluorescence is usually short (10^{-9} to 10^{-7} s, i.e., of the same order as the lifetime of an excited singlet state). In contrast, phosphorescence involves a change in electron spin, and so the transition from the lowest excited triplet state (T_1) to the singlet ground state (S_0) occurs at a low proba-

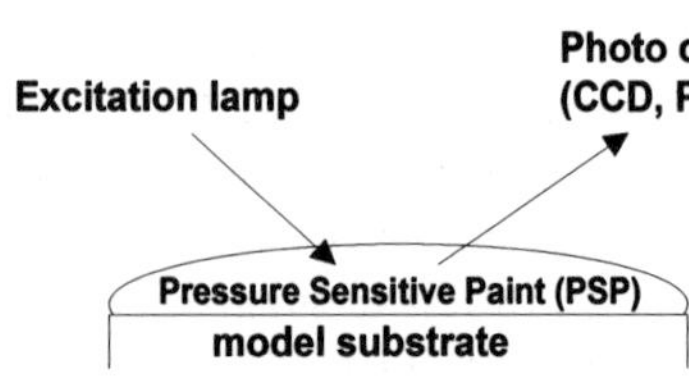

Fig. 1 Concept of pressure sensitive paint (PSP)

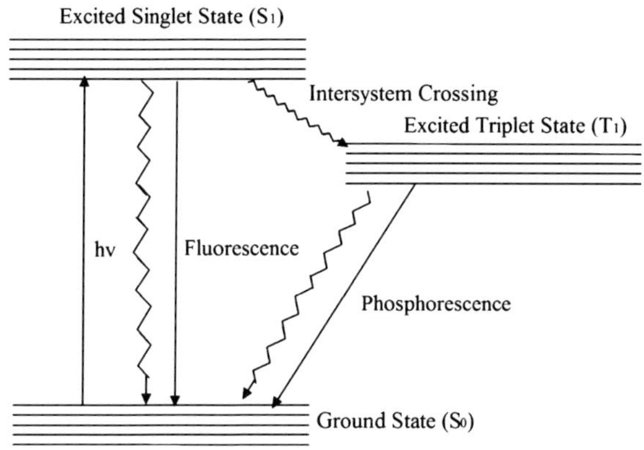

Fig. 2 Jablonsky diagram

bility, and phosphorescence is a long-lived emission (about 10^{-5} to 10 s) [3, 4]. Also, owing to the relative energies of the excited states, phosphorescence occurs at longer wavelengths than fluorescence. Given the fact that molecules in the triplet state have long lifetimes, they are particularly susceptible to interactions with other molecules. Oxygen quenching is diffusion-limited and can be described by the Stern–Volmer relationship [5]. This is valid for both phosphorescence and fluorescence lifetimes and intensities:

$$I_0/I = \tau_0/\tau = 1 + k_q\tau_0[O_2] \tag{1}$$

where, I_0 and I are the luminescence intensities, and τ_0 and τ are lifetimes in the absence of oxygen and in the presence of oxygen (concentration $[O_2]$), respectively. The quenching rate constant, k_q, is dependent on the physical constants of the system and is described by a modified Smoluchowski equation [6–8]:

$$k_q = 4\pi Np(D_A + D_B) \times 10^3 \tag{2}$$

where, N is Avogadro's number, p is a factor related to the probability of each collision causing quenching and to the radius of interaction between the donor and the oxygen. D_A and D_B are the diffusion coefficients for the donor and acceptor, respectively. In general, under defined experimental conditions, k_q can be considered constant. Using Eq. 1, it is possible to make intensity or lifetime measurements of phosphorescence (or fluorescence) and then relate this measurement to the oxygen concentration. Thus, the principle of optical oxygen sensors can be applied to measure the pressure in air, because the concentration of oxygen in air is proportional to pressure. Oxygen pressure measurements using PSP samples are based on quenching phenomenon by oxygen, an active quencher of luminescence. From a standpoint of optical sensing application, the most interesting fact is that the excited singlet and triplet states can be deactivated by quenching processes. The approximation of the luminescence intensity of the PSP was modeled by the Stern–Volmer equation [9]:

$$I_0/I = 1 + K \cdot P \tag{3}$$

where, P is oxygen pressure. I_0 and I are the luminescence intensities in the absence and presence of oxygen, respectively, and K is the Stern–Volmer constant. However, this relationship is generally not practical for measuring I_0 in the wind tunnel environment, since the tunnel would have to be pumped down to a vacuum. Instead of trying to achieve zero oxygen conditions the intensity of luminescence at wind off is used as the reference intensity ($I(ref)$), and the pressure at wind off is considered the reference pressure, $P(ref)$. In practice, this is usually the local barometric pressure. In terms of the Stern–Volmer equation this takes the form of the ratio of the Stern–Volmer for two different pressures. That is:

$$I_0/I_1 = 1 + K \cdot P_1 \text{ at wind off} \tag{4}$$

$$I_0/I_2 = 1 + K \cdot P_2 \text{ at wind on} \tag{5}$$

$$I_1/I_2 = 1 + K \cdot P_2/1 + K \cdot P_1 = 1/1 + K \cdot P_1 + K/1 + K \cdot P_1 P_2 \tag{6}$$

K and P_1 are constants. Thus Eq. 6 may be expressed as:

$$I_1/I_2 = A + K \cdot P_2 \tag{7}$$

where, $A = [1/(1 + K \cdot P_1)]$ and $B = [K/(1 + K \cdot P_1)]$. Since luminescence intensity depends on illumination intensity, values for I_1 are determined for each point on the wind tunnel model at each angle of attack. The values for A and B are then determined from a plot of I_1/I_2 vs. pressure. Equation 7 is a fundamental equation (the modified Stern–Volmer relation) in determining oxygen pressure by PSP.

The conventional PSP consists of sensor probe molecules contained in a transparent oxygen-permeable polymer film matrix (Fig. 3). The oxygen molecules permeate into the polymer film matrix by diffusion. In contrast, PSP consists of sensor molecules directly adsorbed onto a substrate such as an anodic oxidized aluminum plate (Fig. 4). Thus, polymers and probes are important factors for PSP application.

In this chapter, we focus on the properties of polymer matrices and fluorescence probes for PSP in detail.

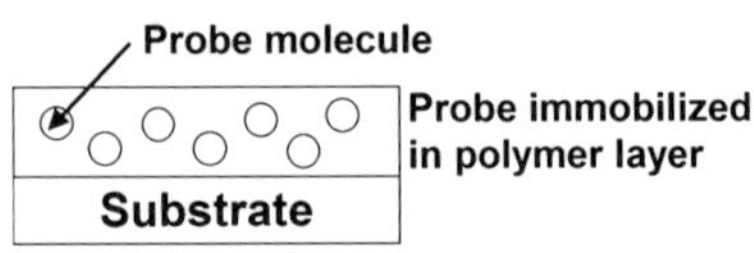

Fig. 3 Polymer-based PSP

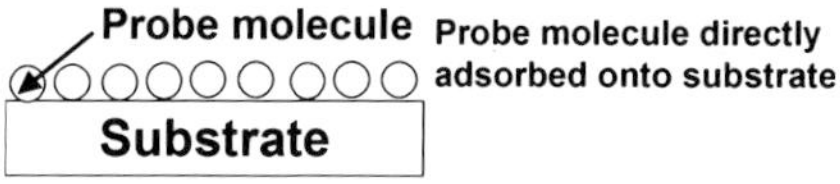

Fig. 4 Chemisorption layer-based PSP

3 Materials for Polymer-Based PSP

Many PSPs are composed of probe dyes, such as polycyclic aromatic hydrocarbons (e.g., pyrene) and coordination compounds (e.g., platinum porphryins and ruthenium(II) polypyridyl complexes) immobilized in various gas permeable polymer films such as silicon polymer, organic glassy polymers (e.g., poly(methylmethacrylate), polystyrene), fluorinated polymers, or cellulose derivatives such as ethyl cellulose [9, 10]. As probe molecules interact with polymer matrices directly, the properties of PSPs strongly depend on the properties of polymer matrices. The oxygen permeability of polymer matrix is an especially important factor for highly sensitive PSP.

In this section, the properties of polymer matrix for PSP are described.

When oxygen permeates a polymer matrix film, the permeation rate is given by the oxygen permeability coefficient P. The dimension of P is cm^3 (STP) cm^{-2} s cm Hg. As the permeation of oxygen molecules through flawless polymer films occurs by consecutive steps of solution of a permeant in the polymer and diffusion of the dissolved permeant, the permeability coefficient is given by $P = DS$. Where D and S are the diffusion constant and the oxygen solubility coefficient, respectively. The usual dimensions for D and S are $cm^2\,s^{-1}$ and cm^3 (STP) cm^{-3} cm Hg^{-1}, respectively. Oxygen permeability, diffusion constants and solubilities for oxygen are listed in Table 1 [11–15].

Table 1 Properties of polymers for PSP

	P [a] ($\times 10^{13}$)	D [b] ($\times 10^{6}$)	S [c] ($\times 10^{6}$)	Ref.
Poly(dimethylsiloxane)	695	40	24	[8–10]
Poly(1-trimethylsily-1-propyne)	7700	47	170	[11]
Polystyrene	2.63	–	–	[8]
Poly(methylmethacrylate)	9	10	8.5	[8]
Poly(vinyl chloride)	0.34	1.2	2.9	[8]
Poly(2,2,2-trifluoroethylmethacrylate)	32	15	0.27	
Ethylcellulose	11.0	0.639	1.73	[8]
Cellulose acetobutyrate	3.56	–	–	[8]
Cellulose acetate	5.85	–	–	[8]

[a] P permeability coefficient, [b] D diffusion constant, [c] S oxygen solubility coefficient

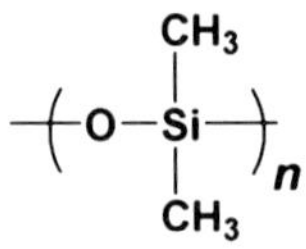

Poly (DMS)

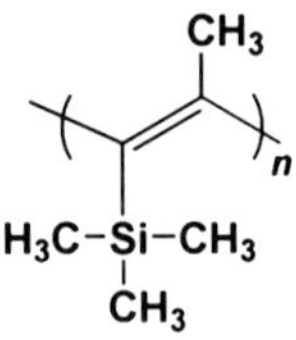

Poly (TMSP)

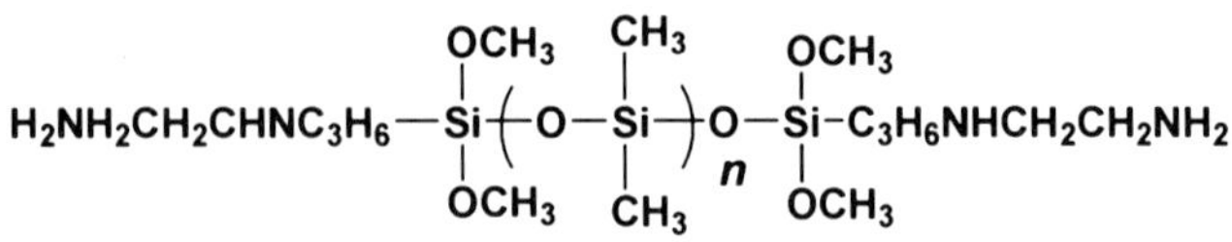

Poly(DMS), amino terminated
GP-134

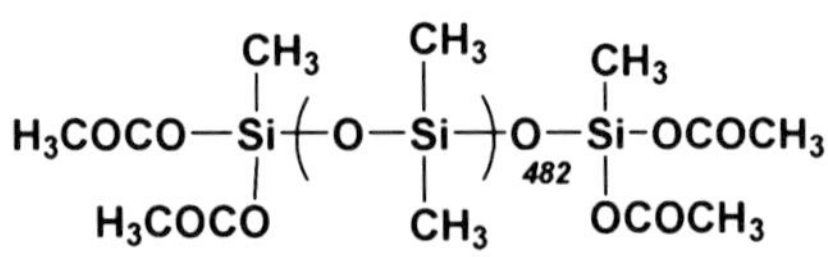

Poly(DMS), methyldiacetoxy terminated
PS 368.5

Fig. 5 Chemical structures of silicone polymers

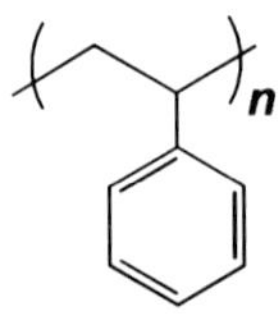

Polystyrene (PS)

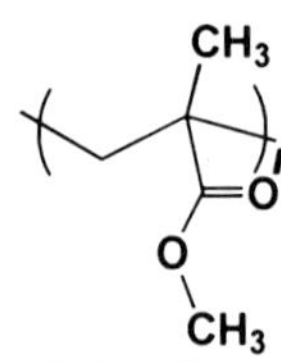

Poly(methyl methacrylate) (PMMA)

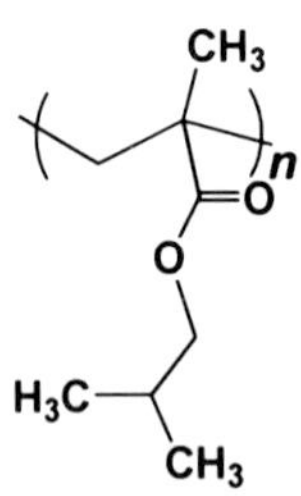

Poly(isobutyl methacrylate) (Poly IBM)

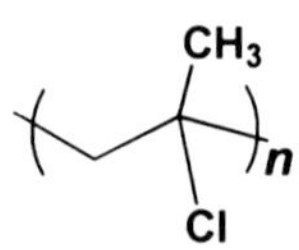

Poly(vinyl chloride) (PVC)

Fig. 6 Chemical structures of organic glassy polymers

The polymers used for PSPs are classified into four categories, i.e., silicon polymer (Fig. 5) , organic glassy polymer (Fig. 6), fluorinated polymer (Figs. 7 and 8), and cellulose derivative polymer (Fig. 9).

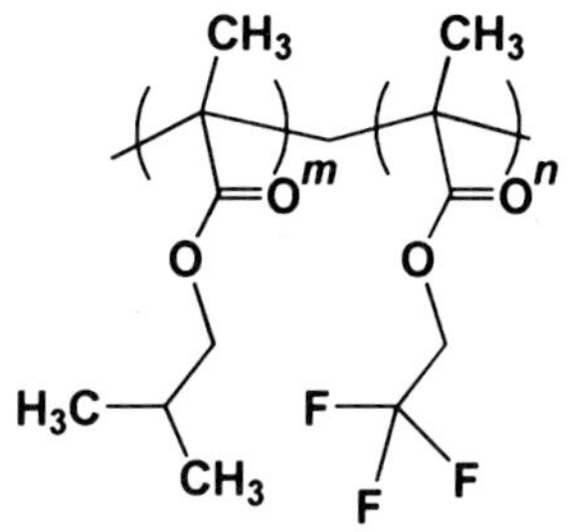

Poly(isobutylmethacrylate-*co*-2,2,2-trifluoroethylmethacrylate)

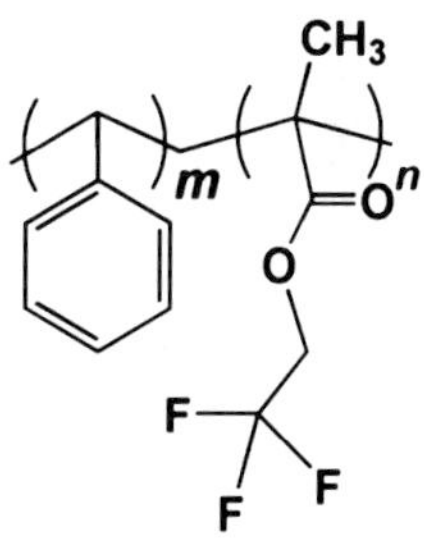

Poly(styrene-*co*-2,2,2-trifluoroethylmethacrylate)

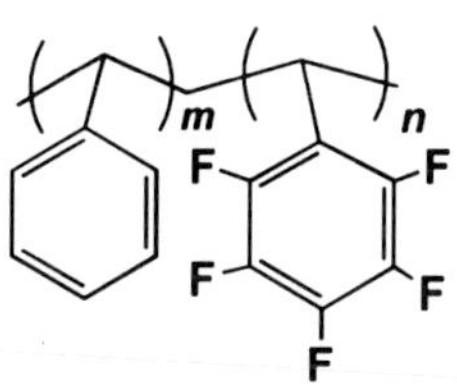

Poly(styrene-*co*-pentafluorostyrene)

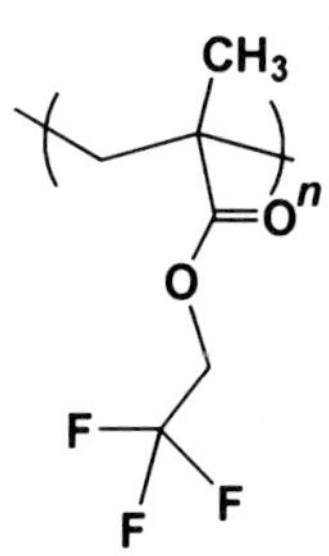

Poly(2,2,2-trifluoroethylmethacrylate)

Fig. 7 Chemical structures of fluorinated polymers

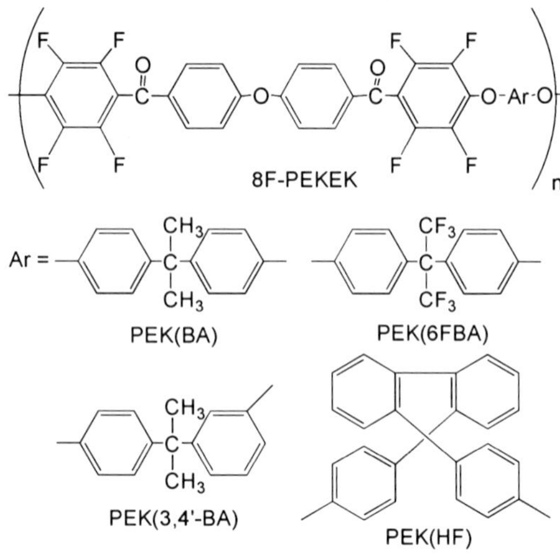

Fig. 8 Chemical structures of fluorinated poly(aryl ether ketone)s

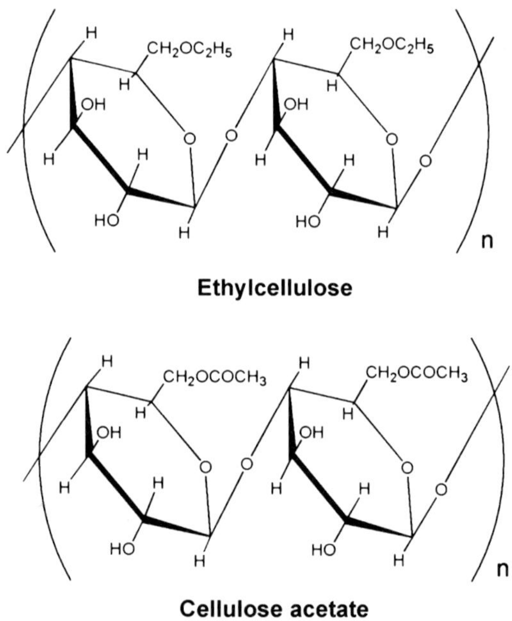

Fig. 9 Chemical structures of cellulose derivatives

3.1
Coordination Compounds as Luminescent Probes for Polymer-Based PSP

The coordination compounds are widely used as luminescent probes for polymer-based PSP. The coordination compounds are classified into two categories. One is the transition metal polypyridyl complex. In this group, transition metals are Ru^{2+}, Os^{+} and Ir^{3+} as shown in Fig. 10. The second group comprises the metalloporphyrins as shown in Fig. 11. Pt^{2+} and Pd^{2+} porphyrins are especially used as PSP probes.

Demas and DeGraff reported the design of highly luminescent transition metal complexes for optical oxygen sensor applications [16]. Table 2 shows the photochemical and photophysical properties of sensing probes using luminescent transition metal complexes in immobilizing polymer films.

In the transition metal polypyridyl complex group, tris(4,7-diphenylphenantroline) ruthenium(II) ($Ru(dpp)_3^{2+}$) is widely used as a probe for a PSP. The luminescence lifetime of $Ru(dpp)_3^{2+}$ is long compared with the other ruthenium(II) polypyridyl complexes [17]. The absorption and emission maxima of $Ru(dpp)_3^{2+}$ are 457 and 610 nm, respectively. The luminescence lifetimes under nitrogen- and air-saturated conditions are ca. 4.0 and 2.0 μs, respec-

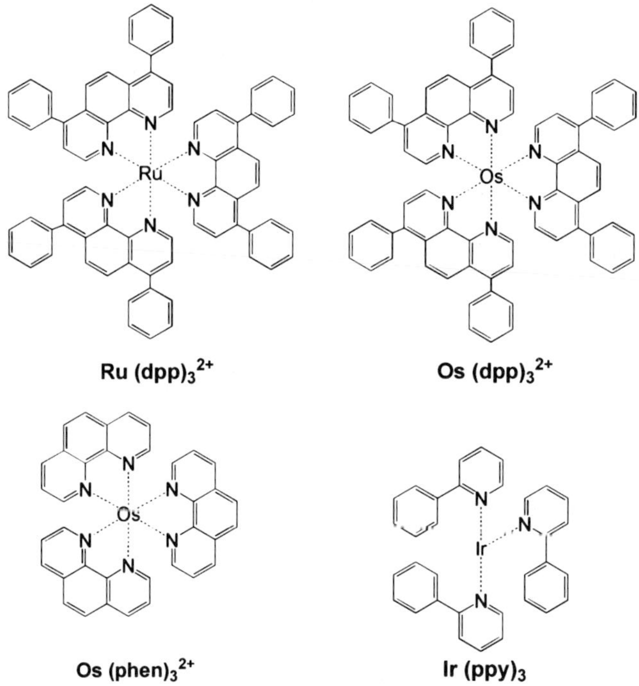

Fig. 10 Chemical structures of transition metal polypyridyl complexes

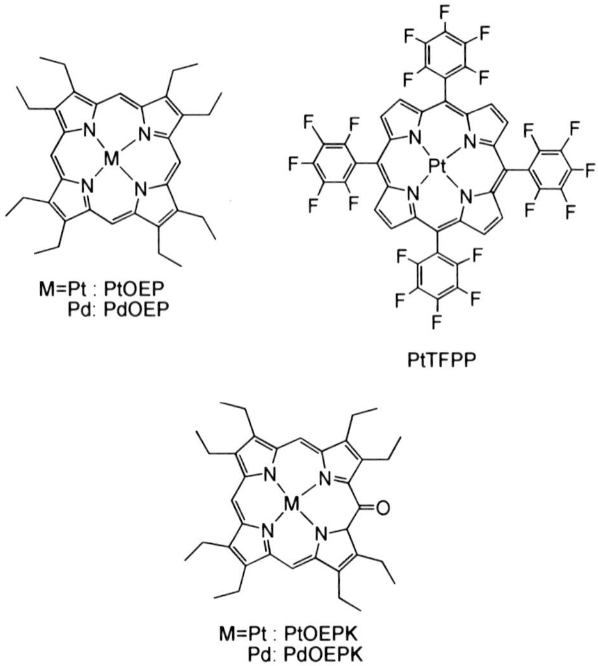

Fig. 11 Chemical structures of Pt^{2+} and Pd^{2+} porphyrins

tively. PSPs based on an optical oxygen sensor using $Ru(dpp)_3^{2+}$ immobilized in various polymer films have been investigated. The I_0/I_{100} values of $Ru(dpp)_3^{2+}$ in various silicone-based polymer films and fumed silica surfaces are listed in Table 2 [18–22]. I_0 and I_{100} indicated the observed fluorescence intensities from a sensing film exposed to argon- and oxygen-saturated conditions, respectively. The I_0/I_{100} values of $Ru(dpp)_3^{2+}$ in silicone and PS films are 4.4 and 1.1, respectively [18, 19]. The I_0/I_{100} value of $Ru(dpp)_3^{2+}$ in diacetoxy siloxane polymer film including silica particles is drastically increased up to 80 [21]. Moreover, many attempts have been made to increase the photostability of Ru(II) polypyridyl complexes and the resulting optical detection performance, as described in [23, 24].

Demas et al. described optical oxygen sensors using analogous osmium(II) complexes that have intense red absorptions and that can be excited with low-cost, high-intensity red diode lasers [25]. The osmium(II) complexes are probably more photochemically robust than ruthenium(II) complexes because of the larger energy gap between emitting state and the photochemically destructive upper d-d state. In Table 2, the photochemical and photophysical properties of osmium(II) tris(1,4-diphenyl-1,10-phenanthroline) ($Os(dpp)_3^{2+}$) and osmium(II) tris(1,10-phenanthroline) ($Os(phen)_3^{2+}$) are indicated as examples of osmium(II) complexes. The luminescence lifetimes of $Os(dpp)_3^{2+}$ and $Os(phen)_3^{2+}$ are 4.6 and 6.0 ns in dichloromethane solution,

Table 2 Transition metal polypyridyl complexes luminescence probes for PSP

Probe	λ_{max}(abs) [nm]	λ_{max}(em) [nm]	I_0/I_{100} in various polymer films	Ref.
$Ru(dpp)_3^{2+}$	337, 457	610	4.4 in silicone film 1.1 in PS film 3.5 in PVC film	[18–20]
			80 in methyl diacetoxy-terminated poly(DMS) (PS 368.5) including silica particle film	[21]
			8.0 in diamino-terminated poly(DMS) (GP-134) film	[21]
			16 in GP-134 including silica particle film	[21]
			25.0 onto fumed silica surface	[22]
$Os(dpp)_3^{2+}$	454, 500, 580, 650	729	4.5 in poly(DMS) film	[25]
$Os(phen)_3^{2+}$	432, 478, 660	720	–	[25]
$Ir(ppy)_3$	376	512	1.2 in PS film	[28]
			15.3 in poly-(styrene-*co*-TFEM)	[28]

respectively. The absorption and emission maximum of osmium(II) complexes are around 500 and 700 nm, respectively. Optical oxygen sensors using various osmium(II) complexes immobilized in poly(DMS) film are being developed [25]. However, the oxygen sensitivity of sensors using osmium(II) complexes immobilized in poly(DMS) film is the same as that of ruthenium(II) complexes in poly(DMS) film [25].

In general, these transition metal polypyridine complexes, ionic dyes, are insoluble in organic polymer film. On the other hand, iridium(III) polypyridine complexes are soluble in organic polymer film and display remarkably strong green luminescence with high quantum yield and have long lifetime ($\tau < 2.0\,\mu s$). Thus, iridium(III) complexes are attractive candidates as novel optical oxygen sensing materials. Di Marco et al. and Vandeloise et al. reported optical oxygen sensors using iridium(III) polypyridine complexes [26–28]. The absorption and emission maximum of iridium(III) tris(2-phenylpyridine anion) ($Ir(ppy)_3$) are 376 and 512 nm, respectively. The luminescence lifetime is 2.0 μs in THF solution. The optical oxygen-sensing properties of various iridium(III) polypyridine complexes immobilized in polymer film are also listed in Table 2. The optical oxygen sensor using $Ir(ppy)_3$ in poly(styrene-*co*-TFEM) film has an especially good sensitivity for oxygen [28].

Of the metalloporphyrin groups, Pt^{2+} and Pd^{2+} porphyrins show strong phosphorescence at room temperature. PtOEP and PdOEP (OEP: octaethylporphyrin) display strong room-temperature phosphorescence with high quantum yields and long lifetimes (ca. 100 μs for PtOEP and ca. 770 μs for PdOEP) [17]. Some optical PSPs based on oxygen sensors using PtOEP or PdOEP have been developed by dispersing them in polymer films. Table 3 lists the photochemical and photophysical properties of Pt^{2+} and Pd^{2+} porphyrins [29–36].

PtOEP is widely used as a PSP probe among the metalloporphyrins. The absorption and emission maximum of PtOEP are 381 and 535, and 646 nm, respectively. For examples of PSPs based on the OEP metal complexes in silicone polymer films, PSPs based on the platinum or palladium OEP immobilized in poly(TMSP) film have been reported. Oxygen permeability of poly(1-trimethylsily-1-propyne) (poly(TMSP)) film is about ten times larger than that of poly(dimethylsiloxane) (poly(DMS)) film, and provides a tough and thin film [14, 15]. The poly(TMSP) film is a porous polymer matrix with a high oxygen permeability and diffusion, indicating that the probe

Table 3 Porphyrins luminescence probes for PSP

Probe	λ_{max}(abs) [nm]	λ_{max}(em) [nm]	I_0 / I_{100} in various polymer films	Ref.
PtOEP	381, 535	646	4.5 in PS film	[36]
			67.3 in poly(isobutyl-methacrylate) (pIBM)	[31, 32]
			82.4 in poly(IBM-*co*-TFEM)	[31, 32]
			87.6 in poly(styrene-*co*-TFEM)	
			24.6 in poly(styrene-*co*-perfluorostyrene)	
			225 in poly(TMSP)	[29, 30]
			2.0 in PVC	[36]
			45.5 in pPEGMA (poly (ethylene glycol)ethyl ether methacrylate)	
			40 in silica glass film	[36]
PdOEP	393, 512, 546	663	11.5 in PS film	[23]

dye in poly(TMSP) film contacts with the oxygen in the gas phase. For example, the I_0/I_{100} values of octaethylporphyrin (OEP) platinum (PtOEP) and palladium (PdOEP) complexes in poly(TMSP) films were 225 and 121, respectively, [29, 30]. Thus, highly sensitive PSPs have been developed using poly(TMSP) films.

PSPs based on OEP metal complexes in various fluorinated polymer films also have been reported [31–35]. To apply the polymer matrix film for PSP, oxygen-permeable polymers with a low diffusion barrier for oxygen and with a high stability against irradiation are desired. A fluoropolymer meets the above requirements. Fluoropolymer films allow high permeability of oxygen. The C – F bond length is short (13.17 nm) and bonding energy is high (116 kcal mol^{-1}) compared with C – H (99.5 kcal mol^{-1}). The electronegativity of fluorine is high, and thus polymers containing fluoro-groups are stable against photo-oxidation. Oxygen affinity is induced by the high electronegativity of fluorine, and the oxygen permeability of the polymers containing fluoro-groups will be large [37–39]. Thus, fluoropolymers are useful for the polymer matrix of PSP. The optical oxygen- sensing system based on the phosphorescence of Pt^{2+} and Pd^{2+} porphyrins immobilized in new various fluoro-copolymer such as poly(isobutylmethacrylate-*co*-2,2,2-trifluoroethylmethacrylate) [31, 32], poly(styrene-*co*-2,2,2-trifluoroethyl-methacrylate) [33, 34] and poly(2,2,2-trifluoroethylmethacrylate) films [40] are reported. Chemical structures are shown in Fig. 7. Use of poly(styrene-*co*-2,2,2-trifluoroethylmethacrylate) increased the values of I_0/I_{100} with increasing 2,2,2-trifluoroethylmethacrylate units, indicating control of PSP properties [34].

A series of new fluorine-containing poly(aryl ether ketone)s (8F-PEKEK (Ar) where Ar is 2-2-bis(4-hydroxyphenyl)-1,1,1,3,3,3-hexafluoropropane (6FBA), 2,2-bis(4-hydroxyphenyl)propane (BA), 2-(4-hydroxyphenyl)-2-(3-hydroxyphenyl) propane (3,4′-BA) or 9,9′-bis(4-hydroxyphenyl)fluorine (HF)), as shown in Fig. 8, have been synthesized [41] and applied to the matrix of PSP using phosphorescence quenching of PtOEP and PdOEP by oxygen [35]. The phosphorescence intensity of PtOEP of PdOEP in 8F-PEKEK(Ar) films decreased with increase in oxygen concentration. For PtOEP in 8F-PEKEK(Ar) film, I_0/I_{100} values are more than 20.0. For PdOEP in 8F-PEKEK(Ar) film, on the other hand, high I_0/I_{100} values of more than 143 were obtained. These results indicate that PtOEP and PdOEP films are useful PSP.

Finally, PSPs based on the OEP metal complexes in various cellulose derivative polymers also have been introduced. Cellulose derivative polymers such as ethyl cellulose, cellulose acetate etc., as shown in Fig. 9, also provide a mechanical strength to thin film. The oxygen permeability, diffusion constant, and solubility for oxygen for typical cellulose derivative polymers are listed in Table 1. Cellulose derivative polymer films with plasticizer tributyl phosphate (TBP) have large oxygen permeability. Cellulose acetate with TBP film is especially widely used for matrices of PSPs. The PSP based on Pt^{2+} or Pd^{2+} porphyrin immobilized in cellulose acetate with TBP is highly sensitive [42].

However, this type of the sensor somewhat lacks photostability under continuous operating conditions. PtOEP shows a decrease in intensity during the initial illumination period, although stable phosphorescence intensity is observed with only minor changes in intensity after hours of illumination. This is attributed to photobleaching of the indicator, leaching of indicator from matrix, and deterioration of matrix itself. However, so far, many attempts have been made to increase the photostability of porphyrin dye and the resulting optical detection performance. Thus, the choice of indicator in a detection material is often the controlling factor in detection characteristics, especially sensor lifetime.

Pt^{2+} fluorinated porphyrin, Pt^{2+} tetrakis(pentafluorophenyl) porphyrin (PtTFPP), Pt^{2+} octaethylporphyrin in corresponding photostable ketones (OEPK) (chemical structures are shown in Fig. 11) have also received much attention as photostable sensing materials. As the electron-withdrawing character of the perfluorophenyl substituents of PtTFPP strongly raises the redox potential and reduces the electron density of the porphyrin ring, PtTFPP is stable against photo-oxidation and photo-reduction [43]. Table 4 shows the photochemical and photophysical properties of PtTFPP. The I_0/I_{100} values of PtTFPP immobilized in PS film PSP is 3.0. The hydrogenation of *exo*-pyrrole bonds in the porphyrin results in substantial red-shifting of both the absorption and emission spectra. Their Pt octaethylporphyrin ketone (PtOEPK) immobilized in polystyrene sensor showed 12% intensity diminishing under continuous illumination for 18 h, detected by absorbance measurement. PtOEPK and PdOEPK display strong room-temperature phosphorescence with a quantum yield of ($\Phi_P < 0.01$) and long lifetime of (ca. 61 μs for PtOEPK and ca. 460 μs for PdOEPK). The PtOEPK is ten times more photostable than that of PtOEP [44, 45].

Table 4 Photostable porphyrins luminescence probes for PSP

Probe	λ_{max}(abs) [nm]	λ_{max}(em) [nm]	I_0 / I_{100} in various polymer films	Ref.
PtTFPP	395, 541	648	3.0 in PS film 15.4 in poly(2,2,2-trifluoroethyl methacrylate)	[43] [40]
PtOEPK	398, 592	758	2.0 in PVC film 20 in PS film	[44, 45]
PdOEPK	410, 603	790	8.0 in PVC film 28 in PS film	[44, 45]

3.2
Poly Aromatic Hydrocarbons as Luminescent Probes for Polymer-Based PSP

Bergman [46] and others [47–51] have used polycyclic aromatic hydrocarbons (PAHs, chemical structures shown in Fig. 12), which are found to be efficiently quenched by oxygen in the 0–40 kPa range. The PAHs are either dissolved in a polymer [48–51], soaked into porous glass [50], or covalently immobilized on a glass support [47]. Peterson et al., by combining Kautsky's

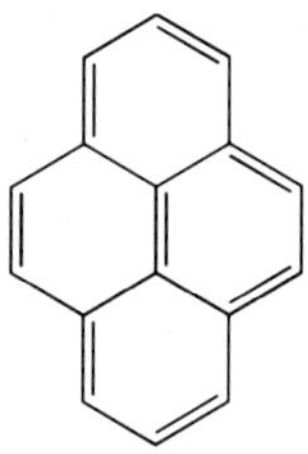

Pyrene

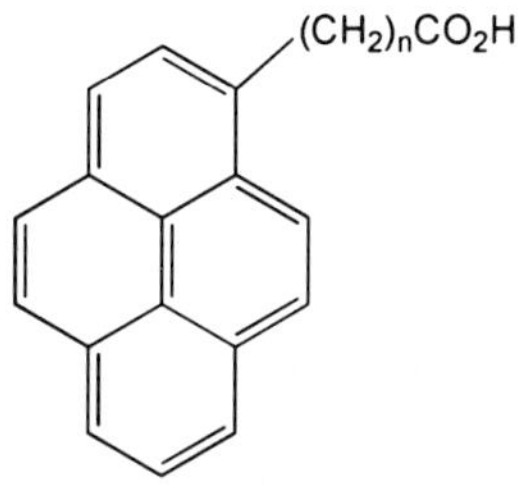

Pyrene-1- carboxylic acid
n=3: Pyrene-1-butylic acid (PBA)
9: Pyrene-1-decanoic acid
10: Pyrene-1-dodecanoic acid

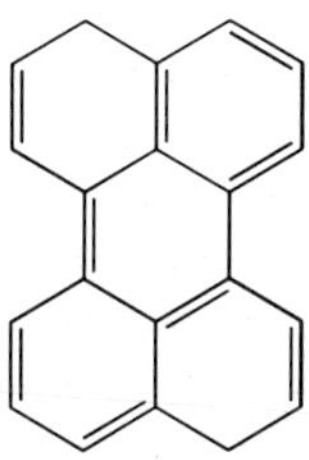

Perylene

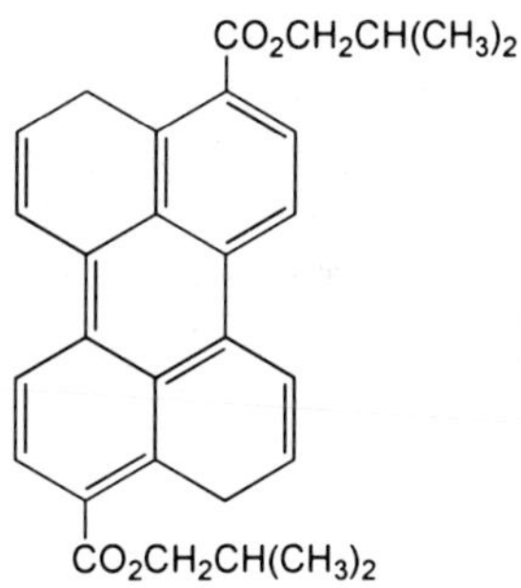

3,9-Perylenedicarboxylic acid diisobutyl ester

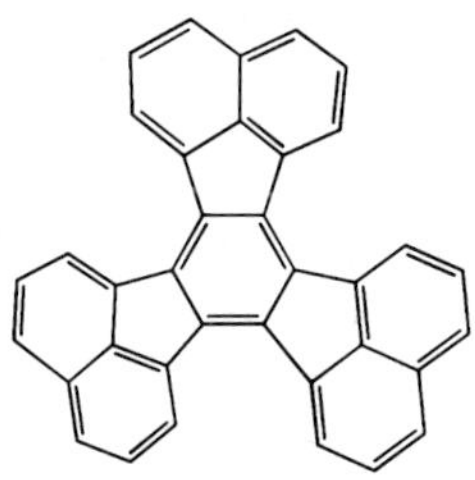

Decacyclene (DCY)

Fig. 12 Chemical structures of polyaromatic hydrocarbons

Table 5 Organic fluorescence probes for PSP

Probe	λ_{max} (abs) [nm]	λ_{max} (em) [nm]	I_0/I_{100} in various polymer films	Ref.
PBA	365	376, 396, 474	1.5 in poly(DMS) film	[56]
Perylene diisobutylate	457	512	1.5 in PS film	[57]
DCY	380	510	7.8 in porous polymer beads	[54]
Pyrene	365	376, 396 474	2.0 in triethoxysilane glass film	[58]

adsorption technique with the sensitivity of the PAHs, developed a fiberoptic sensor for oxygen based on the oxygen-sensitive fluorescence of Amberlite-adsorbed perylene diisobutyrate [52]. Changes in luminescence intensity can therefore be used to monitor oxygen [53].

The decacyclene (DCY) also was used as fluorescence oxygen-sensing probe [54]. The I_0/I_{100} values of PAHs in various polymer films are listed in Table 5 [54–58].

In organic luminescence probes, pyrene-1-butylic acid (PBA) is widely used as a sensing probe. The fluorescence lifetime of PBA is long compared with the other PAHs. The emissions of PBA monomer and of excimer between PBA molecules are quenched by oxygen. The absorption maximum of PBA is 365 nm and the emission maxima are 376 and 396 nm attributed to monomer and 447 nm attributed to excimer emissions. The fluorescence lifetime is 0.2 μs. PSP based on the optical oxygen sensing using PBA immobilized in polystyrene, silicone polymer, or silicon rubber film has been investigated.

4
Materials for Probe Adsorbed Layer-Based PSP

As probe molecules interact with polymer molecules directly in polymer-based PSP, the properties of sensing layers strongly depend on the properties of polymer matrices. To overcome these problems, chemisorption layers and Langmuir–Blogett (LB) membranes have been exploited. These techniques have been paid much attention for the design of solid surfaces with well-defined composition, structure, and thickness for interfacial optical studies [59–62]. The chemisorption layers are formed using spontaneous binding between the thiol group and the metal surface (Au, Ag, or Pt) or between the carboxyl group and the metal oxide surface (Al_2O_3, Fe_2O_3, TiO_2 etc.).

As the chemisorption technique is very convenient, this layer is widely used for optical and optoelectronic devices. Among a number of chemisorption layer techniques, the use of compounds with carboxyl functional group is most prevalent for preparation of the chemisorption layer of probe molecules on the surface of anodic oxidized aluminum. As the probe molecules are arranged on the solid surface directly by using this technique, the chemisorption layer may possess a lower diffusion barrier for oxygen. Thus, highly sensitive devices for PSP can be accomplished by using a chemisorption layer. In this section, the fluorescence probes for PSP based on the chemisorption layer are introduced.

Pyrene-1-carboxylic acid derivatives, as shown in Fig. 13, are widely used as fluorescence probes for PSPs based on a chemisorption layer. PSPs based on the fluorescence quenching of pyrene derivatives with carboxyl groups, pyrene-1-decanoic acid, pyrene-1-dodecanoic acid and PBA chemisorption layers onto alumina plate were developed [63–65]. The ratio I_0/I_{100} was estimated to be 50.7 for pyrene-1-decanoic acid, 44.7 for pyrene-1-dodecanoic acid and 73.4 for PBA, showing that the pyrene-1-butyric acid layer is highly sensitive for oxygen. These samples obeyed Stern–Volmer plots with multisite model and possessed good operational stability. For argon-saturated to oxygen-saturated conditions, response times were 10 s for pyrene-1-decanoic acid, 5.0 s for pyrene-1-dodecanoic acid, and 5.0 s for pyrene-1-butyric acid. For the reverse conditions, response times were 50.0 s for pyrene-1-decanoic acid, 50.0 s for pyrene-1-dodecanoic acid, and 50.0 s for PBA . These are stable sensing layers against irradiation that exhibit minimal decrease (ca. 7%) in initial intensity after continuous irradiation for 24 h.

Porphyrins with carboxylate, as shown in Fig. 13, are also widely used as fluorescence probes for PSP based on the chemisorption layer. This section de-

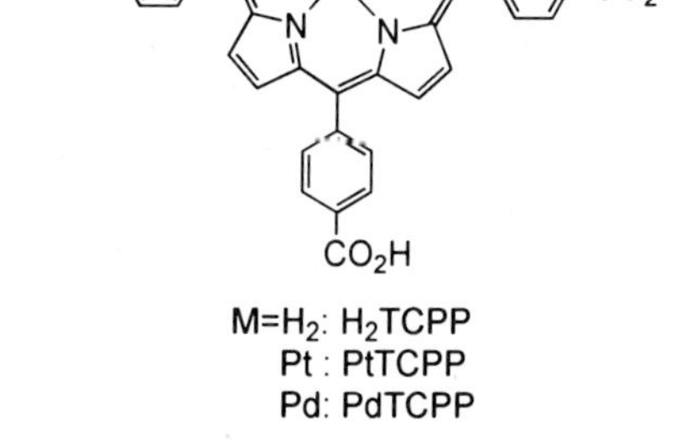

Fig. 13 Chemical structures of tetracarboxyphenylporphyrin metal complexes

Table 6 Probe adsorbed onto anodic oxidized aluminum for PSP

Probe	λ_{max} (abs) [nm]	λ_{max} (em) [nm]	I_0/I_{100}	Ref.
PBA	365	376, 396, 474	73.4	[64]
Pyrene-1-decanoic acid	365	376, 396	50.7	[63, 65]
Pyrene-1-dodecanoic acid	365	376, 398	44.7	[63, 65]
H_2TCPP	418, 514, 548, 591, 646	653	2.58	[67]
			16.5	[67]
PtTCPP	399, 507, 538	665	17.7	[66, 67]
PdTCPP	416, 523	701		

scribes the fabrication of TCPP metal complexes (Pt^{2+} complex, PtTCPP; Pd^{2+} complex, PdTCPP; metal-free complex, H_2TCPP) as luminescence probes for PSP based on the chemisorption films onto alumina plate and their optical properties [54, 55]. The I_0/I_{100} values of PtTCPP, PdTCPP and H_2TCPP films are estimated to be 16.5, 17.7 and 2.58, respectively, indicating that PtTCPP and PdTCPP chemisorption films are highly sensitive devices for PSP [66, 67]. The properties of probe-adsorbed PSP are listed in Table 6.

References

1. McLachlan BG, Bell JH (1995) Exp Thermal Fluid Sci 10:470
2. Demas JN, DeGraff BA, Coleman P (1999) Anal Chem 71:793A
3. Lewis GN, Kasha M (1944) J Am Chem Soc 66:2100
4. Lott PF, Hurtubise RJ (1974) J Chem Edu 51:A315
5. Fischkoff S, Vanderkooi JM (1975) J Gen Physiol 65:663
6. Smoluchowski M (1917) Z Phys Chem 92:129
7. Subczynski WK, Hyde JS (1984) Biophys J 45:743
8. Vanderkooi JM, Maniara G, Green TJ, Wilson DF (1987) J Biol Chem 262:5476
9. Wolfbeis OS (1991) Fiber optic chemical sensors and biosensors vol 2. CRC, Boca Raton FL
10. Lu X, Manners I, Winnik MA (2001) Oxygen diffusion in polymer films for luminescence barometry applications. In: Valeur B, Brochon J-C (eds) New trends in fluorescence spectroscopy: Applications to chemical and life sciences. Springer series on fluorescence methods and applications, vol 1. Springer, Berlin Heidelberg New York, p 229–256
11. Pauly S (1989) Polymer handbook, 3rd edn. Wiley, New York
12. Barrer RM, Chio HT (1966) J Polym Sci C, 10:111
13. Yasuda H, Rosengren K (1970) J Macromol Sci B 14:2839
14. Nakagawa N (1995) Membrane 20:156

15. Masuda T, Isobe E, Higashimura T, Takada K (1983) J Am Chem Soc 105:7473
16. Demas JN, DeGraff BA (1991) Anal Chem 63:829A
17. Kalyanasundaram K (1992) Photochemistry of polypyridine and porphyrin complexes. Academic, New York
18. Carraway R, Demas JN, DeGraff BA, Bacon JR (1991) Anal Chem 63:332
19. Bacon JR, Demas JN (1987) Anal Chem 59:2780
20. Li XM, Wong HY (1992) Anal Chim Acta 262:27
21. Xu W, McDonough RC III, Langsdorf B, Demas JN, DeGraff BA (1994) Anal Chem 66:4133
22. Carraway ER, Demas JN, DeGraff BA (1991) Langmuir 7:2991
23. Fuller ZJ, Bare WD, Kneas KA, Xu W-Y, Demas JN, DeGraff BA (2003) Anal Chem 75:2670
24. Adelt M, Devenney M, Meyer TJ, Thompson DW (1998) Inorg Chem 37:2616
25. Xu WY, Kneas KA, Demas JN, DeGraff BA (1996) Anal Chem 68:2605
26. Vander Donckt E, Camerman B, Hendrick F, Herne R, Vandeloise R (1994) Bull Soc Chim Belg 103:207
27. Di Marco G, Lanza M, Pieruccini M, Campagna S (1996) Adv Mater 8:576
28. Amao Y, Ishikawa Y, Okura I (2001) Anal Chim Acta 445:177
29. Amao Y, Okura I, Shinohara H, Nishide H (2002) Polymer J 34:411
30. Amao Y, Asai K, Okura I, Shinohara H, Nishide H (2000) Analyst 125(11):1911
31. Amao Y, Asai K, Miyashita T, Okura I (1999) Polymer J, 31:1267
32. Amao Y, Miyashita T, Okura I (2000) Anal Chim Acta 421:167
33. Amao Y, Asai K, Miyashita T, Okura I (1999) Chem Lett 1031–1032
34. Amao Y, Miyashita T, Okura I (2000) J Porphyrin Phthalocyanine 5:433
35. Amao Y, Tabuchi Y, Yamashita Y, Kimura K (2002) European Polymer J 38:675
36. Lee S-K, Okura I (1998) Spectrochim Acta A 54:91
37. Yi-Yan N, Felder RM, Koros WJ (1980) J Appl Polym Sci 25:1755
38. Zisma WA (1964) Adv Chem Ser 43:1
39. Pittman AG (1972) Fuluoropolymers. Wiley, New York
40. Amao Y, Miyashita T, Okura I (2001) J Fluor Chem 11:101
41. Kimura K, Tabuchi Y, Yamashita Y, Cassidy PE, Fitch JW III, Okumura Y (2000) Polm Adv Tech 11:757
42. Mills A, Lepre A (1997) Anal Chem, 69:4653
43. Lee S-K, Okura I (1997) Anal Commun 34:185
44. Papkovsky DB, Ponomarev GV, Trettnak W, O'Leary P (1995) Anal Chem 67:4112
45. Hartmann P, Trettnak W (1996) Anal Chem 68:2615
46. Bergmann I (1968) Nature 218:396
47. Wolfbeis OS, Offenbacher H, Kroneis H, Marsoner H (1984) Mikrochim Acta I 153
48. Kroneis HW, Marsoner HJ (1983) Sens Actuators 4:587
49. Lubbers D, Opitz N (1983) Sens Actuators 4:641
50. Cox ME, Dunn D (1985) Appl Optics 24:2114
51. Wolfbeis OS, Posch HE, Kroneis H (1985) Anal Chem 57:2556
52. Peterson JI, Fitzgerald RV, Buckhold DK (1984) Anal Chem 56:62
53. Gehrich JL, Lubbers DW, Opitz N, Hansmann DR, Miller WW, Tusa JK, Yafuso M (1986) IEEE Trans Biomed Eng BME-33 117
54. MacCraith BD, McDonagh CM, O'Keeffe G, Keyes ET, Vos JG, O'Kelly B, McGilp JF (1993) Analyst 118:385
55. Longmuir IS, Knopp JA (1976) J Appl Physiol 41:598
56. Sharma A, Wolfbeis OS (1988) Appl Spect 42:1009
57. Fujiwara Y, Ishikawa Y, Amao Y (2002) Nippon Kagaku Kaishi 2:261

58. Xu W, Schmidt R, Whaley M, Demas JN, DeGraff BA, Karikari EK, Farmer BL (1995) Anal Chem 67:3172
59. Nuzzo RG, Fusco FA, Allara DL (1987) J Am Chem Soc 109:2358
60. Porter MD, Bright TB, Allara DL, Chidsey CED (1987) J Am Chem Soc 109:3559
61. Laibinis PE, Whitesides GM (1992) J Am Chem Soc 114:1990
62. Ulman A (1990) An introduction to ultrathin organic films from Langmuir–Blodgett to self-assembly. Academic, San Diego, CA
63. Fujiwara Y, Amao Y (2002) Sens Actuators B: Chem 85:175
64. Fujiwara Y, Okura I, Miyashita T, Amao Y (2002) Anal Chim Acta 471:25
65. Fujiwara Y, Amao Y (2003) Sens Actuators B: Chem 89:187
66. Amao Y, Asai K, Okura I (1999) Anal Commun 36:179
67. Amao Y, Okura I (2000) Analyst 125:1601

Springer Ser Chem Sens Biosens (2005) 3: 323–341
DOI 10.1007/5346_011

Published online: 27 September 2005

Optical Sensing of Enantiomers

Maura Kasper · Stefan Busche (✉) · Günter Gauglitz

Institute of Physical and Theoretical Chemistry, Eberhard Karls University,
Auf der Morgenstelle 8, 72076 Tübingen, Germany
stefan.busche@ipc.uni-tuebingen.de

Abstract The optical sensing of enantiomers is a current topic. During the last decade many different applications for optical sensing of enantiomers have been reported in the literature. The principles of distinction of enantiomers using amide and cyclodextrin phases, molecularly imprinted polymers and fluorescence sensors are depicted in this chapter. Label-free methods, like surface plasmon resonance and reflectometric interference spectroscopy, can be adopted for polymer-based chiral amides, cyclodextrines and molecularly imprinted polymers. These materials derived from chromatographic methods are used for enantiomeric separation in the aqueous phase as well as in the gaseous phase. Calixarene and 1,1′-binaphthyl fluorophores as well as fluorescent cyclodextrins could be established as sensing materials for fluorescence measurements. Examples for the most commonly used methods—enantioselective fluorescence quenching or enhancement—are presented.

1
Introduction

Since Pasteur separated crystalline sodium ammonium tartrate manually in 1848, many researchers have worked on the subject of enantiomeric separation. In 1939 Henderson and Rule fully separated derivatives of camphor by column chromatography using lactose as a stationary phase material [1]. Gil-Av et al. [2] were able to separate amino acid derivatives on a polysiloxane-based stationary phase by gas chromatography (GC) in 1966. Since then many approaches for a successful distinction between enantiomers have been developed for capillary GC and liquid chromatography [3]. It is still a current topic for researchers: searching for "chiral separation" with SciFinder [4] results in 812 hits and searching for "chiral recognition" leads to 285 hits for the year 2003 only.

Sensors for the detection of enantiomers are of great interest, as so far the on-line monitoring of production processes and medical diagnostics using standard chemical analytical methods is not possible. Quite often only one enantiomer of a chiral compound is actually a bioactive therapeutic. Therefore a proper analysis of the final product is essential. Currently, this involves separation techniques like liquid chromatography, GC and capillary electrophoresis, and determination of enantiomeric purity with circular dichroism and specific rotation. These are all off-line procedures and therefore no real-time analysis can be performed. Sensing devices for the distinction of different enantiomers would be a much cheaper, faster and easier-to-use alternative for this task, amenable to automation.

However, only one theoretical plate (one single absorption and desorption process) is available for sensors, while discrimination in chromatographic approaches results from an accumulation of some thousands of subsequent theoretical plates. Chiral discrimination with selective and sensitive gas sensors and sensors for the aqueous phase proved a great challenge. With sensing devices, a fast and reversible monitoring of the chiral compounds of interest should be possible. Modern microsensors work at room temperature, which is an advantage compared with GC. During the last few years many different approaches to the sensing of and with enantiomers have been realised. The unambiguous chiral discrimination of several types of analytes by different types of optical sensors was demonstrated [5–7].

This chapter deals with the different separation mechanisms of chiral discrimination which are applied for optical sensors. Several types of optical sensors based on enrichment of analyte molecules in thin polymer films and fluorescence sensors were introduced for sensing of enantiomers in gaseous and aqueous media.

2
Principles of Discrimination of Enantiomers

In this section an overview of the numerous methods and principles for the discrimination of enantiomers is given. First, the interaction principles of the polymer-based methods adapted from chromatographic procedures are illustrated. The discrimination of enantiomers was achieved some decades ago by using different types of stationary materials, like cyclodextrins or polymer-bonded amide selectors. These stationary-phase materials have successfully been appointed for label-free optical sensing methods like surface plasmon resonance (SPR) or reflectometric interference spectroscopy (RIfS). Furthermore, various successful applications to optical spectroscopy of the well-established method of fluorescence measurements for the discrimination of enantiomers are described.

2.1
Amide Phases

One of the main principles for chiral separation used in modern capillary GC is the bonding of the optically active compounds via hydrogen bridges to a stationary-phase material. Feibush and Gil-Av [8] suggested a rapid and reversible formation of association complexes between carbonyl and amide functions of selector and selectand. The formation of diastereomeric associates yields complexes of different stability, depending on the relative configuration. The introduction of dipeptide and diamide phases leads

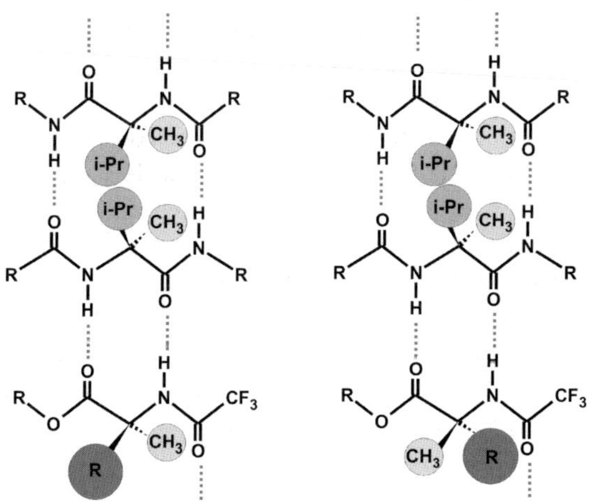

Fig. 1 Association of *N*-trifluoroacetyl (*N-TFA*) amino acid derivatives and diamide phases (*left* L,L′; *right* L,D′)

to two more hydrogen bonds for the distinction of amino acids. In Fig. 1 a selectand molecule is connected via hydrogen bonds to a diamide selector with a 5 + 5 = 10-ring according to Beitler and Feibush [9].

Many spectroscopic and structural investigations have confirmed a model which assumes an association leading to different ring sizes depending on the arrangement of the chains in the stationary-phase material. The observed enantioselectivity can be explained by the different arrangements of the isopropyl groups in the diamide phase. In accordance with experiments, the combination of L-selector and L-amino acid is stabler than L-selector and D-amino acid [10]. Beside hydrogen bonds, other enantiospecific interactions, like dipole–dipole interactions, can lead to the discrimination of enantiomers [11].

Owing to the high thermal and long-term stability, amide phases bonded to polysiloxane are regarded as sensitive materials for the sensoric approach to enantiomeric separation. An important factor for the resolution of the stationary material is the number of dimethylpolysiloxane units between the chiral moieties. More than 200 different amide phases have been synthesised and applied for the discrimination of amino acids, lactate esters and many other substances [10].

In this chapter two different kinds of phase materials are presented: the first is a chiral diamide selector bonded to a polysiloxane matrix; the other selector system is a calix[4]arene with chiral residues which is attached to a polysiloxane backbone (Sect. 2.4). These systems were used in direct optical methods based on a change in the refractive index or the optical thickness of a transparent polymeric layer.

2.2 Cyclodextrin Phases

Cyclodextrins, which can be considered as empty capsules of different sizes, have the ability to entrap various organic molecules in their central cavity. The molecular intercalation of chiral guests is in many cases enantioselective owing to the inherent molecular asymmetry of cyclodextrins; this has been known for more than four decades [12]. Cyclodextrins are cyclic oligosaccharides consisting of six to twelve glucose units; the most important contain six (α), seven (β) or eight (γ) glucose units (Fig. 2). They are produced by biodegradation of starch.

The native cyclodextrin is a truncated cone having an axial cavity with primary hydroxyl groups around its narrower rim and secondary hydroxyl groups on the opposite, wider rim. The ability of cyclodextrins to distinguish between enantiomers is based on several phenomena. They have numerous chiral centres (five in every glucose unit) with different orientation and distance from the neighbouring atoms. Their twisted shape explains why the cyclodextrins show better chiral recognition than linear oligoglucosides. The

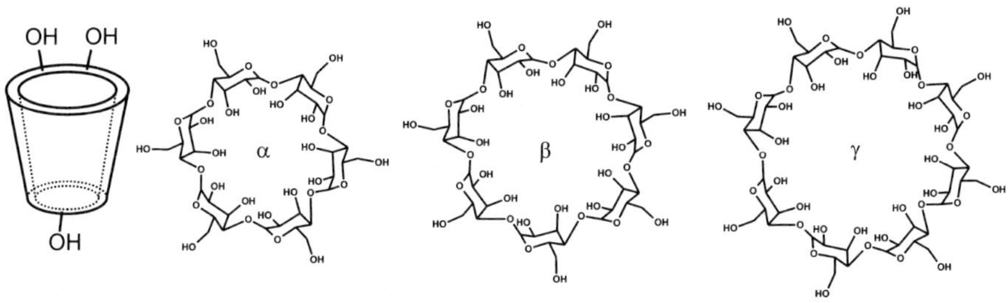

Fig. 2 *Left* Scheme of a cyclodextrin; *right* structures of α-, β- and γ-cyclodextrin

recognition spectra can be broadened by substituents which add more chiral centres to the structure. Cyclodextrins show chiral selectivity with rigid, cyclic, mostly aromatic compounds as well as with aliphatic enantiomers. As mechanisms for the separation of enantiomers, the molecular entrapment mechanism and the interaction of hydrogen bonding are discussed in the literature [10, 13]. Covalent bonds between guest and host molecules are excluded.

Cyclodextrins can be used as fluorescence sensors or as hosts for direct optical detection with methods like SPR.

2.3 Molecularly Imprinted Polymers

The poor chemical and physical stability of biomolecules is a great disadvantage and frustrates their use as detector molecules in harsh environments. Biomimetic receptor systems are an interesting alternative approach because of their ability to bind target molecules with affinities and specificities equivalent to those of natural receptors. Molecular imprinting in synthetic polymers is a technique which is being increasingly adapted in optical sensor technology [14]. Binding sites are generated during the imprinting process, and they often have affinities and selectivities approximating those of antibodies. Owing to their highly cross-linked polymeric nature, they are intrinsically stable and robust, facilitating the application in extreme environments.

In the molecular imprinting technique, a cross-linked polymer matrix is formed around a target analyte (the template). The precursor mixture contains a functional monomer which can interact with the template molecule by covalent or non-covalent bonding. After the polymerisation process, the functional groups are held in position by the polymer backbone and the template molecule is removed. The residual binding sites are complementary to the target molecules in size and shape.

The great potential of this technology has been recognised recently; there is now strong development towards the use of molecularly imprinted polymers (MIPs) as recognition elements in sensors [15].

2.4
Fluorescence Sensors

Application of photoluminescence to chiral separation has been studied for over two decades. As enantioselective fluorescence sensors are potentially useful for the rapid assay of the composition of chiral substrates, the development of molecule-based sensors is receiving growing research attention [5]. They generally consist of a fluorophore and a binding site; by introducing chirality to the binding site, the resulting sensor can carry out enantioselective recognition. Numerous signalling modes to distinguish enantiomers by fluorescence measurements are available: enhancement or quenching of fluorescence by interaction of dyes with analyte molecules are the commonest. Both methods benefit from different stabilities of the enantiomer–receptor complexes formed by intercalation.

For enhancement of fluorescence, molecules with intramolecular fluorescence quenching of photoinduced electron transfer (PET) of lone pairs are used. When interacting with the target molecule, this quenching will be inhibited and therefore the fluorescence can be turned on. The 1,1′-binaphthyl macrocycles are most extensively applied for this method; they provide multiple chiral functional groups as binding sites for analytes such as α-hydroxycarboxylic acids, amines and even amino acid derivatives [5] (Fig. 3, left).

Calixarenes are "cup-like" cyclooligomers formed via a phenol–formaldehyde condensation. As they have a rigid conformation providing an upper and a lower rim, which can both be functionally modified, various derivatives with adapted properties for chiral recognition can be prepared [16] (Fig. 3, right). The fluorescence emission of modified calix[4]arenes can be quenched by the complexation with the respective analyte molecules. An aryl ring is a necessary feature of the analyte as for non-aromatic substances no quenching can be observed. Therefore, the guest molecules commonly contain a hydrogen-bonding site and a chiral centre immediately adjacent to an aryl ring.

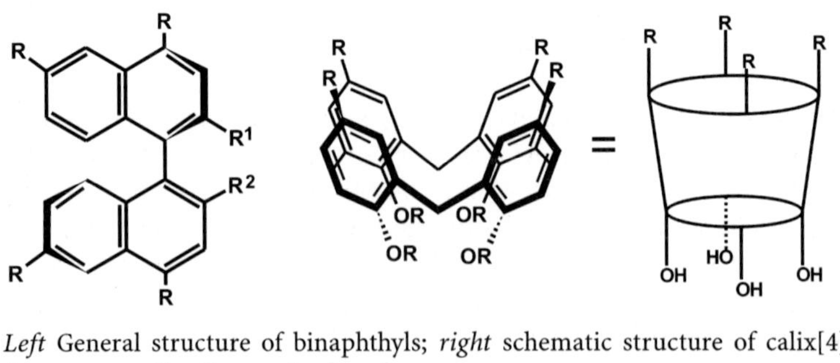

Fig. 3 *Left* General structure of binaphthyls; *right* schematic structure of calix[4]arene containing macrocyclic oligomeric phenols

3 Optical Sensing Applications for Enantiomers

3.1 Polymer-Based Sensors

For the application of label-free optical transduction principles like SPR or RIfS, a chiral receptor bound to a transparent polymer layer is required. As various types of these polymers have already been applied to chromatographic separation processes, a substantial wealth of knowledge was achieved during the last few decades. Stationary materials like bonded amide selectors or cyclodextrins were adopted as sensor coatings. Several different applications of these materials in various fields of interest have been reported in the literature [17].

3.1.1 Chiral Amides

The high potential of chiral amide-containing selectors for the application in optical sensing devices was shown during the last few years. These selective materials have been used for a sensoric approach to monitor enantiomeric purities of amino acid derivatives and lactate esters. On-line monitoring of a production process is possible. In this section two applications of chiral amide-based sensors are presented. The implementation of chiral discrimination by chiral amide selectors with optical and piezoelectrical gas-sensing devices has been reported recently [18]. Two optical methods (SPR and RIfS) were used to discriminate both enantiomers of several analytes by a calix[4]arene bound to a polysiloxane backbone (S. Busche, M. Kasper, A. Ruderisch, V. Schurig, G. Gauglitz, unpublished results).

3.1.2 Chiral Discrimination by Chirasil-Val

Both enantiomers of Chirasil-Val receptors have been applied to optical and piezoelectrical sensors [20]. In this chapter only the result of the optical transduction principle is presented.

RIfS was used to characterise changes in the optical thickness Δ(nd) of a polymeric sensitive layer. RIfS is based on interference effects in thin transparent films. By interaction of analyte molecules with the sensitive layer, a swelling takes place and so the interference pattern is changed [21].

The chiral selector is derived from the well-known chiral chromatographic stationary-phase Chirasil-Val [22] and both enantiomers of the receptor have been applied as coatings in a sensor array. The chiral selector octyl-Chirasil-Val [*N*-(2-methylpropanoyl)-*S*/*R*-valine *tert*-butylamide] contains chiral peptide residues for enantiomeric recognition and non-chiral lipophilic side chains (Fig. 4). Solutions of these polymers were spin-coated on glass transducers for RIfS. To recognise artefacts caused by fluctuations of the gas stream or contaminations of the *S*- and *R*-analytes, polydimethylsiloxane layers without chiral selectors were included in the sensor array.

All sensors were exposed to analyte gas simultaneously at constant temperature. The sensors were exposed alternately to dry air and analyte vapour. The sensor responses are defined as the signal differences between analyte equilibrium and baseline (dry air). Both enantiomers of methyl and ethyl lactate and both enantiomers of *N*-trifluoroacetylalanine methyl ester (*N*-TFA-Ala-OMe) were used as chiral analytes.

In Fig. 5, the relative sensor signals of one *S*-sensor and one *R*-sensor and an additional non-chiral SE-30 sensor upon exposure to enantiomeric mixtures of *N*-TFA-Ala-OMe are shown. The sensitive layers display stronger interactions between the *S*-analyte and the *S*-polymer or the *R*-analyte and the *R*-polymer than between the respective *R*-/*S*-pairs.

The discrimination factors α (α is the ratio of the two sorption equilibrium signals towards both enantiomers) of methyl lactate by *R*- and *S*-sensors and the α^{-1} values to demonstrate the inversion symmetry are shown in Fig. 6, right. The results are in complete accordance with gas chromatographic inves-

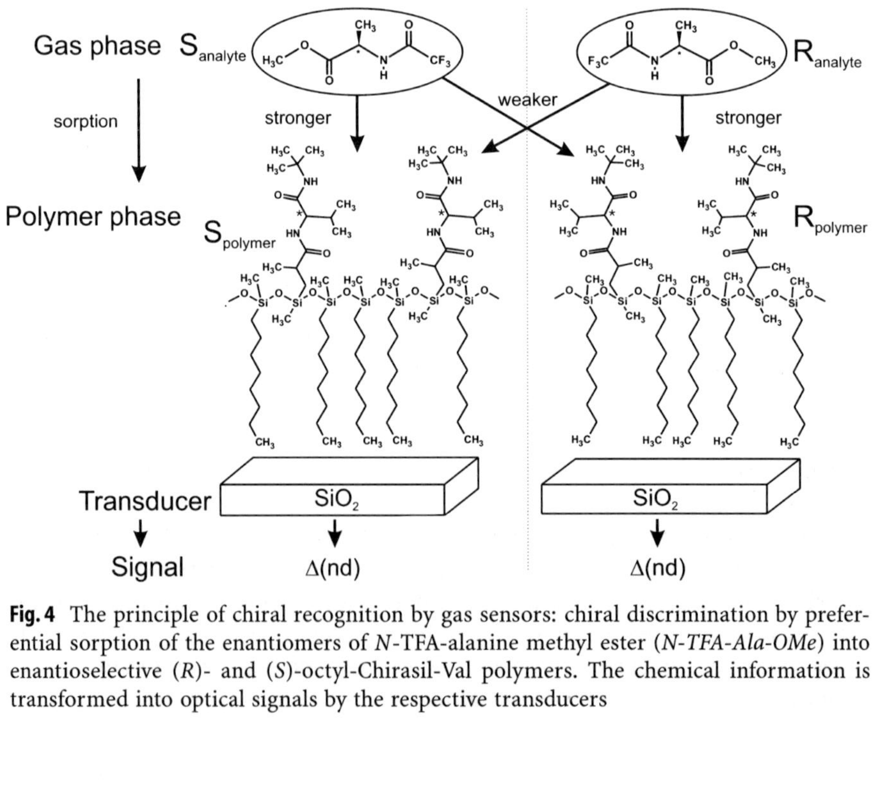

Fig. 4 The principle of chiral recognition by gas sensors: chiral discrimination by preferential sorption of the enantiomers of *N*-TFA-alanine methyl ester (*N-TFA-Ala-OMe*) into enantioselective (*R*)- and (*S*)-octyl-Chirasil-Val polymers. The chemical information is transformed into optical signals by the respective transducers

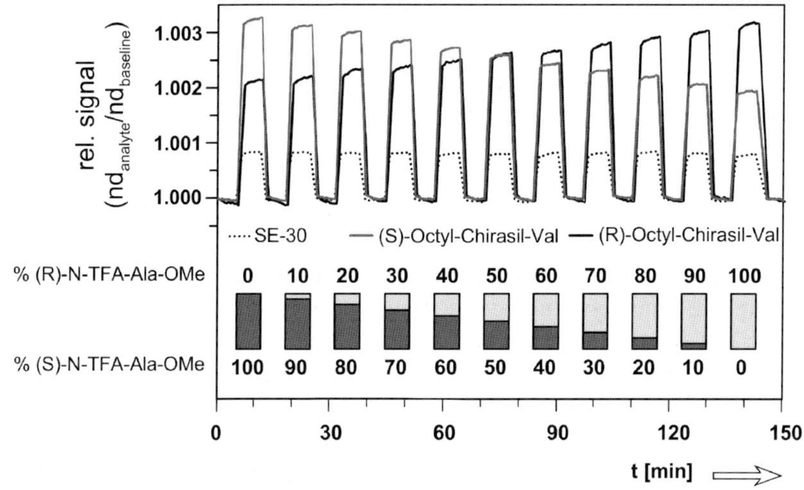

Fig. 5 Relative reflectometric interference spectroscopy sensor signals (normalised with regard to the layer thickness without analyte gas) of two chiral sensors (*S*-sensor, *grey line*; *R*-sensor, *black*) and an additional SE-30 sensor (*points*) upon exposure to mixtures of different enantiomeric composition (in percent) of *N*-TFA-Ala-OMe [18]

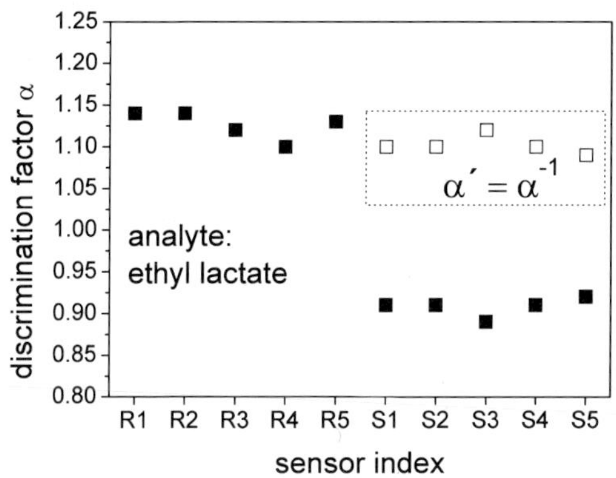

Fig. 6 Discrimination factors of methyl lactate obtained by five different *R*- and *S*-sensors. For the *S*-sensors, the α' values are also shown to demonstrate the inversion symmetry [18]

tigations. The signals of the chiral sensors upon exposure to racemic mixtures are identical and show medium intensity.

The enantiomeric composition of methyl lactate ester mixtures was measured by an array of *S*- and *R*-sensors and the data were evaluated by chemometric tools (multiple line regression, partial component analysis and principal component regression). The ability to distinguish the enantiomeric composition of various mixtures of *N*-TFA-Ala-OMe or lactates quantitatively with octyl-Chirasil-Val was demonstrated [18].

3.1.3
Discrimination with Optical Chirasil-Calix Sensors

Recently the separation of enantiomers by RIfS and SPR using calixarenes with chiral amide residues was demonstrated [19]. Chirasil-Calix (Fig. 7) is well known from capillary GC as a stationary-phase material because of its good thermal and long-term stability. The separation of amino acid derivatives and lactic esters was widely studied [23].

Changes of the refractive index of a sensitive layer are monitored by SPR. By the uptake of analyte molecules into the polymeric film, the refractive index is changed and the resonance wavelength of the surface plasmons is also changed. There are excellent publications available on the theoretical basis and the numerous applications of SPR [25–27].

Both enantiomers of methyl and ethyl lactate ester and several amino acid derivatives were successfully discriminated by both optical transduction principles. In Fig. 8 the results of SPR measurements of both enantiomers of the valine esters with L-Chirasil-Calix are shown. In accordance with gas chromatographic investigations, the L-enantiomer of the amino acid derivatives shows stronger interaction with an L-Chirasil-Calix sensitive layer. This leads

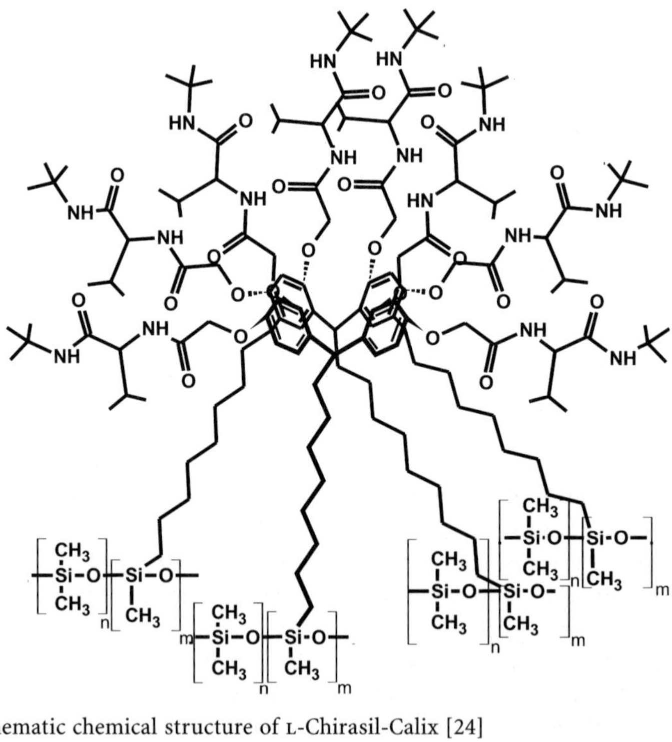

Fig. 7 Schematic chemical structure of L-Chirasil-Calix [24]

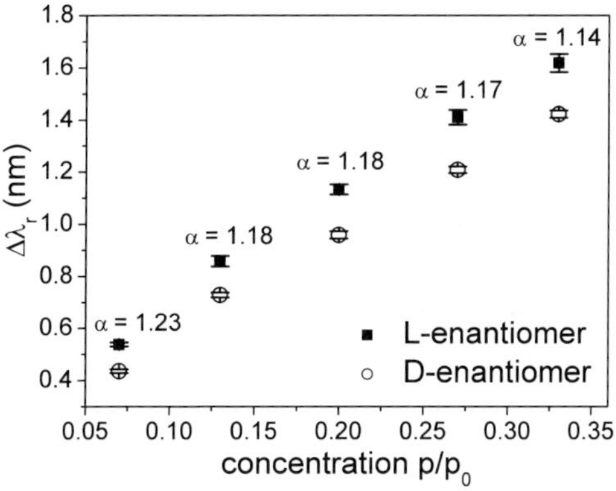

Fig. 8 Calibration curves of valine ester with Chirasil-Calix using surface plasmon resonance (*SPR*)

to higher response signals for both transduction principles. For racemic mixtures the response lies between the respective signals of the pure enantiomers.

Regarding the response signals of the lactate esters, the D-enantiomers showed stronger interaction with the L-polymer of Chirasil-Calix. This is in agreement with the elution sequence in GC.

3.1.4
Cyclodextrins

The importance of stereochemistry for biological activity and for chiral drugs is well known. For chiral inhalation anaesthetics, differences of the narcotic action of the individual enantiomers were discussed. Therefore, a fast and cheap sensing device for on-line monitoring of anaesthetic compounds is highly desirable. Several inhalation anaesthetics were separated by GC first [28]; afterwards the enantioselective complexation in chiral host–guest systems was investigated by optical and electrical transducers [7]. Chiral inhalation anaesthetics can be separated into enantiomers with large chiral separation factors α on modified cyclodextrins. Gas chromatographic investigations have shown no separation of halodiether B [1,1,1,3,3-pentafluoro-2-(fluoromethoxy)-3-methoxypropane] with α-cyclodextrin, whereas β-cyclodextrin shows a relatively high separation factor, and with γ-cyclodextrin (Lipodex E) the highest known chiral separation factor in GC was achieved.

SPR spectroscopy and RIfS were used as optical transduction principles for the detection of halogenated diethers as chiral guests with modified cyclodextrin as the stationary phase [7].

Lipodex E [octakis(3-*O*-butanoyl-2,6-di-*O*-*n*-pentyl)-γ-cyclodextrin] dissolved in a polysiloxane matrix (SE-54) was used as a chiral receptor. The binding strength of the analyte molecules depends on the interaction mechanism and the forces (hydrogen bonds and dispersion forces) between receptor and target molecules.

Both enantiomers of enflurane [2-chloro-1-(difluoromethoxy)-1,1,2-trifluoroethane], isoflurane [2-chloro-2(difluoromethoxy)-1,1,1-trifluoroethane] and desflurane [2-(difluoromethoxy)-1,1,1,2-tetrafluoroethane] were used as chiral analytes.

All sensors of the array were exposed simultaneously to the test gases. Dry air was used as a carrier gas. In Fig. 9 the sensor responses of Lipodex E to both enantiomers of halodiether B in a concentration range from 0 to 140 $\mu g\,l^{-1}$ is shown. The interaction between the cyclodextrin-recognition units and the *S*-enantiomer is stronger than that with the *R*-enantiomer.

The calibration curves for the pure *R*- and *S*-enantiomers of halodiether B with SPR are shown in Fig. 10. A significant and reproducible difference upon exposure to the optical antipodes was observed as the response of the chiral sensors to the chiral compounds. Enantioselectivity factors α can be determined by dividing the respective signal heights. For a concentration of 20 $\mu g\,l^{-1}$, the α value for SPR is 9.6 ($\pm$0.7). The α values decrease with increasing concentrations, because fewer complexation sites are available for guest molecules.

Investigation with the two optical transduction principles mentioned before confirmed the preferential enrichment of the *S*-enantiomer resulting in more than ninefold higher signals, which is the highest separation factor found for chiral separation on cyclodextrins up to now. The good agree-

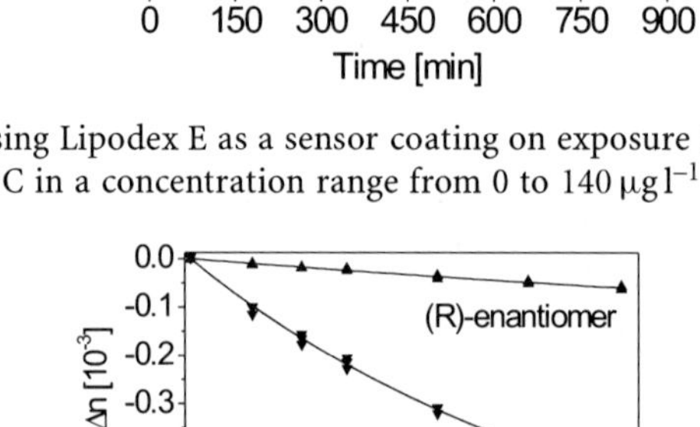

Fig. 9 SPR signals using Lipodex E as a sensor coating on exposure to the enantiomers of halodiether B at 30 °C in a concentration range from 0 to 140 $\mu g\,l^{-1}$

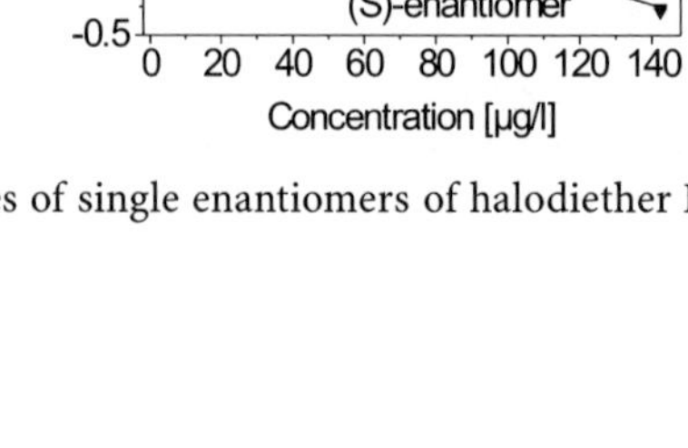

Fig. 10 Calibration curves of single enantiomers of halodiether B for SPR with Lipodex E

ment between the optical transduction methods and GC is explained by the robustness against structural changes of the surface. The enantioselective complexation can be separated from the non-enantioselective sorption into the polymer backbone by applying a model based on Langmuir-type and Henry-type sorption.

3.1.5 Molecularly Imprinted Polymers

MIPs with benzamidine-containing binding sites were used for RIfS measurements in aqueous media. Stoichiometric non-covalent binding with a dicarboxylic acid was realised. These kinds of MIPs are good candidates for application in chemosensors because of their fast and reversible interaction between the imprinted matrix and the target analyte [29]. In Fig. 11 the uptake and release of an analyte molecule in the preformed MIP is shown. The template molecules are chelated by hydrogen bridges including an ion pair.

Both enantiomeric forms of 2,3-di-*O*-benzoyltartaric acid were used as optically active templates. *N*,*N*′-diethyl-4-vinylbenzamindine and a large excess of ethylene dimethacrylate were used as co-polymer and cross-linker. The MIP layers were exposed to different concentrations of the two antipodes in water. The interaction of the target analytes with the MIP layers was observed directly by RIfS [30]. In Fig. 12 the calibration functions for enantiomeric forms of the tartaric acids are shown.

A linear dependence of the signal on the concentration is found. By linear regression, the slope of this calibration function (the so-called sensitivity

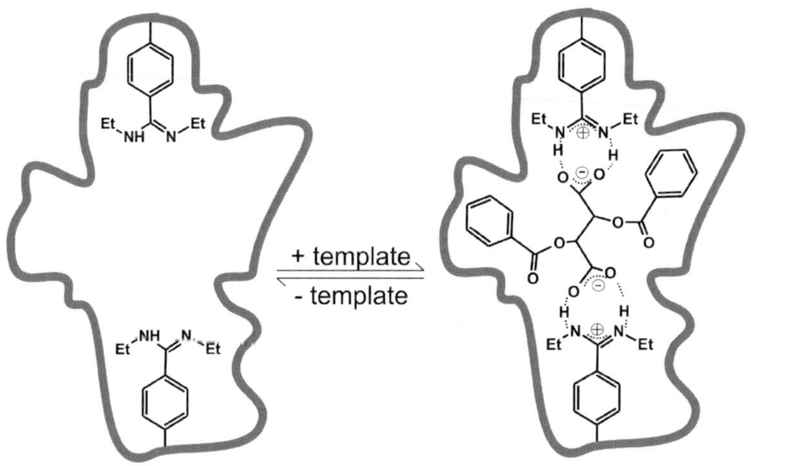

Fig. 11 Uptake and release of templates by the imprinted polymer. On the *left* the free cavity is shown. On the *right*, the template is bound by two chelated hydrogen bridges including an ion pair

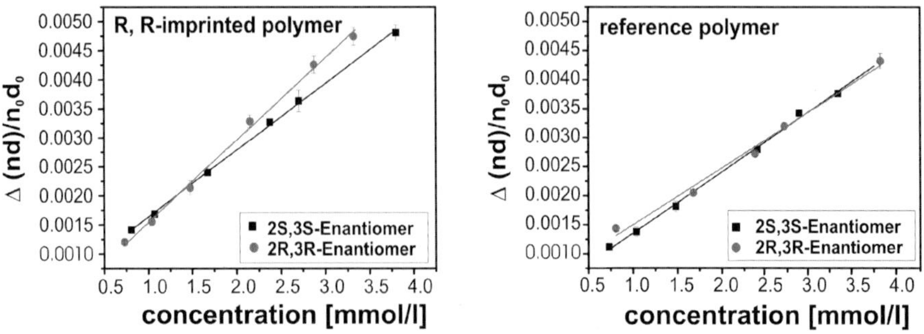

Fig. 12 Comparison of the calibration functions for both enantiomeric forms of the tartaric acid examined. On the *left*, the signals for an (*R*,*R*)-imprinted polymer are shown. On the *right*, the results for the reference polymer can be seen

S) can be determined. The imprinted polymers show differences in their sensitivities towards the enantiomers of the analytes (Fig. 12, left), whereas the non-chiral reference polymer shows no difference between the two enantiomers (Fig. 12, right). Therefore, a chiral discrimination by the imprinted polymers is proved; this can be described by a separation factor α (Table 1). The separation factor is defined here as the ratio of the sensitivity of the template to the sensitivity of the antipode.

Table 1 Separation factors α for all polymers

Polymer	α
(*S*,*S*)-imprinted	1.19
(*R*,*R*)-imprinted	1.23
Reference polymer	0.98

For the (*S*,*S*)-imprinted polymer a separation factor of 1.19 was achieved, while for the (*R*,*R*)-imprinted polymer an α value of 1.23 was found. The reference polymer shows nearly no chiral discrimination ($\alpha = 0.98$) [30].

The combination of the high stability, the fast response and good enantiomeric discrimination ability makes MIPs promising materials for chiral chemosensors in the aqueous phase. Further improvement can be expected, as new simple methods for spin-coating were reported recently [31].

3.2 Fluorescence Sensors

Numerous modes of signalling of fluorescence for the distinction of optically active compounds are known. For an excellent overview of all methods and applications of fluorescence of organic molecules in chiral recognition see Pu [5]. Herein, examples for the most commonly used methods—enantioselective fluorescence quenching or enhancement—will be presented.

3.2.1
Calixarene Fluorophores

The quantitation of undesired enantiomers in drug raw materials is one of the objectives of the pharmaceutical industry. Several calixarene derivatives were investigated as fluorescence sensors for chiral pharmaceutical compounds. The mechanism of these examples is based on different fluorescence quenching of the calixarenes by the two enantiomeric forms of a specific analyte.

A significant ability to discriminate between chiral amines based on the quenching of *S*-di-2-naphthylprolinol fluorescence emission was reported by Diamond et al. [32]. *R*-Phenylethylamine (PEA) was seen to have a much greater efficiency as a quencher than the *S*-enantiomer. L- and D-norephedrine, which have structural conformation similarities to PEA, were also observed to have an enantiomeric selectivity. The mechanism of chiral recognition is proposed to be a combination of hydrogen bonding and 3D chirally restricted space.

The effect of the relative spacing between the naphthyl signalling groups and the 3D chiral distribution of binding sites within the calixarene cavity was investigated by Diamond et al. recently [33]. Therefore two selectors were used: one with these binding sites and the chiral groups separated from the naphthyl groups by an additional ether group (Fig. 13, compound **1**) compared with the other selector (compound **2**). Phenylglycinol and phenylalaninol were used as target guests, whereby phenylalaninol has the chiral binding sites separated from the aryl group by a methylene spacer.

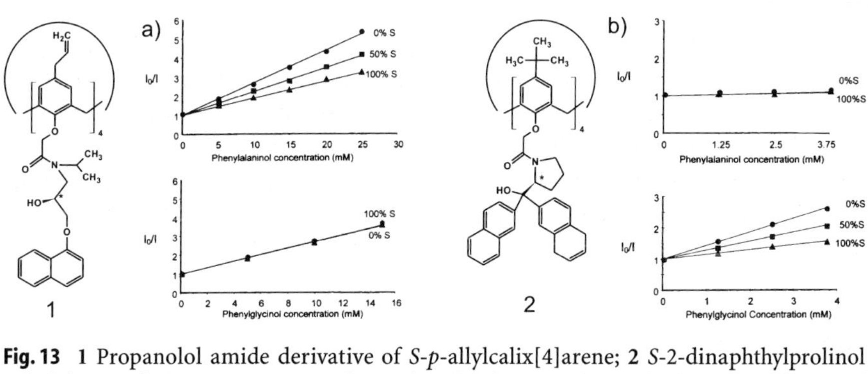

Fig. 13 **1** Propanolol amide derivative of *S*-*p*-allylcalix[4]arene; **2** *S*-2-dinaphthylprolinol calix[4]arene derivative. **a** Stern–Volmer plots for the quenching of **1** by *S*-phenylalaninol and *S*-phenylglycinol in percent; **b** Stern–Volmer plots for the quenching of **2** by *S*-phenylalaninol and *S*-phenylglycinol in percent [33]

As can be seen in the Stern–Volmer plots, for the quenching of the fluorescence of **1** upon addition of phenylalaninol and phenylglycinol, respectively (Fig. 13, left), the selector exhibits significant ability to distinguish between the enantiomers of phenylalaninol, but not of phenylglycinol. For *S*-dinaphthylprolinol calix[4]arene, an excellent ability to discriminate between the enantiomers of phenylglycinol can be found, but there is no significant quenching effect for both enantiomers of phenylalaninol. So **2** can successfully discriminate the two enantiomers of phenylglycinol, whose chiral centre is immediately adjacent to its aromatic ring. In the presence of a methylene spacer between the chiral centre and the aromatic ring, a successful discrimination is not possible. In contrast, for selector **1**, an effective discrimination is only realisable when the analyte has a spacer at its disposal. Furthermore, these receptors enable an estimation of the enantiomeric composition with better than 5% accuracy on the basis of a single fluorescence measurement, if the total concentration of both enantiomers is known.

3.2.2 Fluorescent Cyclodextrins

Cyclodextrins (Sect. 2.2), have the ability to include various organic molecules in their central cavities. Chromophores have been linked to cyclodextrins to build spectroscopic sensors for organic molecules. Many fluorescent cyclodextrins have been prepared to be used as molecular sensors in solution.

Corradini et al. [34] prepared β-cyclodextrin based sensor molecules containing binding sites for Cu(II) derived from *R*- and *S*-phenylalanine, respectively, and a dansyl fluorophore. By complexation with Cu(II), the fluorescence of the dansyl group is quenched, whereby the quenching of the *S*-enantiomeric complex was greater. Fluorescence enhancement was achieved by addition of amino acids since coordination of amino acids to Cu(II) displaced the sulphonamide of the dansyl group. The best enantioselective fluorescence response was found for interaction of the *S*-selector with D- or L-proline, respectively [ef = $(I_D - I_0)/(I_L - I_0) = 3.89$]. For the *R*-selector reversed enantioselectivities of a much smaller amount were found. Histidine caused the greatest fluorescence enhancement but with an almost negligible enantioselectivity. Further work with different selector arms was done and a model for the recognition process based on the formation of ternary diastereomeric complexes was proposed [35].

Dansylglycine-modified cyclodextrin (DnsC4-β-CD) was found to be capable of being immobilised on a cellulose membrane and of acting as a fluorescence sensor for enantiomers [36]. DnsC4-β-CD decreased its fluorescence intensity upon exposure to guest molecules. This result implicates a useful application of a cellulose membrane as a supporting material for various chromophore-modified cyclodextrins. Disposable sensor chips or paper may be developed for molecular detection.

3.2.3
1,1′-Binaphthyl Fluorophores

Binaphthyl-based fluorescence sensors are interesting receptors for carrying out enantioselective recognition. Multiple chiral functional groups of the macrocycles availed as fluorescence sensors can be used for the detection of the different groups of chiral compounds. The design and synthesis of a chiral bisbinaphthyl-based selector for the enantioselective recognition of α-hydroxycarboxylic acid was described by Pu et al. [37]. The interaction of mandelic acid as a chiral guest with the chiral binaphthyl selector in benzene was investigated. A significant fluorescence enhancement due to the suppressed PET quenching was observed when protonation by the acid occurred. This fluorescence enhancement was found to be highly enantioselective. The interaction of the *S*-enantiomer of mandelic acid with the *S*-1,1′-binaphthol derivative was found to increase the fluorescence signal 2.87 times, while the *R*-mandelic acid increased the intensity only 1.75 times. This leads to an enantiomeric fluorescence difference ratio $ef = (I_S - I_0)/(I_R - I_0)$ of 2.49. When determining the interaction of both enantiomers with the *R*-selector, a mirror-image relationship can be found; therefore, the observed differential fluorescence enhancement for the two enantiomers is a result of chiral recognition by the respective selector. The fluorescence signal of the receptors can be significantly amplified up to 22 times by introducing dendritic branches to the chiral receptor unit.

Recognition of α-amino acids is more challenging than recognition of α-hydroxycarbocyclic acids, as the primary amine group in an amino acid can quench fluorescence of the fluorophores; still, the carboxylic group can enhance fluorescence. Recently, a bisbinaphthyl-based fluorescence sensor for the recognition of α-amino acid derivatives was reported by Pu et al. [38]. An *N*-carbonyl group was introduced into the amino acid substrate in order to delocalise the lone pair of the amine group. This inhibits fluorescence quenching and converts the zwitterionic amino acid into a neutral molecule. Both enantiomeric forms of the bisbinaphthyl macrocycle selector were synthesised. Multiple chiral functional groups provide efficient binding sites for organic molecules. Experiments with D- and L-*N*-benzyloxycarbonylphenylglycine demonstrated an increase in the fluorescence intensity of more than 4 times for one enantiomer of an N-protected phenyl glycine, while the other enantiomer does not cause significant fluorescence enhancement. A mirror-image relationship between the responses of the two enantiomers of the selector was found, which proves the assumption of chiral recognition.

The fluorescence sensor can easily be used to determine the enantiomeric composition of amino acid derivative mixtures because of the linear relationship of the concentration and the fluorescence signal.

4 Conclusions and Outlook

Regarding the scope of applications for enantiomeric discrimination by optical sensors, it is becoming clear how important this field of investigation is. Nowadays, chiral investigations are conducted in the design and production of pharmaceuticals, in medical monitoring as well as in environmental probing. Development of fast and cheap sensors would be desirable. Numerous promising advances towards sensor applications for the discrimination of enantiomers have been reported in the literature.

Basic concepts of interaction have been refined as new models could be established. The fields of applications were broadened. Currently, the detection of enantiomers by fluorescence measurements and by direct optical methods are well-investigated topics. By introducing various functional groups to established skeletal structures, highly specific recognition of enantiomers is likely to be achieved. Those methods based on recognition by polymer-bonded selectors will continue to profit from advances in modern chromatographic methods.

References

1. Henderson GM, Rule HG (1939) J Chem Soc 1568
2. Gil-Av E, Feibush B, Charles-Sigler R (1966) Tetrahedron Lett 1009
3. Ward TJ (2002) Anal Chem 74:2863
4. SciFinder Scholar, version 2004. American Chemical Society
5. Pu L (2004) Chem Rev 104:1687
6. Bodenhöfer K, Hierlemann A, Seemann J, Gauglitz G, Christian B, Koppenhoefer B, Göpel W (1997) Nature 387:577
7. Kieser B, Fietzek C, Schmidt R, Belge G, Weimar U, Schurig V, Gauglitz G (2002) Anal Chem 74:3005
8. Feibush B, Gil-Av E (1970) Tetrahedron 26:1361
9. Beitler U, Feibush B (1976) J Chromatogr 123:149
10. Schreier P, Bernreuther A, Huffer M (1995) Analysis of chiral organic molecules—methodology and applications. de Gruyter, Berlin
11. Stölting K, König WA (1976) Chromotographia 9:331
12. Cramer F, Dietsche W (1959) Chem Ber 92:378
13. Snopek J, Smolková-Keulemansová E, Cserháti T, Gahm KH, Stalcup A (1996) Cyclodextrins in analytical separation methods. In: Lehn JM, Atwood JL (eds) Comprehensive supramolecular chemistry: cyclodextrins. Elsevier, Amsterdam, p 515

14. Díaz-García ME, Badía R (2004) Molecularly imprinted polymers for optical sensing devices. In: Narayanaswamy R, Wolfbeis OS (eds) Optical sensors—industrial, environmental and diagnostic applications. Springer, Berlin Heidelberg New York, p 35
15. Haupt K, Mosbach K (2000) Chem Rev 100:2495
16. McMahon G, O'Malley S, Nolan K, Diamond D (2003) ARKIVOC vii:23
17. Maier NM, Franco P, Lindner W (2001) J Chromatogr A 906:3
18. Bodenhöfer K, Hierlemann A, Seemann J, Gauglitz G, Christian B, Koppenhoefer B, Göpel W (1997) Anal Chem 69:3058
19. Busche S, Kasper M, Ruderisch A, Schurig V, Gauglitz G; in preparation
20. Albert JA, Lewis NS, Schauer CL, Sotzing GA, Stitzel SE, Vaid TP, Walt DR (2000) Chem Rev 10:2595
21. Gauglitz G, Brecht A, Kraus G, Nahm W (1993) Sens Actuators B 11:21
22. Frank H, Nicholson GJ, Bayer EJ (1977) Chromatogr Sci 15:174
23. Pfeiffer J, Schurig V (1999) J Chromatogr A 840:145
24. Ruderisch A, Pfeiffer J, Schurig V (2001) Tetrahedron Asymmetry 12:2025
25. Reather H (1983) Surface plasmons on smooth and rough surfaces and on gratings. Springer tracks in modern physics, vol 111. Springer, Berlin Heidelberg New York
26. Homola J, Yee SS, Gauglitz G (1999) Sens Actuators B 54:3
27. Homola J (2004) Surface plasmon resonance biosensors for food safety. In: Narayanaswamy R, Wolfbeis OS (eds) Optical sensors—industrial, environmental and diagnostic applications. Springer, Berlin Heidelberg New York, p 145
28. Meinwald J, Thompson WR, Pearson DL, König WA, Runge T, Francke W (1991) Science 251:560
29. Wulff G, Schönfeld R (1998) Adv Mater 10:957
30. Nopper D, Lammershop O, Wulff G, Gauglitz G (2003) Anal Bioanal Chem 377:608
31. Schmidt RH, Mosbach K, Haupt K (2004) Adv Mater 16:719
32. Grady T, Harris SJ, Smyth MR, Diamond D (1996) Anal Chem 68:3775
33. Lynam C, Jennings K, Nolan K, Kane P, McKervey MA, Diamond D (2002) Anal Chem 74:59
34. Pagliari S, Corradini R, Galaverna G, Sforza S, Dossena A, Marchelli R (2000) Tetrahedron Lett 41:3691
35. Corradini R, Paganuzzi C, Marchelli R, Pagliari S, Sforza S, Dossena A, Galaverna G, Duchateau A (2003) Chirality 15:S30
36. Tanabe T, Touma K, Hamasaki K, Ueno A (2001) Anal Chem 73:3126
37. Xu MH, Lin J, Hu QS, Pu L (2002) J Am Chem Soc 124:14239
38. Lin J, Li ZB, Zhang HC, Pu L (2004) Tetrahedron Lett 45:103

Springer Ser Chem Sens Biosens (2005) 3: 343–365
DOI 10.1007/5346_012

Published online: 22 September 2005

Optical Sensors for Ions and Protein Based on Digital Color Analysis

Yoshio Suzuki[1,3] · Koji Suzuki[1,2] (✉)

[1]Cooperation for Innovative Technology and Advanced Research in Evolutional Area (CITY AREA), Kanagawa Academy of Science and Technology, 3-2-1 Sakato, Takatsu-ku, Kawasaki, 213-0012 Kanagawa, Japan

[2]Department of Applied Chemistry, Faculty of Science and Technology, Keio University, 3-14-1 Hiyoshi, Kohoku-ku, Yokohama, 223-8522 Kanagawa, Japan
suzuki@applc.keio.ac.jp

[3]*Present address:*
National Institute of Advanced Industrial Science and Technology, 1-1-1 Higashi, Tsukuba, 305-8562 Ibaraki, Japan
suzuki-yoshio@aist.go.jp

Abstract A novel colorimetric method, digital color analysis (DCA), was proposed using a digital color analyzer and was applied to various quantitative analyses using chromaticity coordinates and suitable sensors for visual colorimetry based on the characteristics of human visual perception by virtual simulations based on digital color information. On the basis of DCA, we developed a visual colorimetric sensor for Li^+, NH_4^+ and protein determination by the mixing of two kinds of lipophilic dyes, whose optimum mixing ratio was obtained by virtual simulation based on DCA. Using DCA, a linear relation calibration curve can be obtained over the wide range of Li^+, NH_4^+ and protein concentrations and this indicated a different color change with increase of the concentrations of these substances. These color sensors based on DCA have great potential for this analytical purpose.

1 Introduction

Absorption spectrometry is a traditional method used for the measurement of various chemical substances and makes it possible to carry out visual colorimetry allowing easy measurements. Conventional absorption spectrophotometry is the measurement of numerical values such as that of absorbance to carry out qualitative and quantitative analysis. In such cases, if the spectra obtained are complicated, the determination often becomes difficult. However, even if the spectral changes are quite complicated, our eyes recognize them simply as color changes. Determination utilizing the colors themselves is a perceptual method instead of simple absorption spectrophotometry.

In our group, digital color analysis (DCA) was proposed, which was based on the variety of the color information for applications with an analytical purpose [1, 2]. In this method, the observed colors are treated as digital information using a digital color analyzer. The concept of DCA consists of the analysis and utilization of "color" information, where a digital color analyzer is used as a means of utilizing "colors". The digital color analyzer used in this study (COLORTRON) is hand-held size and is capable of measuring colors quite easily. It can rapidly transform color information into numerical values by processing the data with a personal computer which is connected to the color analyzer. This analyzer can easily determine colors, calculate the tristimulus values (X,Y,Z) with the spectral data obtained, and the XYZ values can be converted into a variety of numerical color data [3, 4]. In the case of calculating the tristimulus values, the characteristics of a standard observer's eyes and the spectral data of a light source are included in the calculation process as shown in Fig. 1. The basic tristimulus values can be further transferred into several-figured color data such as HSB, RGB,and $L^*a^*b^*$, or Q_xQ_y as complementary color data. These color coordinates have their own unique chromaticity diagrams. By transforming color information into numerical values, colorimetric analysis that has been used only as semiquantitative analysis can serve as an accurate method. Moreover, even if the spectrum is quite complicated to analyze, it is possible to make an accurate determination without the spectral data by the color changes, on the basis of color difference or digital color library data for the color obtained. DCA can produce calibration curves by converting the spectrum into tristimulus values and plotting the color on a chromaticity diagram. DCA can also create a color library for

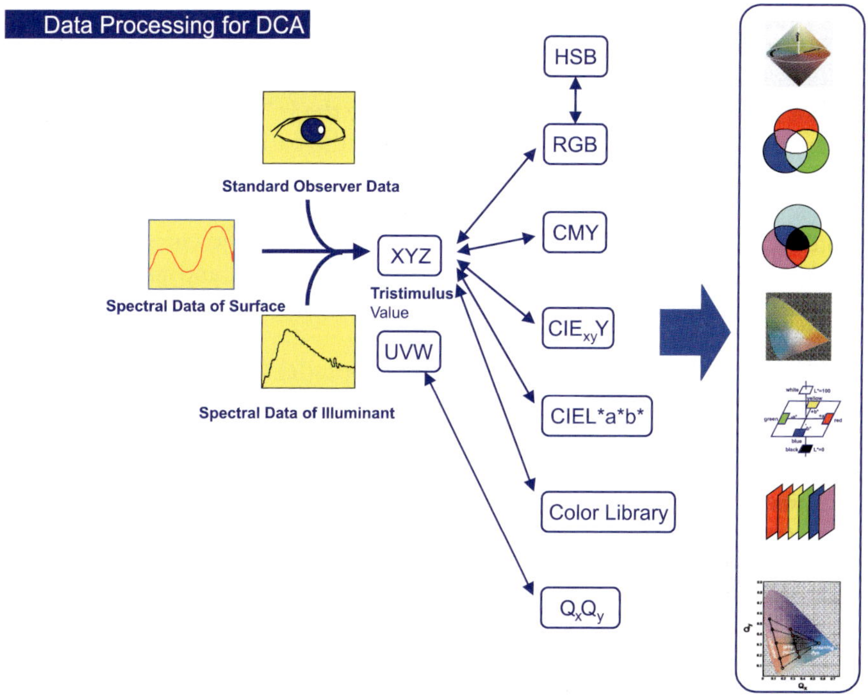

Fig. 1 Concept and data-processing diagrams for digital color analysis

direct visual calibration and simulate the optimum color changes based on limited experimental color data using a personal computer with suitable software.

In this review, Li^+, NH_4^+, and protein color sensors were treated with DCA analysis for the visual determination using several major advantages for its utilization in the calibration/determination of an analyte using color data as follows: (1) calibration/determination based on numerical color data, (2) calibration/determination based on color library data, (3) calibration/determination based on chromaticity diagrams, and (4) suitable visual calibration/determination based on color simulating calibration which can be used for the design of an optimum color-based optode.

2 Li^+ Ion Sensors Based on Digital Color Analysis

Some lithium salts such as Li_2CO_3 are used to treat manic-depressive illness and hyperthyroidism. The manic-depressive therapeutic Li^+ concentration range in blood is 0.75–1.25 mM, and it is known that more than 1.5 mM Li^+ is toxic and that 3.5–4.5 mM Li^+ is lethal. Because of the toxicity of excess lithium salt and the requirement to maintain the concentration of Li^+ within the narrow effective therapeutic range, it is necessary to control the Li^+ concentration in the human body [5, 6]. Although Li^+ can be determined by blood analysis, it is preferable to determine the Li^+ concentration in saliva because it is more easily extracted and allows the patients to monitor the Li^+ concentration themselves [7]. The Li^+ ion normally exists in saliva at a concentration twice that in the blood; the effective therapeutic Li^+ range in saliva is 1.5–2.5 mM. Monitoring the concentration of a specific clinically relevant substance is important for controlling overprescription and for confirming that a patient actually took the medicine.

In this section, we introduce a mechanism of dry chemistry and attempt to construct a film-type sensor (membrane optode) which needs only dropping of the sample onto the optode film device that includes all the reagents necessary for analysis. For this purpose, a PVC membrane film optode containing a lithium ionophore, an indicator dye, and a lipophilic additive was prepared, and its response characteristics were investigated. For visual determination, an attempt was made to produce the color change passing through a colorless gray point, which can be performed by screening the dye [8, 10]. Such a color change is easy to recognize visually at a certain point or narrow level of analyte concentration. When the sensor changes from one color to its complementary color, passing through the gray point at nearly 10^{-3} M Li^+, it is possible to efficiently evaluate whether the optimum therapeutic Li^+ concentration is maintained or if Li^+ is in excess over the toxic limit.

2.1 Theory

2.1.1 Response Mechanism for the Ion Optode

The response mechanism of the Li^+ optode constructed in this study is based on the ion extraction/exchange from the aqueous phase (sample solution) to the organic phase (optode membrane) as shown in Fig. 2. The ion-pair extraction/exchange can be described by the following equation:

$$i^+_w + S_o + DH^+R^-_o \rightleftarrows Si^+R^-_o + D_o + H^+_w \,, \qquad (1)$$

where i^+ represents the cation to be extracted (analyte, Li^+ in the present case), and H^+ represents a proton. S, DH^+R^-, Si^+R^-, and D represent the neutral ionophore, the ion-pair of a protonated dye and a lipophilic anionic additive, the ion-pair of an ion-ionphore complex and a lipophilic anionic additive, and the deprotonated dye, respectively. Subscripts o and w represent the organic phase and the water phase, respectively. When the neutral ionophore S complexes with the analyte cation, the cationic dye DH^+ deprotonates and becomes neutral. Subsequently, the ion–ionophore complex and the

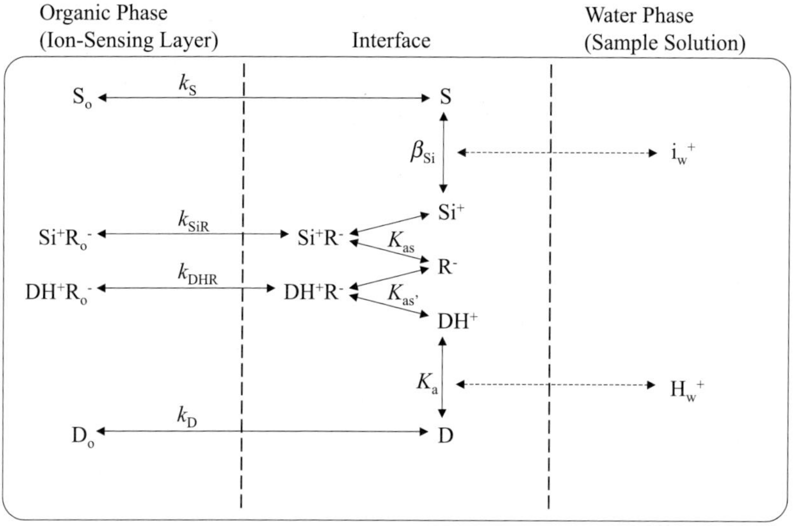

Fig. 2 The model of ion extraction/ion exchange for an optode based on a neutral ionophore and a lipophilic cationic dye in relation to the ion-optode response mechanism i^+ cation to be extracted, H^+ proton, S neutral ionophore, R lipophilic cationic additive, D color-changeable dye. The subscripts o and w represent the organic phase and the water phase, respectively)

lipophilic anion R^- form an ion-pair in the organic phase. The total extraction constant, K, is written as follows:

$$K = \frac{a_{\mathrm{SiR_o}} a_{D_o} a_{\mathrm{H_w^+}}}{a_{i_w^+} a_{\mathrm{S_o}} a_{\mathrm{DHR_o}}}, \tag{2}$$

where a denotes the activity of each chemical species. On the basis of all of the equilibrium constants and the reactions in the model shown in Fig. 2, the analyte activity,′ $a_{i_w^+}$, is given by Eq. 3:

$$a_{i_w^+} = \frac{a_{\mathrm{SiR_o}} a_{D_o} a_{\mathrm{H_w^+}}}{a_{\mathrm{S_o}} a_{\mathrm{DHR_o}} K} = \frac{\left(a_R^{\mathrm{tot}} - a_D^{\mathrm{tot}} + a_{D_o}\right) a_{D_o} a_{\mathrm{H_w}^+}}{\left(a_S^{\mathrm{tot}} - a_R^{\mathrm{tot}} + a_D^{\mathrm{tot}} - a_{D_o}\right)\left(a_D^{\mathrm{tot}} - a_{D_o}\right) K}, \tag{3}$$

where a^{tot} denotes the total activity of each chemical species. Because the absorbance, A, is expressed as Eq. 4 according to Bouget–Beer's law, Eq. 3 is rewritten as Eq. 5:

$$A = \frac{\left(a_R^{\mathrm{tot}} - a_D^{\mathrm{tot}} + a_{D_o}\right) a_{\mathrm{H_w^+}}}{\left(a_S^{\mathrm{tot}} - a_R^{\mathrm{tot}} + a_D^{\mathrm{tot}} - a_{D_o}\right)\left(a_D^{\mathrm{tot}} - a_{D_o}\right)}, \tag{4}$$

$$a_{i_w^+} = \frac{\left(\varepsilon b a_R^{\mathrm{tot}} - \varepsilon b a_{RD}^{\mathrm{tot}} + A\right) A_{\mathrm{H_w^+}}}{\left(\varepsilon b a_S^{\mathrm{tot}} - \varepsilon b a_R^{\mathrm{tot}} + \varepsilon b a_D^{\mathrm{tot}}\right)\left(\varepsilon b a_D^{\mathrm{tot}} - A\right)/K}, \tag{5}$$

where ε and b represent the molar absorption coefficient and the optical path length, respectively. Equation 5 represents the general relationship between the ion activity for the analyte (i^+) in the water phase and the absorbance value (A) of the optode. For the evaluation of the optode response, introduction of the normalized absorbance, α, is useful, where α is defined by the ratio of the activity of the deprotonated dye to that of the total amount of dye as expressed by Eq. 6:

$$\alpha = \frac{a_{D_o}}{a_t^{tot}} = \frac{A - A_0}{A_1 - A_0} , \tag{6}$$

where A_0 and A_1 represent the absorbance values of the dye in the protonated form and in the completely deprotonated form, respectively.

2.1.2
Theoretical Equations for Color Mixing on the Basis of Q_xQ_y Coordinates

Q_xQ_y coordinates are complementary xy chromaticity coordinates as shown in Fig. 3. These are analogues of the xy coordinates which are calculated from tristimulus XYZ values in order to plot color points in a two-dimensional diagram [11]. In this case, Q_xQ_y values are calculated by the same procedure adopted in the case of xy coordinates, but the transmission spectra $T(\lambda)$ are replaced with the absorption spectra $A(\lambda)$. The equations for Q_xQ_y are given as follows:

$$Q_x = \frac{U}{U + V + W} , \tag{7}$$

$$Q_y = \frac{V}{U + V + W} , \tag{8}$$

$$U = k \int_{380}^{780} A(\lambda) P(\lambda)\bar{x}(\lambda)\mathrm{d}\lambda , \tag{9}$$

$$V = k \int_{380}^{780} A(\lambda) P(\lambda)\bar{y}(\lambda)\mathrm{d}\lambda , \tag{10}$$

$$W = k \int_{380}^{780} A(\lambda) P(\lambda)\bar{z}(\lambda)\mathrm{d}\lambda , \tag{11}$$

$$k = \frac{100}{\int_{380}^{780} P(\lambda)\bar{y}(\lambda)\mathrm{d}\lambda} , \tag{12}$$

where $x(\lambda)$, $y(\lambda)$, and $z(\lambda)$ are the color-matching functions. $A(\lambda)$ and $P(\lambda)$ are the absorbance values of the sample and the intensities of the illuminant as a function of the wavelength, respectively.

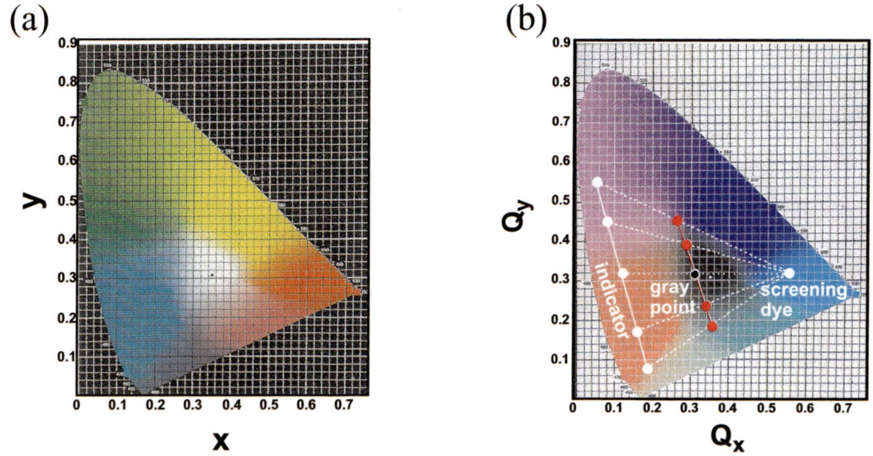

Fig. 3 *xy* and Q_xQ_y chromaticity coordinates

The color of the dye is presented as one specific color point in the Q_xQ_y chromaticity diagram, and the Q_xQ_y coordinates are independent of the dye concentration. When two colors are mixed, the color point corresponding to the subtractive mixture will always lie on the line joining these two colors. The location of the color point depends on the mixing ratio.

For the mixture of the color-changeable dye (indicator) and the inert dye (screening dye) to exhibit grayness during the color change of the indicator, an experimental condition satisfying the following equation is required:

$$V_{\mathrm{ind},r} J_{\mathrm{ind}} m_{\mathrm{ind}} + V_{\mathrm{s},r} J_{\mathrm{s}} m_{\mathrm{s}} = 0 \tag{13}$$

where m represents the volume of solution where the concentration of the solution has to be fixed, r represents x or y, and the subscripts ind and s denote the indicator and screening dye, respectively. V and J values are given by

$$V_r = Q_r - G_r \tag{14}$$

and $$J = \frac{X + Y + Z}{X_{\mathrm{n}} + Y_{\mathrm{n}} + Z_{\mathrm{n}}}\,, \tag{15}$$

in which G_r indicates the coordinates of the gray point and J is the optical concentration. $Q_{\mathrm{ind},r}$ represents a certain color point during the color transition of the indicator. Assuming that $m_{\mathrm{ind}} = 1$ in Eq. 13, the volume of the screening dye m_{s} to be added to the indicator can be calculated by solving the simultaneous Eqs. 16 and 17:

$$\begin{cases} V_{\mathrm{ind},x} J_{\mathrm{ind}} + V_{\mathrm{s},x} J_{\mathrm{s}} m_{\mathrm{s}} = 0\,, & (16) \\ V_{\mathrm{ind},y} J_{\mathrm{ind}} + V_{\mathrm{s},y} J_{\mathrm{s}} m_{\mathrm{s}} = 0\,. & (17) \end{cases}$$

In general, when only one screening dye is used, these simultaneous equations cannot be solved unless the color points of the indicator, the screening dye, and the gray point lie exactly on the same line in the Q_xQ_y diagram. However, treating $V_{\mathrm{ind},r}$ and $V_{\mathrm{s},r}$ as the vectors on the Q_xQ_y chromaticity diagram in which G_r is an original point, we should be able to find m_s by searching for the m_s value where the size of the sum of the vectors $(V_{\mathrm{ind},r}\,J_{\mathrm{ind}} + V_{\mathrm{s},r}\,J_\mathrm{s}m_\mathrm{s})$ expressed by Eq. 18 becomes minimum:

$$f(m_\mathrm{s}) = \left[(V_{\mathrm{ind},x}\,J_{\mathrm{ind}} + V_{\mathrm{s},x}\,J_\mathrm{s}m_\mathrm{s})^2 + (V_{\mathrm{ind},y}\,J_{\mathrm{ind}} + V_{\mathrm{s},y}\,J_\mathrm{s}m_\mathrm{s})^2\right]^{1/2} . \tag{18}$$

By calculating m_s using Eq. 18, the color change that passes through the gray point can be obtained. Q_xQ_y, J, and m_s values can be calculated, for instance, by the Mathematica software (Wolfram Research, Champaign, IL, USA) with a personal computer.

2.1.3 Optimization of the Mixing Ratio of the Dyes

A simulation of the color variations for various mixing ratios of the dyes was performed in order to select the most suitable color variation. When the absorption spectra are obtained at a certain concentration ($C_{\mathrm{dye},0}$) of the dye, $L^*a^*b^*$ values at various concentrations of the dye (C_{dye}) can be calculated. The absorption spectra $A(\lambda)$ at C_{dye} are given by the following equation:

$$A(\lambda) = (C_{\mathrm{dye}}/C_{\mathrm{dye},0})A_0(\lambda) , \tag{19}$$

where A_0 (λ) is the absorption spectra at $C_{\mathrm{dye},0}$. On the other hand, the relationship between transmittance T and absorption A is

$$T = 10^{-A} . \tag{20}$$

Therefore, the tristimulus values XYZ can be expressed as Eqs. 21–23 using the absorption spectra data [12]:

$$X = k \int_{380}^{780} 10^{-A(\lambda)} P(\lambda)\bar{x}(\lambda)\mathrm{d}\lambda , \tag{21}$$

$$Y = k \int_{380}^{780} 10^{-A(\lambda)} P(\lambda)\bar{y}(\lambda)\mathrm{d}\lambda , \tag{22}$$

$$Z = k \int_{380}^{780} 10^{-A(\lambda)} P(\lambda)\bar{z}(\lambda)\mathrm{d}\lambda . \tag{23}$$

Here, k is given by Eq. 24:

$$k = \frac{100}{\int_{380}^{780} P(\lambda)\bar{y}(\lambda)\mathrm{d}\lambda} . \tag{24}$$

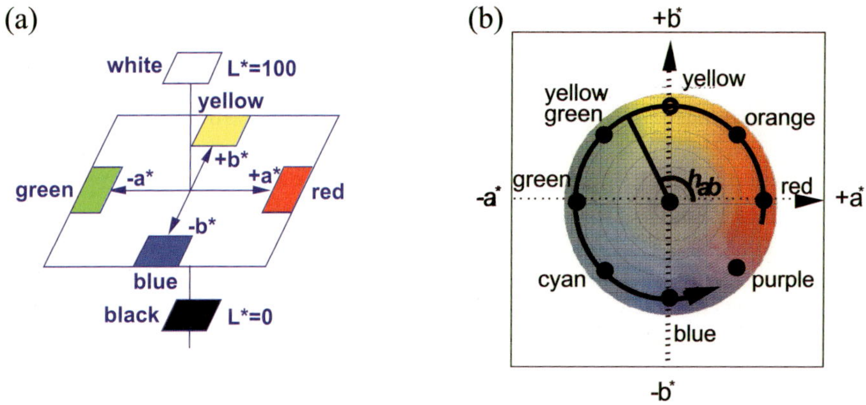

Fig. 4 $L^*a^*b^*$ color space (**a**) and color circle and hue angle (**b**) shown on an a^*b^* chromaticity diagram

The XYZ values at C_{dye} are obtained by substituting Eq. 19 into Eqs. 21–23. The $L^*a^*b^*$ values, the color space of which proposed by CIE, the color circle, and the hue angle are shown in Fig. 4, at C_{dye} can be calculated using Eqs. 25–27 from the XYZ values:

$$L^* = 116\left(\frac{Y}{Y_n}\right)^{1/3} - 16\,, \tag{25}$$

$$a^* = 500\left\{\left(\frac{X}{X_n}\right)^{1/3} - \left(\frac{Y}{Y_n}\right)^{1/3}\right\}, \tag{26}$$

$$b^* = 200\left\{\left(\frac{Y}{Y_n}\right)^{1/3} - \left(\frac{Z}{Z_n}\right)^{1/3}\right\}, \tag{27}$$

where X_n, Y_n, and Z_n are the tristimulus values of a completely diffuse and reflective surface (standard white plate of magnesium oxide); $X_n = 96\,422$, $Y_n = 100\,000$, and $Z_n = 82\,521$. The color difference ΔE^*_{ab} is expressed as follows:

$$\Delta E^*_{ab} = \sqrt{(\Delta L^*)^2 + (\Delta a^*)^2 + (\Delta b^*)^2}\,. \tag{28}$$

The $L^*a^*b^*$ color space is a visually uniform color space and is quite useful for designing devices suitable for visual colorimetry. The angle originating from the positive direction of the a^* axis expresses the hues themselves and is defined as the hue angle, h_{ab}, expressed by Eq. 29 [13]:

$$h_{ab} = \tan^{-1}(b^*/a^*)\,. \tag{29}$$

When we evaluate the color variation qualities of the sensors, it is important to evaluate both a color difference (ΔE) and a hue angle variation (Δh_{ab}). As an example of an investigation using h_{ab}, Tlaczala and Bartecki [14, 15] studied the solvatochromism for metal ion complexes using h_{ab}. They studied the relationship between the acceptor number (E_T values) of solvents and the h_{ab} values. In general, solvatochromism refers to variations of absorbance spectra influenced by solvents. The maximum absorption wavelength shifts as a result of the electronic energy levels changing under the influence of solvent polarities. Hues, chromaticity, and brightness as parameters expressing colors are related to the λ_{max} values, the sharpness of peaks, and the absorbance values of the spectra. It is concluded that the h_{ab} values are useful for evaluating color variations of solvatochromic dyes induced by solvent polarity variations.

2.2 Construction of a Li^+ Sensing Film Optode

Two types of Li^+ sensing optode based on DCA were developed; one was constructed from TTD14C4, KD-M11, KD-S1, TFPB, and NPOE (film 1), where the abbreviations refer to the Li^+ ionophore, color-changeable dye, screening dye, lipophilic anionic additive, and membrane solvent, respectively (Fig. 5), in a PVC membrane; and the other optode was made from TTD14C4, KD-M11, KD-C4, TFPB, and NPOE (Film 2) [1, 2, 16, 17]. KD-C4 (Fig. 5) was a color-changeable cationic dye, the pK_a values of which were different from

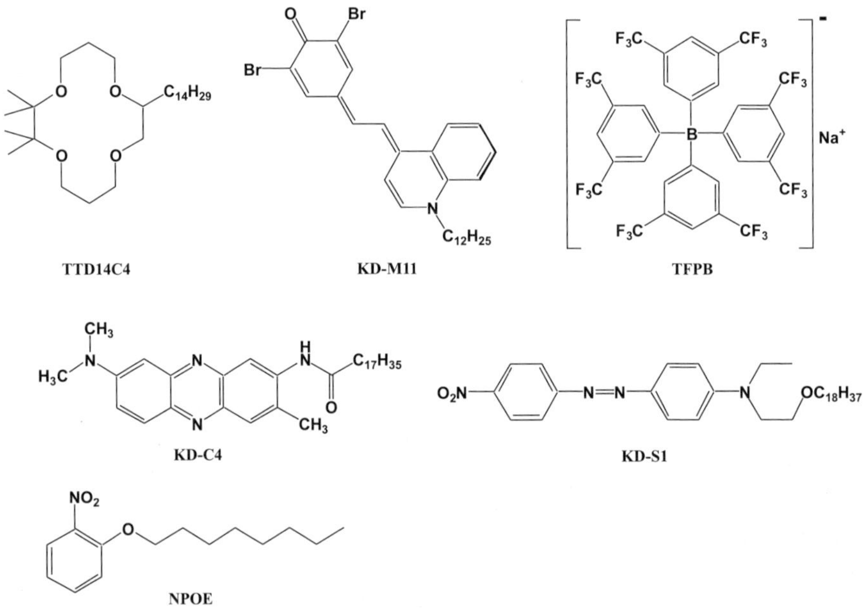

Fig. 5 Chemical structures of the compounds incorporated in the Li^+ sensing optode membrane

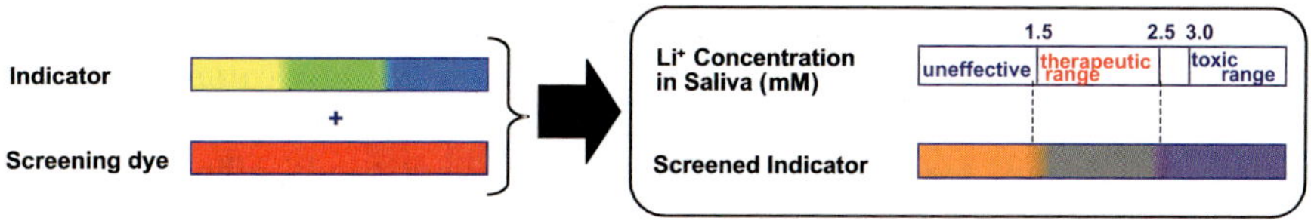

Fig. 6 Theoretical color change passing through a colorless gray point by mixing two dyes under optimum conditions

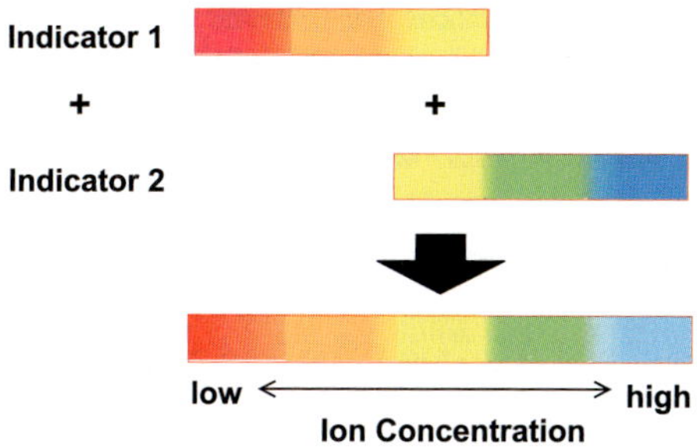

Fig. 7 Schematic epresentation of the multicolor creation by mixing two indicators

those of KD-M11. The former Li^+ optode created colorless gray in the process of color change as a function of Li^+ concentration under the optimum mixing ratio of KD-M11 and KD-S1, the process of which was able to be simulated to obtain the Q_xQ_y and J values from the calculation of the spectral data using Mathematica software, as illustrated in Fig. 6. The latter made it possible to carry out the multicolor visual colorimetry over a large concentration range of Li^+ under the optimum mixing ratio of KD-M11 and KD-C4, which indicated a different color change, yellow → green → blue and red → orange → yellow with increasing pH, respectively, from that for the simulation using $L^*a^*b^*$ color coordinates, the schematic representation of which is shown in Fig. 7.

2.3 Response of Optode Film 1 to Li^+ Concentration

Figure 8 shows the color change of film 1 with and without KD-S1 for the response to various Li^+ concentrations measured by the digital color analyzer. Film 1 without KD-S1 showed a color change from yellow to blue via green as a function of Li^+ concentration. On the other hand, film 1 containing KD-S1 had a color change passing through the gray point at 10^{-3} M Li^+. KD-M11 was yellow in its protonated form and turned blue in its deprotonated form. When the quantity of the protonated form of KD-M11 equals that of the deprotonated form, KD-M11 is green. To obtain the gray point, a red dye is adequate because red is the complementary color of green. KD-S1 (λ_{max} = 497 nm) did not show a spectral change for solutions of various pH, and the response

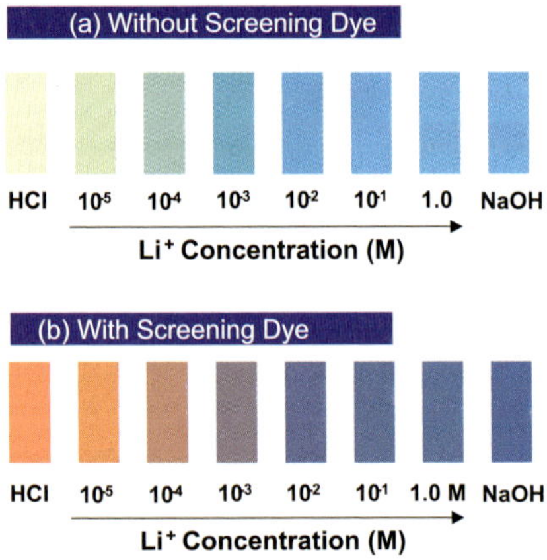

Fig. 8 Color variations in the Li^+ solutions (10^{-5}–1.0 M, pH 9) obtained by the Li^+-sensitive optode without (**a**) and with (**b**) the screening dye

curve of the optode containing both KD-M11 and KD-S1 was identical with that of the optode containing only KD-M11. As a result, KD-S1 was suitable for the screening dye.

The Q_xQ_y coordinates of the membranes containing KD-M11, KD-S1, and KD-M11 + KD-S1 for 10^{-5}–1.0 M Li^+ solutions are shown in Fig. 9. We tried to draw calibration curves using Q_xQ_y, HSB, RGB, CMY, and CIE $L^*a^*b^*$ tristimulus color spaces based on digital color data. The calibration curves obtained based on any tristimulus color spaces except for the Q_xQ_y coordinate gave a complicated line, and only the Q_xQ_y coordinate showed a good straight line as shown in Fig. 9. The calibration curve of film 1 passed a gray point near

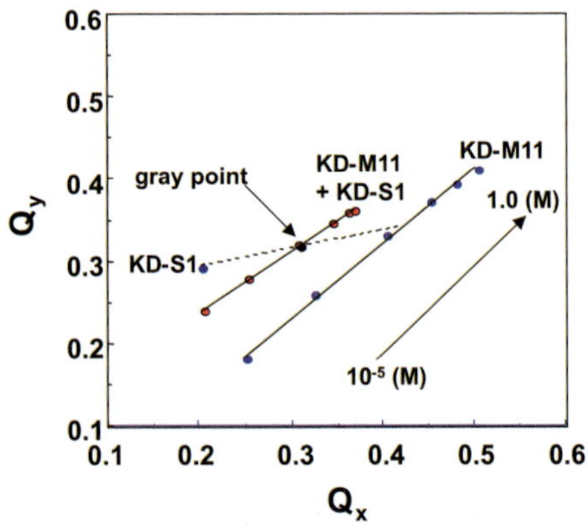

Fig. 9 Q_xQ_y chromaticity coordinate of the optode sensors with and without the screening dye as a function of Li^+ concentration

10^{-3} M Li^+, and the mixing ratio of these two dyes could be calculated from Q_xQ_y values at the 10^{-3} M Li^+ point and the KD-S1 color point in the Q_xQ_y chromaticity diagram. The color difference $\%\Delta E^*$ values in the range from 10^{-4} to 10^{-2} M Li^+ was 52.1 for KD-M11 and 55.8 for KD-M11 + KD-S1 in the optode, respectively. The $\%\Delta E^*$ value of the optode containing KD-M11 + KD-S1 was slightly higher than that of the optode containing only KD-M11. It is difficult to discriminate a slight color chage from yellowish-green to bluish-green via green, whereas it is easy to discriminate between chromatic and achromatic color. In this case, the optode film for clinical use that allows for easy visual detection at concentrations near 10^{-3} M Li^+ was successfully developed.

2.4 Response of the Optode Film 2 to Li^+ Concentration

The responses of the optode film containing the individual dye KD-C4 or KD-M11 toward Li^+ were investigated, and the corresponding colors observed by the COLORTRON are shown in Fig. 10. The KD-C4-based optode responded in the range from 10^{-6} to 10^{-2} M while changing the color from red to yellow. On the other hand, the optode film containing KD-M11, whose pK_a values are higher than those of KD-C4, responded in the range from 10^{-4} to 1.0 M with a color change from yellow to blue via green. Each dye responded over 4 orders of magnitude to the ion concentration. From this result, we assumed that a desirable color variation will be obtained when the two dyes having different pK_a values are mixed under optimum conditions.

To prove this hypothesis, the $L^*a^*b^*$ values were calculated at each concentration of 10^{-6} – 1 M Li^+ when the mixing ratio of KD-C4 to KD-M11 was 9/2, 3/1, 2/1, and 3/2 (mol/mol), respectively. The appearance of the color variations was expressed by plotting the values on the a^*b^* coordinate as shown

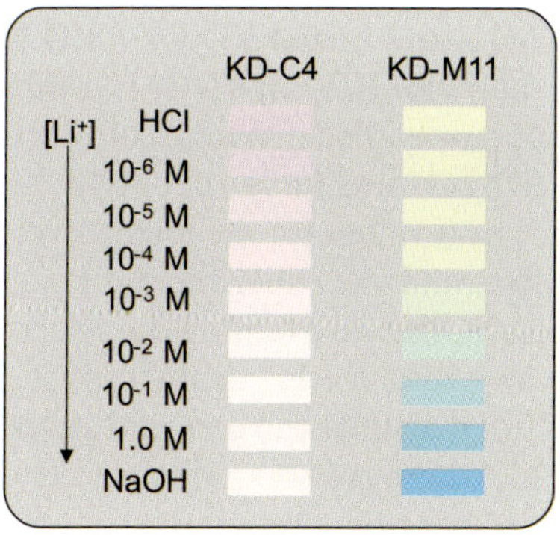

Fig. 10 Color variations of the Li^+ optode membranes containing KD-C4 and KD-M11, respectively

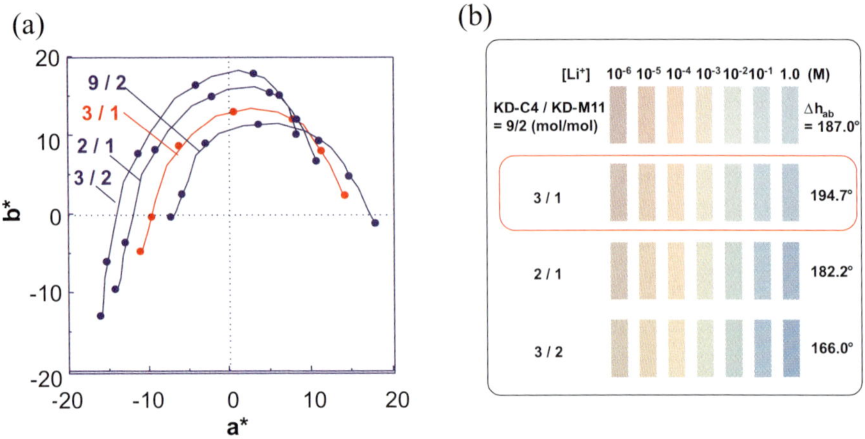

Fig. 11 The a^*b^* values for the simulated color variation of the optode containing the mixed dyes (KD-C4/KD-M11) in different mixing ratios (**a**) and the corresponding simulated color variations (**b**)

in Fig. 11a. Figure 11b shows the membrane colors by the simulation and hue angle variations (Δh_{ab}) that were calculated in the range from 10^{-6} to 1 M Li^+. As a result, it was shown that Δh_{ab} was largest when the mixing ratio of KD-C4 to KD-M11 was 3/1; the optode of the mixing ratio can be useful for visual colorimetry for Li^+ concentrations. It indicates that the individual absorbances of KD-C4 and KD-M11 for λ_{max} were approximately equal when the mixing ratio of KD-C4 to KD-M11 was 3/1.

On the basis of the simulation results, the actual response of the optode film containing the mixture of the two dyes under optimum conditions was investigated. The color change as a function of the Li^+ concentration was monitored using COLORTRON, and the results are shown in Fig. 12a. The hue angle variation of the mixed dyes was larger than that of an individual dye, and the color variation was clear and visually distinguishable. The color difference (ΔE) for individual KD-C4, KD-M11, and mixed KD-C4 + KD-M11 was 30, 44, and 54, respectively. The Δh_{ab} values for KD-C4, KD-M11, and KD-C4 + KD-M11 were 98°, 112°, and 187°, respectively. The dynamic concentration range for KD-C4, KD-M11, and KD-C4 + KD-M11 was $10^{-6}-10^{-2}$ M, $10^{-4}-1.0$ M, and $10^{-6}-1.0$ M, respectively. These numerical values for the mixed dyes were larger in all cases than for an individual dye, and the experimental results were in good agreement with the simulation results. To examine the selectivity of this optode film containing mixed indicator dyes for Li^+ against other alkali metal ions (Na^+, K^+), $\Delta E/\Delta E_{total}$ was plotted against the concentration of each metal ion. The data are shown in Fig. 12, where ΔE was calculated from the beginning of the color variation on the a^*b^* coordinates to the point of each respective ion concentration, and ΔE_{total} expresses the maximal possible color difference of the optode film measured from the completely protonated (HCl, pH 2) to the fully

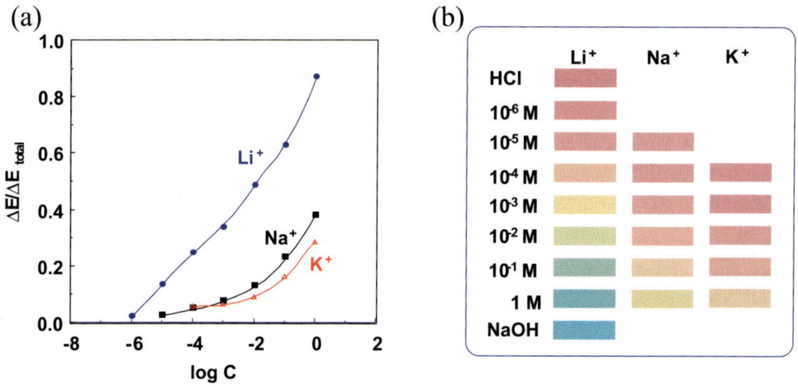

Fig. 12 Color change of the optode membrane for Li^+, Na^+, and K^+ solutions using COLORTRON (**a**) and the correlation between the ion concentration ($\log C$) and $\Delta E/\Delta E_{\text{total}}$ (**b**)

deprotonated (NaOH, pH 11) state. As a result, the selectivity for Li^+ over Na^+ and K^+ was about 1000 and 5000 times higher, respectively, and the multicolor was observed for only Li^+ as shown in Fig. 12b. The color variation of the two dyes had no offset, and the multicolor variation was realized over a wide dynamic range of the ion concentration. Therefore, it was possible to develop an advantageous optode device for visual colorimetry by mixing different dyes after selecting the optimal combination using characteristics such as the color variations and pK_a difference of the indicators.

3 NH_4^+ Ion Sensor Based on Digital Color Analysis

Ammonium ions (NH_4^+) in rivers arise from the degradation of proteins and organic compounds containing nitrogen from sewage water, human waste, and industrial wastewater. Easy determination methods of ammonium ion are needed for screening the wastewater. For example, the standard value for the total of ammonium, nitrite, and nitrate nitrogen is 100 mg/mL in wastewater in Japan. The indophenol method has generally been used for the determination of ammonium ion. This method takes over 10 min to complete the reaction and show a blue color. The reaction does not display a change in the hue but shows a change in the color strength. The reaction solution is toxic and needs waste disposal. On the other hand, NH_4^+ sensors have advantages such as quick response and reversibility. Almost all ammonia gas and ammonium ion sensors have been developed using nonactin, as an ammonium ionophore or a gas-permeable membrane constructed from colorimetric or fluorometric indicators [18–22].

NH_4^+ optodes were developed using an ammonium ionophore and indicators which were made by our group, and the linear calibration graphs were produced by DCA, which converts the color information of the sensors into numerical values (Q_xQ_y and $L^*a^*b^*$ values) and optimizes the color of the sensors to be suitable for human visual perception based on the Li^+ sensor described before.

3.1 Construction of NH_4^+ Sensing Film Optode

The NH_4^+ sensing film optode was fabricated from TD19C6, KD-M13, K-TCPB, and NPOE under optimum conditions, which correspond to the NH_4^+ ionophore, the color-changeable dye of pK_a 7.9, the lipophilic anionic additive, and a membrane solvent, respectively, included in a PVC membrane [23, 24]. These chemical structures are shown in Fig. 13. KD-M13 becomes yellow in its protonated form and turns blue in the deprotonated form. When the quantity of the protonated form of the dye equals that of the deprotonated form, the mixture becomes green.

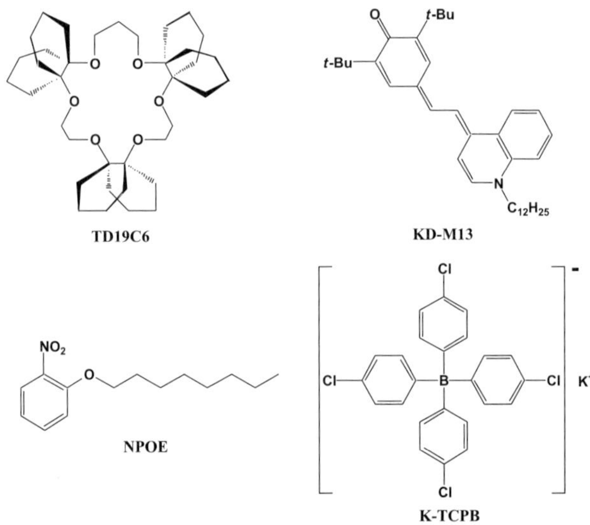

Fig. 13 Chemical structures of the compounds incorporated into the NH_4^+ optode film sensor

3.2 Response of the NH_4^+ Film Optode

On the basis of the ion-exchange mechanism, the color change of the ion-sensing film occurs according to the ammonium ion concentration change.

The ion-sensing film optode responds to NH_4^+ on the basis of the ion-pair extraction/exchange mechanism as shown in the following reaction [25–27]:

$$NH_4{}^+{}_w + S_o + DH^+R_o{}^- \rightarrow NH_4{}^+SR_o{}^- + D_o + H^+{}_w\,, \qquad (30)$$

where S, DH^+R^-, $NH_4^+SR^-$, D, and H^+ represent a neutral ionophore, the ion-pair of a protonated dye and a lipophilic anionic additive, the ion-pair of an NH_4^+-ionophore complex and a lipophilic anionic additive, and the deprotonated dye and proton, respectively. The subscripts o and w represent the organic phase and the water phase. When the complexation of the neutral ionophore with NH_4^+ takes place, the cationic dye deprotonates and becomes neutral, which causes the color change of the optode membrane.

To investigate the response of the optode membrane to NH_4^+, its absorption spectrum was monitored as a function of the NH_4^+ concentration after immersing the freshly prepared PVC film in 0.1 M aqueous HCl to protonate the dye. The absoption data are shown in Fig. 14. The absorption maximum shifted from 475 to 610 nm, and the color of the optode membrane changed from yellow (10^{-6}–10^{-5} M) → light yellow (10^{-4} M) → light green (10^{-3} M) → cyan (10^{-2} M) → blue (10^{-1} – 1 M). The absorption spectra have a clear isosbestic point.

After the membrane had been removed from the glass cell, it was placed on a white Teflon plate to measure the color of the membrane using a digital color analyzer in the reflection mode, and the typical resulting color of the membrane is indicated in Fig. 15. The Q_xQ_y coordinates of this membrane for 10^{-6}–1.0 M NH_4^+ solutions (pH 9) are shown in Fig. 16. The Q_xQ_y coordinate obtained by the optode has a linear calibration curve between (10^{-6} and 10^{-1} M NH_4^+, similarly to the Li^+ optode.

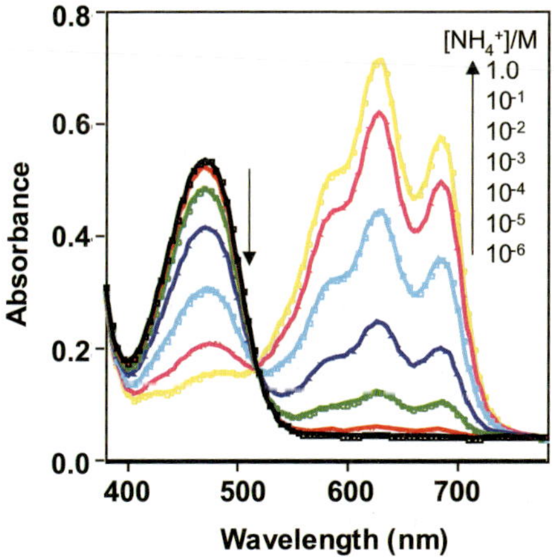

Fig. 14 Absorption spectra of the optode sensor before and after the addition of NH_4^+ solution

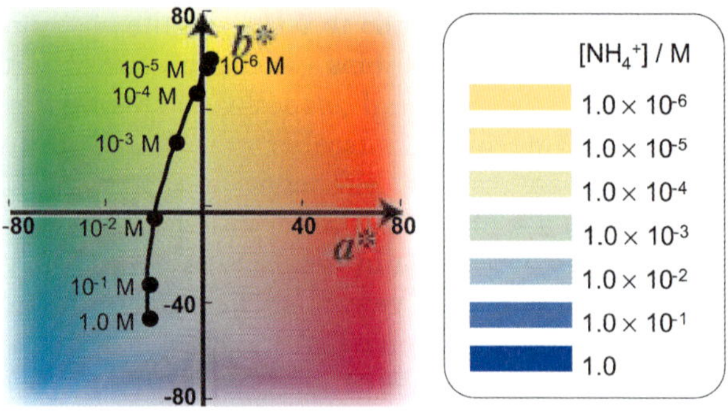

Fig. 15 Color images of the NH_4^+ sensing optode membrane

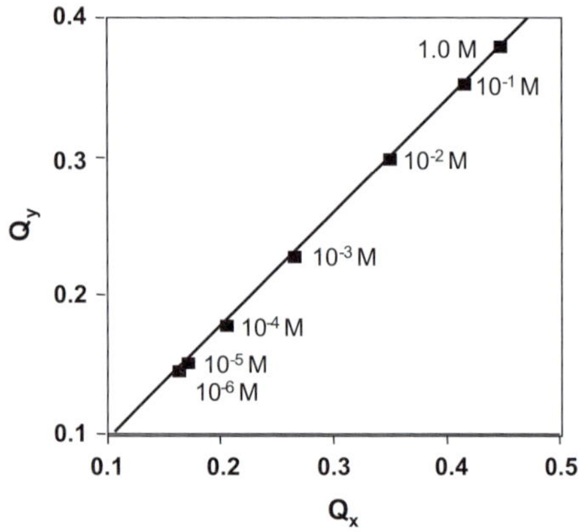

Fig. 16 Q_xQ_y chromaticity coordinates of the NH_4^+ optode

To examine the response time to NH_4^+, ammonium solutions were dropped on this optode membrane, and then the time-course dependence of the Q_x and Q_y values was monitored. As a result, within 10 s the Q_x and Q_y values became saturated.

The reproducibility test of the sensor response to NH_4^+ was carried out, and the results are shown in Fig. 17a. A good relative standard deviation (RSD) of the Q_x and Q_y values was obtained; the RSDs of the Q_x values were 2.08, 0.5, and 0.35% and those of the Q_y values were 2.02, 1.28, and 0.44%, which corresponded to the buffer (NH_4^+ free), 10^{-3} M and 10^{-1} M NH_4^+ solutions.

For the evaluation of the optode selectivity, the normalized absorbances were calculated and plotted against NH_4^+, Na^+, and K^+ concentration, (Fig. 17b). The selectivity of this NH_4^+ optode against Na^+ and K^+ is about 1000 and 10 times greater, respectively. This selectivity against foreign substances was in good agreement with the results of the ion-selective electrode using TD19C6.

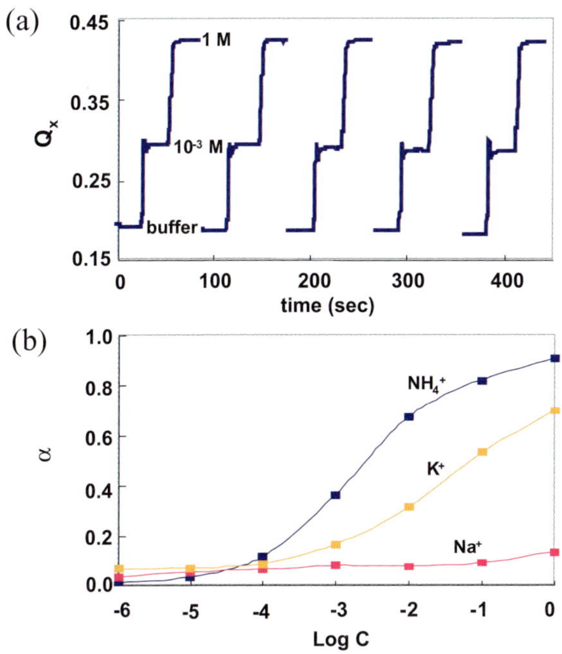

Fig. 17 The reproducibility test (**a**) and the ion selectivity (**b**) of the NH_4^+ sensing film optode

In this section, a visual and colorimetric NH_4^+ sensing film was described. The observed color gradation was suitable for human visual perception over a wide dynamic range. This NH_4^+ sensing film is useful for routine semiquantification of NH_4^+.

4 Protein Sensor

Various colorimetric reagents which are pH indicators that change their color in a weak-acid region were developed to monitor the proteins in human urine. These reagents have sulfonyl groups or carboxyl groups, and bind to the N-terminus in the protein to produce the color change by halochromism. Tetrabromophenol blue, which is one of the dyes used to detect the urinary total protein, is included in the easy and rapid sensing paper to introduce a mechanism of dry chemistry, and this sensing paper has been widely used not only by the specialists in this field but also by many people [28, 29]. However, it is not so sensitive around a criterion value in spite of the good color variation and hue at high protein concentration as shown in Fig. 18. On the other hand, eosin B is more sensitive than tetrabromophenol blue [30, 31];

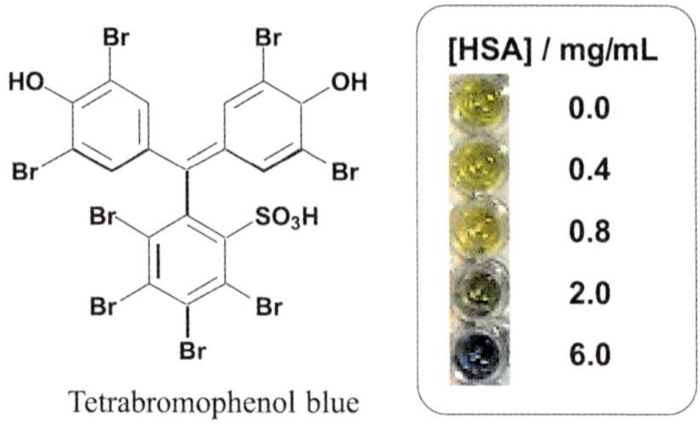

Fig. 18 Chemical structure of tetrabromophenol blue and the color change after the reaction with human serum albumin (*HSA*)

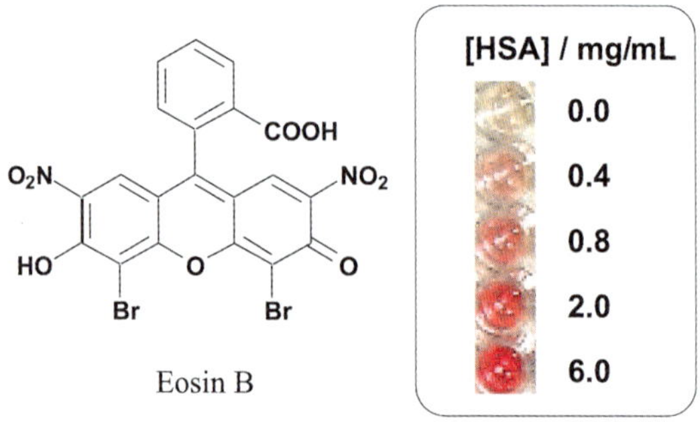

Fig. 19 Chemical structure of eosin B and the response to HSA

however, the color variation is poor owing to the simple bathochromic effect with increasing protein concentration (Fig. 19).

To make use of their merits and to erase their drawbacks, we carried out the multicolor visual colorimetry over a large concentration range of protein under the optimum mixing ratio of tetrabromophenol blue and eosin B using $L^*a^*b^*$ color coordinates based on DCA.

4.1 Simulation of the Color Change of the Mixed Dyes in Response to Human Serum Albumin

The $L^*a^*b^*$ values were calculated at various human serum albumin (HSA) concentration in the 0–12 mg/mL range when the mixing ratio of tetrabromophenol blue to eosin B was 1/0, 4/1, 2/1, 1/1, 1/2, 1/3, and 1/4 (mol/mol), respectively. The appearance of the color variations was expressed by plotting the values on the a^*b^* coordinate as shown in Fig. 20a. Figure 20b shows the membrane colors by the simulation. As a result, the color difference value, ΔE, was largest when the mixing ratio of tetrabromophenol blue to eosin B was 1/3, which can be useful for visual colorimetry of HSA.

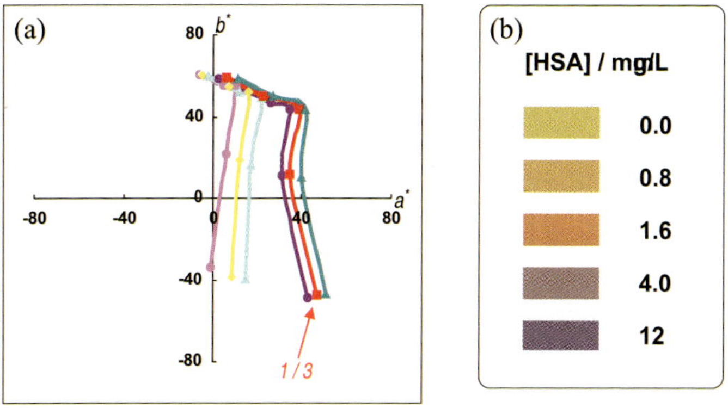

Fig. 20 Color variation of the mixed dyes (tetrabromophenol blue and eosin B) on the a^*b^* color coordinate for the response to the various concentrations of HSA (**a**) and their color sample (**b**) by simulation

4.2 Actual Color Change of the Mixed Dyes in Response to Human Serum Albumin

On the basis of the simulation results, the actual response of the mixture of the two dyes under optimum conditions was investigated for HSA, and the color samples by simulation and by actual color change are shown in Fig. 21a. The color change as a function of the HSA concentration was monitored using COLORTRON, and the results on the a^*b^* coordinates are shown in Fig. 21b. The color changed from yellow to purple via green in the concentration range from 0 to 6.0 mg/mL. The actual results of both color samples and a^*b^* coordinates were in good agreement with those of the simulation results.

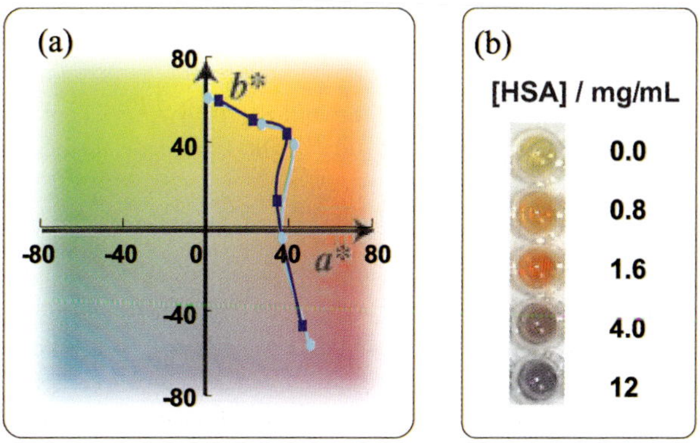

Fig. 21 Simulation and actual color changes on the a^*b^* color coordinate (**a**) and color samples of the simulation and actual color change (**b**) in response to different HSA concentrations

Furthermore, the colors of these two dyes did not interfere each other, and it is possible to measure the multicolor variation and a wide range of the protein concentration owing to the optimum selection of the functional indicators.

5 Conclusions and Future Direction

We have demonstrated that DCA deals with a wide variety of analytical information and simulates color changes to create optimum visual sensors. To take advantage of useful DCA, we have developed optical sensors for Li^+, NH_4^+, and protein based on DCA.

By transforming the color information into numerical values, it is possible to make an accurate determination without the spectral data using the color changes on the basis of the color difference or digital color library data even if the spectrum is complicated to analyze. DCA creates calibration curves by transforming the spectral data into the tristimulus values and plotting the color on a chromaticity diagram, which makes it possible to simulate the optimum color changes by combining suitable multiple dyes on the basis of characteristics such as color variations and pK_a values of the indicators using a personal computer with appropriate software.

The optical sensors described in this review are suitable for easy, rapid, and widely used measurements and are expected to be applicable to many fields, such as environmental analysis, clinical examinations, and in situ process checks.

Acknowledgements We thank Mrs. Etsuko Hirayama and Dr. Hiroaki Okabe for their ambitious engagements in DCA analysis.

References

1. Hirayama E, Sugiyama T, Hisamoto H, Suzuki K (2000) Anal Chem 72:465
2. Suzuki K, Hirayama E, Sugiyama T, Yasuda K, Okabe H, Citterio D (2002) Anal Chem 74:5766
3. COLORTRON user manual (1994) Light Source Computer Images, Larkspur, CA
4. Prasad KMMK, Rahhem S, Vijayalekshmi P, Sastri CK (1996) Talanta 43:1187
5. Frezzotti A, Gambini AMM, Coppa G, Sio GD (1996) Scand J Clin Lab Invest 56:591
6. Gorham JD, Walton KG, McClellan AC, Scott MG (1994) Ther Drug Monit 16:277
7. Bodman V, Arter T, Maseiewicz F, Dychko D, Schaeffer J, Winterkorn R (1992) Clin Chem 38:1049
8. Reilly CN, Flaschka HA, Laurent S, Laurent B (1960) Anal Chem 32:1218
9. Flaschka H (1961) Talanta 8:342
10. Bosch E, Casassas E, Izquierdo A, Roses M (1984) Anal Chem 56:1422
11. Flaschka H (1960) Talanta 7:90

12. Pauli H (1976) J Opt Soc Am 66:866
13. Japanese Industrial Standard, color, Z8729 (1996) Japanese Standard Association
14. Tlaczala T (1997) Pol J Chem 71:823
15. Tlaczala T, Bartecki A (1997) Monatsh Chem 128:225
16. Suzuki K, Yamada H, Sato K, Watanabe K, Hisamoto H, Tobe Y, Kobiro K (1993) Anal Chem 65:3404
17. Hisamoto H, Tohma H, Yamada T, Yamaguchi K, Siswanta D, Suzuki K (1998) Anal Chim Acta 373:271
18. Ozawa S, Hauser PC, Seiler K, Tan SSS, Morf WE, Simon W (1991) Anal Chem 63:640
19. West SJ, Ozawa S, Seiler K, Tan SSS, Simon W (1992) Anal Chem 64:533
20. Seiler K, Morf W, Rusterholz B, Simon W (1989) Anal Sci 5:557
21. Wang E, Zhu L, Ma L, Patel H (1997) Anal Chim Acta 357:85
22. Strömberg N, Hulth S (2001) Anal Chim Acta 443:215
23. Suzuki K, Siswanta D, Otsuka T, Amano T, Ikeda T, Hisamoto H, Yoshihara R, Ohba S (2000) Anal Chem 72:2200
24. Hisamoto H, Tohma H, Yamada T, Yamaguchi K, Siswanta D, Suzuki K (1998) Anal Chim Acta 373:271
25. Hisamoto H, Miyashita N, Watanabe K, Nakagawa E, Suzuki K (1995) Sens Actuators B 29:378
26. Suzuki K, Tohda K, Tanada Y, Ohzora H, Nishihama S, Inoue H, Shirai T (1989) Anal Chem 61:382
27. Suzuki K (1998) J Syn Org Chem Jpn 59:291
28. Sasaki M, Pugia MJ, Parker DR, Kuromoto K, Furukawa I, Konishi I (1999) J Clin Lab Anal 13:246
29. Pugia MJ, Lott JA, Profitt JA, Cast TK (1999) J Clin Lab Anal 13:180
30. Waheed AA, Gupta PD (1996) J Biochem Biophys Methods 33:187
31. Waheed AA, Gupta PD (2000) J Biochem Biophys Methods 42:125

Subject Index

Printing: Krips bv, Meppel
Binding: Stürtz, Würzburg